Storying the Ecocatastrophe

How do writers and artists represent the climate catastrophe so that their works stir audiences to political action or at least raise their environmental awareness without, however, appearing didactic? *Storying the Ecocatastrophe* attempts to answer this question while interrogating the potential of narrative to become a viable political force. This collection of essays achieves this by examining the representational strategies and ideological goals of contemporary cultural productions about climate change. These productions have been created across different genres, such as the traditional novel, dance performance, solarpunk, economic report, collage, and space opera, as well as across different languages and cultures. The volume's twelve chapters demonstrate that rising temperatures, erratic weather, extinction of species, depletion of resources, and coastal erosion and flooding are an effect of our abusive relationship with nature. They also show that our use of nuclear power, extraction of natural resources, and extensive farming, including heavy reliance on pesticides, intersect with intrahuman violence, as fleshed out by heteropatriarchy, racism, (neo)colonialism, and capitalism. They finally argue that human activity has indirectly contributed to other contemporary crises, namely the migrant crisis and the spread of contagious diseases such as Covid-19.

Helena Duffy (MSt Oxon, PhD Oxford Brookes) is Professor of French at the University of Wrocław, Poland. Her research on the cultural representations of World War II and the Holocaust has resulted in the publication of the monographs *World War II in Andreï Makine's Historiographic Metafiction* (2018) and *The Holocaust in French Postmodern Fiction* (2022). She has edited issues of *French Forum*, *Holocaust Studies, Journal of Holocaust Research*, and *Eastern European Holocaust Studies*, and, with Avril Tynan, has co-edited a collection of essays, *Trauma, Ethics, Hermeneutics* (2024).

Katarina Leppänen is Professor of Intellectual History and the Head of the Department of Literature, Intellectual History and Religion at the University Gothenburg, Sweden. Her ongoing project deals with the importance of literary transnational, international, and regional exchanges in the Nordic and Baltic countries in the early twentieth century. Titled 'Cosmopolitan and Vernacular Dynamics in World Literature,' the project is funded by the Swedish Foundation for Humanities and Social Sciences.

Routledge Studies in World Literatures and the Environment
Series Editors: Scott Slovic and Swarnalatha Rangarajan

Reading Contemporary Environmental Justice
Narratives from Kerala
R. Sreejith Varma

(Eco)Anxiety in Nuclear Holocaust Fiction and Climate Fiction
Doomsday Clock Narratives
Dominika Oramus

Decolonial Animal Ethics in Linda Hogan's Poetry and Prose
Toward Interspecies Thriving
Małgorzata Poks

The Imagined Arctic in Speculative Fiction
Maria Lindgren Leavenworth

Desertscapes in the Global South and Beyond
Anthropocene Naturecultures
Edited by Sushila Shekhawat, Rayson K. Alex and Swarnalatha Rangarajan

Reading Madeleine L'Engle
Ecopsychology in Children's and Adolescent Literature
Heidi A. Lawrence

Storying the Ecocatastrophe
Contemporary Narratives about the Environmental Collapse
Edited by Helena Duffy and Katarina Leppänen

For more information about this series, please visit: www.routledge.com/Routledge-Studies-in-World-Literatures-and-the-Environment/book-series/ASHER4038

Storying the Ecocatastrophe

Contemporary Narratives about the Environmental Collapse

Edited by Helena Duffy and Katarina Leppänen

Routledge
Taylor & Francis Group
NEW YORK AND LONDON

First published 2024
by Routledge

605 Third Avenue, New York, NY 10158
and by Routledge
4 Park Square, Milton Park, Abingdon, Oxon, OX14 4RN

Routledge is an imprint of the Taylor & Francis Group, an informa business

ISBN: 978-1-032-72694-6 (hbk)
ISBN: 978-1-032-72697-7 (pbk)
ISBN: 978-1-032-72695-3 (ebk)

DOI: 10.4324/9781032726953

Typeset in Sabon
by Apex CoVantage, LLC

Contents

Notes on contributors

Katarzyna Bojarska is Assistant Professor in the Department of Cultural Studies at SWPS University in Warsaw and President of the NGO 'View: Foundation for Visual Culture' which she co-founded and where she is the editor of the international, open access, online academic journal *View: Theories and Practices of Visual Culture*. She has received numerous grants and awards, including Fulbright, Horizon 2020, and individual and collective grants from the National Centre for Science, Poland. Her research interests include cultural memory, gender and memory, trauma, and visual culture studies, as well as contemporary arts. Her book, titled *Wydarzenia po Wydarzeniu. Białoszewski – Richter – Spiegelman* (*Events after the Event. Białoszewski – Richter – Spiegelman*), was published in 2013. Bojarska is the editor of Ernst van Alphen's *Krytyka jako interwencja: sztuka, pamięć, afekt* (*Criticism as Intervention. Art, Memory, Affect*) (2019). She has also translated and published works of other thinkers, including Achille Mbembe's *Necropolitics* (2018), Michael Rothberg's *Multidirectional Memory* (2016), and Susan Buck-Morss's *Hegel, Haiti and Universal History* (2013), as well as works by Cathy Caruth, Marianne Hirsch, Lauren Berlant, Kristin Ross, Mignon Nixon, Susan Schuppli, Benjamin H. D. Buchloh, and Ariella Aïsha Azoulay. Bojarska has presented her work at numerous international conferences and taught international students in both Europe and the United States. She is an active art critic and member of the AICA.

Chloé Bour--Lang is a Fulbright alumna and a PhD candidate at the University of Strasbourg, France. Her research work concentrates on the influence of gendered behaviours on the environment, and specifically on the treatment of nonhuman animals in contemporary American literature, with a focus on the ecopoetics of spectrality.

Georgiana Bozîntan is a PhD candidate at Babeş-Bolyai University in Cluj-Napoca, Romania. She holds a BA in Norwegian and French languages and literatures, and an MA in francophone literature. Her research lies within the field of ecocriticism, and she is currently working on a thesis about representations of climate change in contemporary Norwegian fiction.

Helena Duffy (MSt Oxon, PhD Oxford Brookes) is Professor of French at the University of Wrocław, Poland. In her career, she lectured at the University of Hull, the University of Queensland, and the University of Turku, Finland. She held the prestigious Marie-Curie Individual Fellowship (Royal Holloway) and the Fernandes Fellowship (University of Warwick), and served as Visiting Professor at the Université Blaise-Pascal in France. Her research on the cultural representations of World War II and the Holocaust has resulted in the publication of the monographs *World War II in Andreï Makine's Historiographic Metafiction* (Brill, 2018) and *The Holocaust in French Postmodern Fiction* (Legenda, 2022). She has edited issues of *French Forum*, *Journal of Holocaust Research*, and *Eastern European Holocaust Studies*, and, with Avril Tynan, has co-edited collection of essays *Trauma, Ethics, Hermeneutics* (Legenda, 2024). She is currently preparing a monograph on the representations of the forest in Holocaust literature and cinema, and, with David Tollerton, is co-editing a special issue of *Holocaust Studies* dedicated to the links between genocide and ecocide.

Hannah Jocelyn is a PhD candidate in American studies at Rutgers University, in Newark. She holds an Honors BA in English literature from Concordia University, an MA in humanities from York University, and an MA in social thought from New York University. She has contributed a chapter to *Critical Insights: American Short Story* (2015), and her work has been published in *The Journal of American Studies*. For ten years she has worked in the Foundry Department of *The New Yorker* magazine.

Markku Lehtimäki, PhD, is Professor of Comparative Literature at the University of Turku, Finland. His fields of expertise are narrative theory, visual culture, ecocriticism, and American literature. His research projects include *The Changing Environment of the North: Cultural Representations and Uses of Water* (2017–2021), funded by the Academy of Finland, and *The Novel's Knowledge: Changing Roles of Author and Book in Society* (2022–2024), funded by the Kone Foundation. He is the author of two monographs, *The Poetics of Norman Mailer's Nonfiction: Self-Reflexivity, Literary Form, and the Rhetoric of Narrative* (Tampere University Press, 2005) and *Sofi Oksasen romaanitaide: Kertomus, etiikka, retoriikka* (*The Art of Sofi Oksanen's Novels: Narrative, Ethics, Rhetoric*, Suomalaisen Kirjallisuuden Seura, 2022) and co-editor of several books, most recently *Visual Representations of the Arctic: Imagining Shimmering Worlds in Culture, Literature and Politics* (Routledge, 2021) and *Cold Waters: Tangible and Symbolic Seascapes of the North* (Springer, 2022).

Katarina Leppänen is Professor of Intellectual History and the Head of the Department of Literature, Intellectual History and Religion at the University Gothenburg, Sweden. Her doctoral thesis was on the feminist Elin Wägner and the European feminist movement in the interwar era, especially focusing on issues of alternative social orders and ecological feminist ideas. Leppänen's research covers question of women's independent nationality, education for internationalism, and trafficking in women in the interwar era. Her ongoing project deals with the importance of literary transnational, international, and regional exchanges in the Nordic and Baltic countries in the early twentieth century. Titled 'Cosmopolitan and Vernacular Dynamics in World Literature,' the project is funded by the Swedish Foundation for Humanities and Social Sciences.

Xin Liu is Senior Lecturer at the Center for Gender Studies, Karlstad University, Sweden. She has published articles in journals such as *Feminist Review*, *Australian Feminist Studies*, *Journal of Environmental Media*, *Parallax*, *Catalyst: Feminism, Theory, Technoscience*, *MAI: Feminism & Visual Culture*, *Feminist Encounters: A Journal of Critical Studies in Culture and Politics*, *Media Theory Journal*, and *Agenda: Empowering Women for Gender Equality*. Her recent research projects are located at the intersection of feminist theory, environmental humanities, critical race studies, science and technology studies, social theory and digital media research.

Faeze Rezaii, MA, is a doctoral researcher in comparative literature at the University of Turku, Finland. Faeze's research interests include environmental literature and film, science fiction, ecofiction, and spatial studies. In her forthcoming monograph, she investigates the mediations of water in transplanetary science fiction literature and cinema. In her research, she focuses on the imaginaries of water, fluid geographies, and narratives of ice in science fiction vis-à-vis the Anthropocene, agency, ideology, and ecological suffering. Faeze has taught courses on ecocriticism, science fiction, and speculative fiction at the University of Turku. She is also a fiction writer, mainly of short stories.

Elizabeth Tavella is a scholar of critical animal studies and environmental humanities whose work builds on critical theory, decolonial methods, and queer ecology to investigate interlinked systems of oppression and liberation histories from an artistic and cultural perspective. Tavella is the creative writing and arts editor of *Ecozon@: European Journal of Literature, Culture and Environment*, and co-editor of a multimedia digital publication titled *Animated Wor(l)ds: Language and Relationality for Multispecies Kinship*. Recent publications include chapters in two edited volumes, *Interrogating the Boundaries of the Nonhuman: Literature, Climate Change, and Environmental Crises* (Lexington Book, 2022) and *Animals and Race* (Michigan State University Press, 2023). Tavella has held a postdoctoral fellowship at the University of Chicago, with affiliations in the Department of Romance Languages and Literatures and the Center for the Study of Gender and Sexuality, and teaching appointments at the School of the Art Institute of Chicago and UIC. Tavella's research has been funded by the Culture and Animals Foundation and the National Endowment for the Humanities.

Sébastian Thiltges holds a PhD in comparative literature and is currently a postdoctoral researcher at the University of Luxembourg (2018–2023). After completing a dissertation on novelistic descriptions of silent landscapes (2013), he turned to ecocriticism and cultural ecology in two postdoctoral projects. While the first project, conducted at the Université du Mans (France), is concerned with ecology in children's literature, the second, conducted at the Universität des Saarlandes (Germany), addresses nature and the environment in literatures from Luxembourg. Thiltges has published numerous articles and co-edited the books *Eco-graphies: écologie et littératures pour la jeunesse* (with N. Prince, Presses universitaires de Rennes, 2018), *Kulturökologie und ökologische Kulturen in der Großregion* (with Ch. Solte-Gresser, Peter Lang, 2020), *Peut-on tout leur dire? Formes de l'indicible en littérature jeunesse* (with T. Raus, Presses universitaires de Bordeaux, 2023), and an online issue of *Cultural Express* 'Faceless Evil in Popular Culture' (with O. Kohns, 2021). Thiltges is currently president of the *Luxembourgish Association for Comparative Literature* (SLLGC) and president of the jury for the Luxembourgish literary award *Prix Servais*.

Spring Ulmer is the author of *Benjamin's Spectacles* (2009), *The Age of Virtual Reproduction* (2009), and *Bestiality of the Involved* (2002). She is a recent recipient of a NEA in Translation and the winner of the 2016 Willis Barnstone Translation Prize. Her scholarship has appeared in the following anthologies: *Post-Racial America?: An Interdisciplinary Study*; *Comparative Feminism, Postmodernism, Postcolonialism: Gender and Sexual Identity in Contemporary Turkish Literature and Culture*, and *Photography and Cinema, 50 Years of La Jetée*. She teaches in the English Department at Middlebury College in Middlebury, Vermont, USA.

Introduction

Storying the ecocatastrophe: from doom-and-gloom scenarios to messages of hope

Helena Duffy

Storying the Ecocatastrophe: Contemporary Narratives about the Environmental Collapse is an outcome of the work carried out by an international study group, 'Narrative and Violence,' which operated between 2020 and 2022 under the generous auspices of the Nordic Summer University (NSU). Founded in 1950, the NSU is an independent, non-hierarchical, and democratic academic institution which for the past seventy years has provided a platform for discussions and collaboration of researchers and practitioners mainly, albeit not only, from the Nordic and Baltic region. Recently, the summer symposia of NSU became an opportunity for its delegates to observe the ramifications of human activity for the non-human world. In July 2018, on the Swedish island of Fårö, temperatures soared up to 36°C. The reputedly freezing Baltic became covered in a thick coat of blue-green algae which are partially caused by a combination of hot temperatures and overfertilisation. In 2019 and 2020, NSU was unable to hold in-person meetings because of COVID-19. Like global warming, the pandemic has been identified as a symptom of the Anthropocene, as is called the geological epoch during which humanity has noticeably impacted nature. According to some, due to a rapid expansion of the human realm, the Anthropocene will also be the 'era of pandemics' (Chakrabarty, 2021: 324, 327).[1] In other words, 'Human beings are the ultimate cause of pandemics' because 'deforestation, agricultural intensification, urbanization, and ecosystem disruption bring people into contact with wildlife and their potentially zoonotic pathogens' (Morens et al., 2020: 4). If 'human-caused ecological pressures' expose us to animal diseases, 'human technology and behavior are spreading those pathogens even more widely and quickly' (Quammen, 2012: 40).

The penultimate workshop of the study group 'Narrative and Violence,' held in October 2021, was dedicated to stories that thematise the devastating outcomes of the humans' controlling, exploitative, and abusive treatment of natural environment. As well as extended versions of some of the papers presented at the workshop, contributions from scholars external to

DOI: 10.4324/9781032726953-1

the group constitute a collection of essays whose twofold aim is to ponder the representational strategies employed in narratives about the ecocatastrophe and to comment on the ethical and ideological agendas informing these narratives. Correspondingly, the twelve essays engage in a discussion of the intersectionality of the environmental goals of the analysed narratives with other political concerns, such as a critique of heteropatriarchy, racism, (neo)colonialism, and capitalism, and a promotion of democracy, equity, and solidarity, including between humans and non-humans. What sets the present volume apart from similar studies is its cross-cultural and cross-generic approach to the problem of the environmental collapse. Moreover, as well as in symptoms of global warming, *Storying the Ecocatastrophe* is interested in the broader signs of humanity's world-building project on Earth during the Anthropocene. These include extraction of natural resources, extensive use of pesticides, nuclear power, water and air pollution, and industrial farming.

Indeed, what usually comes into the spotlight when we speak of the Anthropocene are rising temperatures and extreme and erratic weather caused by the greenhouse gasses that our homes, institutions, factories, and means of transport have been emitting since the onset of the industrial revolution. On the day I am writing these words, the front page of *The Guardian* announced that in 2022 twelve European countries had broken monthly temperature records as the continent registered the hottest ever summer (de Hoog and Kirk, 2023). The increase in temperatures across the globe, which the scientists predict will reach between three and five degrees by 2100, will engender further droughts, tropical cyclones, heatwaves, and floods, such as those that have already been affecting the world. Among other foreseeable outcomes of global warming are coastal erosion, desertification, ocean acidification, crop failures, and forest fires, which will lead in turn to inadequate water and food supplies, spread of contagious diseases, extinction of some species of non-human animals, and proliferation of others, such as flies, mosquitoes, cockroaches, and rats. These phenomena risk to render large sways of Earth uninhabitable, causing famine and malnutrition, creating millions of climate refugees, and, consequently, provoking social, cultural, and, ultimately, political crises in the areas that will accommodate the displaced people. Drawing a parallel between the economic slump of the interwar period and the anticipated depletion of resources, some scholars fear that the encroaching ecocatastrophe may cause genocides similar to the Nazi Holocaust of Europe's Jews (Snyder, 2015; Bartov, 2023). Its other potential repercussions are the collapse of democracy and spread of totalitarianism, or even human extinction (Carrington, 2022).

On the same day as it published the news of broken temperature records, *The Guardian* ran a story about some seven hundred climate activists (including Greta Thunberg herself) who, in the hope of stopping the opening of a coal mine, had been occupying the German village of Lützerath in North Rhine-Westphalia (Connolly, 2023). As part of the plans to start mining in a farming village dating back to the twelfth century, nine hundred inhabitants were evacuated, a nineteenth-century church was deconsecrated and destroyed, and several wind turbines were dismantled. The controversial decision, taken under Germany's previous administration (headed by Angela Merkel's Christian Democratic Union), has surprisingly not been overturned by Germany's recently elected government, despite the participation of the Greens in the ruling coalition.

The determination of *Lützi Bleibt* [Lützi Stays], as is called the movement opposing the coal mine, proves the galvanising impact of the news of climate change that features almost daily in our media. The protesters squatting in Lützerath's abandoned houses claim to be ready to risk their lives in the imminent confrontation with the police. In others, the news of the looming eco-disaster may incite less extreme forms of engagement than those adopted by *Lützi Bleibt*, Extinction Rebellion, and Insulate Britain.[2] They may, for example, choose to travel by train rather than to fly or drive, replace a steak with tofu, or abstain from flicking a light switch. School children may join the Fridays for Future movement, initiated in 2018 by the then fifteen-year-old Thunberg who, holding a homemade placard reading 'Skolstrejk för klimatet' [School strike for climate], sat in front of the Swedish parliament for three weeks in protest against the inertia of politicians regarding environmental change. Since then, many pupils around the globe have been skipping Friday classes to demonstrate against the fossil fuel industry. To society's growing ecological awareness testifies the public opinion poll showing that about 50 per cent of Americans are concerned about water and air pollution. Some 75 per cent worry about the consequences of acid rain, nuclear waste, and global warming, and therefore applaud environmental action (de Steiguer, 2006: 1).

However commendable in many people's eyes, climate protests evidence that the same news is having little bearing on government climate policy. In the United Kingdom, the Conservative administration led by Rishi Sunak is even currently trying to introduce legislation that would enable the police to disperse protesters and take legal action against them, a move that has been described as absurd and authoritarian (Reicher, 2023). The rise of populist, extreme-right, and reactionary politics observed in many countries worldwide has not helped the environmental cause. While in America Donald Trump acolyte Lauren Boebert has derided the green agenda as 'America last' (Scheckner), British Home Secretary Suella Braverman has called climate protesters 'tofu-eating wokerati' (Syal, 2022). Similarly, right-wing Spanish politicians have dismissed the United Nations' climate agenda as 'cultural Marxism' (Fanjul, 2022). On the international arena, coordinated efforts to mitigate the degradation of the environment have been made since 1992, when foundations were first laid for an annual climate summit organised under the banner of the UN. However, as Hubert Zapf points out, 'the reaction of international politics has remained deplorably weak and inappropriate in the face of escalating threat of climate change' (2016: 53). Both climate activists and climate experts have deemed the recent two Conferences of the Parties (COPs) ineffective against the environmental crisis.[3] In Thunberg's words, the Glasgow COP was a failure or, as she disparagingly branded it, 'greenwashing,' since it was trying to 'solve the crisis with the same methods that got us into it in the first place.' Thunberg also called the summit 'a PR event where leaders are giving beautiful speeches and announcing fancy commitments and targets, while behind the curtains governments of the Global North countries are still refusing to take any drastic climate action' (Meredith, 2021). The war in Ukraine, which began in March 2022, has made it easier for politicians to abstain from making further pledges or even to break the existing commitments. The energy and food crises, both exacerbated by Russia's aggression, have displaced the impending ecocatastrophe at the top of the political agenda. It needs adding that, apart from the weakening of the worldwide resolve to stop global warming in its tracks, Moscow's attack on Kyiv has been responsible for enormous environmental damage. Its impact includes emission of thirty-three million tons of greenhouse gasses (the equivalent of adding sixteen million cars to the UK's roads for two years), the destruction of precious animal and plant life, and the pollution of water (Rannard, 2022).

Like activists, scientists, and journalists, novelists, poets, and artists have been warning us against the consequences of the anthropogenic environmental devastation with the aim of stirring us to action that may help to preserve our planet as we know it. Adam Trexler believes that, as well as our daily lives, the Anthropocene, a term the critic uses interchangeably with climate change, has inevitably influenced cultural production (2015: 9), and Timothy Clark considers literature as capable of being a 'significant political force' (2019: 3). One sign of the growing engagement of culture in combatting the climate crisis is the emergence of the genre of 'climate fiction' (or cli-fi), as in 2007 freelance news reporter Dan Bloom called 'novels, short stories and movies that . . . deal with climate change and global warming concerns' (Thorpe and Bloom, n.d.). In 2012, Bloom himself published a novella, *Polar City Red*, which, set in 2017 in Alaska and featuring a group of climate refugees, is an example of cli-fi. In 2013, several journalists, including Rodger Glass writing in *The Guardian* and Carolyn Kormann writing in *The New Yorker*, acknowledged the presence of a corpus of novels that '[set] out to warn readers of possible environmental nightmares to come' and, hence, an emergence of a new genre (Glass, 2013). In fact, the novels with some cli-fi traits had begun appearing as early as the 1950s (Kormann, 2013; Bell, 2021: 103). Among those are Arthur C. Clarke's *The Sands of Mars* (1951), John Christopher's *The Death of Grass* (1956), J.G. Ballard's *The Drowned World* (1962), Ursula K. Le Guin's *The Lathe of Heaven* (1972), and Arthur Herzog's *Heat* (1976). As exemplified by Christopher's novel, in which a virus kills off grass species including wheat and rice, causing a massive famine, early cli-fi novels were dominated by apocalyptic rhetoric intended on revealing what the climate apocalypse would look like and on purveying doom as a way of raising awareness (Buell, 2005: 18–20).

However, climate fiction has flourished only in the last twenty years, with renowned English-language novelists such as Margaret Atwood, Jeanette Winterson, Nathaniel Rich, Barbara Kingsolver, Annie Proulx, and Ian McEwan broaching the topic. For example, in Atwood's *The Handmaid's Tale* (1985), a region of North America is being threatened by pollution and environmental destruction. In *Oryx and Crake* (2003), which can be regarded as the earlier novel's sequel, these problems have escalated into a worldwide ecological disaster. Many of the novels by Kingsolver, who enjoys the reputation of 'a proponent of a strongly ecological worldview,' are preoccupied with the struggle of rural families and communities in conservative and deprived North America, and with 'the way in which such dramas play out against wider issues of environmentalism' (Johns-Putra, 2019: 156). Kingsolver's seventh and best-selling novel *Flight Behaviour* (2012) follows an unhappily married woman who, on the way to meet her lover, stumbles upon a swarm of monarch butterflies. Having been thrown off their customary migratory course by unsettled weather, the butterflies are being threatened by extinction. Younger authors have also approached climate change. This is evidenced by Rivka Galchen's *Atmospheric Disturbances* (2008), Sarah Hall's *The Carhullan Army* (2007), Sarah Moss's *Cold Earth* (2009), and Tim Winton's *Dirt Music* (2001) (Trexler, 2015: 9). Galchen's award-winning debut novel exposes climate change scepticism by following a character who confuses climate with the daily experience of weather anomalies.

The environmental crisis has also been the topic of other genres. This is illustrated by the passionate support of poets for the '10:10' campaign to reduce carbon emissions (Griffiths, 2017: 1) as well as by the publication of several volumes of verse and of a plethora of individual works (Agrawal, 2020; Simpson, 2021).[4] As for theatre and performance, Western theatrical imagination has been, according to Una Chaudhuri, 'haunted' by 'the ticking time bombs of ecological disaster' (1994: 23). For Lisa Woynarski, theatre and performance can indeed 'upend reductive narratives and images, embodying and performing contradictions, erasures and imaginative possibilities' (2020: 2). Jennifer Fay in turn believes that film is the ideal medium for conveying and warning against climate change. This is because cinema, which was born with the Anthropocene, occasions the creation of 'artificial worlds, unnatural and inclement weather, and deadly environments':

> The Anthropocene is to natural science what cinema . . . has been to human culture. It makes the familiar world strange to us by transcribing the dimensionalities of experience into celluloid, transforming and temporally transporting humans and the natural world into an unhomely image.
>
> (2018: 4, 3)

In fact, two films are often cited as the seminal representations of climate change (Perkowitz, 2010: 222–223). These are Roland Emmerich's hugely successful feature film *The Day after Tomorrow* (2004), which was watched in cinema by twenty-one million Americans, and Davis Guggenheim's documentary *An Inconvenient Truth* (2006), which features the former vice president Al Gore. Guggenheim's film won the Academy Award for Best Documentary Feature and has become the third-highest-earning documentary ever. As the present volume confirms, the environmental emergency has indeed been a burning concern across both literary genres and different art forms.

Some, however, still regret that so few writers should dramatise the environmental crisis (Kormann, 2013). Writing in 2019, Amitav Gosh noted that the ecological catastrophe had a significantly less conspicuous presence in literary fiction than in the media and non-fictional writings. Even authors who, like Paul Kingsnorth and Arundhati Roy, demonstrate their commitment to climate action, steer clear of the topic in their fiction. Gosh goes on to observe that climate change literature has not been treated seriously by critics and academics, who tend to relegate it to science fiction 'as though in the literary imagination climate change were somehow akin to extraterrestrials or interplanetary travel' (2019: 103). Another reason for the relatively small number of works that fit the rubric of cli-fi may be the resistance of conventional literary forms to the topic just as traditional historiography struggles to accommodate the reality of the Anthropocene (Chakrabarty, 2009: 197). This can be puzzling considering that for centuries literature has responded to all sorts of crises, including epidemics, wars, and natural disasters. In this case, Gosh concludes, our culture's failure to address the approaching environmental meltdown may be symptomatic of a broader failure of imagination. It could also be a sign of the enduring complicity of western culture with the forces of imperialism and capitalism, a complicity that manifests itself through the promotion by this culture of irresponsible lifestyles made up of reckless consumption and air travel.

If Gosh's concern about the underdevelopment of climate fiction and, by extension, of environmental literary criticism by comparison with gender or postcolonial studies would have been justified a few years ago (Buell, 2005: 1), the last two decades have seen the publication of several studies dedicated specifically to ecological literature. Apart from Gosh's own *The Great Derangement: Climate Change and the Unthinkable* (2017), these studies include Adam Trexler's *Anthropocene Fictions: The Novel in a Time of Climate Change* (2015), Antonia Mehnert's *Climate Change Fictions* (2016), Gregers Andersen's *Climate Fiction and Cultural Analysis: A New Perspective on Life in the Anthropocene* (2020), Adeline Johns-Putra's *Climate Change and the Contemporary Novel* (2019), Erin James and Eric Morel's edited volume *Environment and Narrative: New Directions in Econarratology* (2020), Pieter Vermeulen's *Literature and the Anthropocene* (2020), Nicole Rogers's *Law, Fiction and Activism in the Time of Climate Change* (2021), James Craig Holte's *Climate Change in Popular Culture: A Warming World in the American Imagination* (2022), and Erin James's *Narrative in the Anthropocene* (2022).

The close focus on fiction of the many of the listed studies has most likely been motivated by the novel's believed 'capacity to interrogate the emotional, aesthetic, and living experience of the Anthropocene' (Trexler, 2015: 8–9). According to Clark, the novel is particularly well suited to the topic of ecocatastrophe, as 'it is free to trace all imaginable scenarios and to survey how prejudice, personal background, cultural assumptions, scientific research and the complacencies of day-to-day life all form part of how people engage with or evade environmental questions' (2019: 78). Similarly, in Glass's opinion, 'finely constructed, intricate narratives help us broaden our understanding and explore imagined futures, encouraging us to think about the kind of world we want to live in' (2013). Finally, Andersen posits climate fiction as

> a vital supplement to the reports published by the United Nations' Intergovernmental Panel on Climate Change (IPCC) because, by depicting humans in worlds resembling those forecasts by the IPCC, climate fiction provides speculative insights into how it might be to feel and understand in such worlds. And these are basically insights we as contemporary humans cannot obtain anywhere else.
>
> (2020: 1)

The aims and rationale of the volume

The existing critical studies of climate fiction not only privilege a single genre, but usually limit their corpus to anglophone production. In contrast, *Storying the Ecocatastrophe* departs from a monographic approach in terms of the texts it analyses. Apart from the United States and the United Kingdom, the narratives examined in this volume come from Finland, Poland,

Norway, France, Luxembourg, the Asia-Pacific region, and the Democratic Republic of Congo. This cultural diversity, which has been emphasised by the editorial choice to include quotations in the original languages (as well as in the English translation), has been dictated by several factors. Firstly, ecocatastrophe has a global character, even if its symptoms vary depending on the location. To provide some examples, some parts of Africa have been gripped by severe drought, while Sudan and Nigeria have been experiencing flash floods which sweep away homes and businesses by the hundreds. Similarly, at the time when Australia, California, and Northern Europe were battling horrific forest fires, Southeast Asia was ravaged by typhoons and floods. Secondly, the multicultural perspective adopted by the present volume has been driven by the need to examine not only how the Anthropocene has affected different places around the world, but also how local communities have understood, symbolised, and responded (or failed to respond) to the adversity they face. These differences are hardly unexpected considering that even terms such as 'nature' and 'environment' are 'globally varied and contentious' (Clark, 2019: 137). Thirdly, behind the volume's multicultural scope is the fact that in different places of the planet, anthropogenic impact on nature has been entangled with other oppressive forces. Depending on the specific region, these forces may be patriarchy, racism, capitalism, hegemonic masculinity, (neo)colonialism, and social inequity. Fourthly and finally, by examining the work of authors coming from outside the anglophone zone, *Storying the Ecocatastrophe* contributes to the cultural decolonisation – or rather, de-Americanisation – of ecocriticism (Clark, 2019: 137).

As diverse as the origins of the narratives analysed by this volume are the genres these narratives represent. Apart from the realist novel, the individual chapters examine young adult (YA) fiction, the post-exotic novel, ecopoetry, popular science books, solarpunk, the economic report, photography, space opera, and dance performance. The rationale of such generic diversity is to foreground the intricate connection between the narrative form and the climate emergency dramatised by the texts examined in this collection of essays. According to Ursula Kluwick, 'the public perceptions of climate change are intrinsically tied to narrative strategies; whether or not we are aware of it, when we read and hear about climate change, we are offered not just facts and models but stories' (2014: 504). Analogically to the widely believed inadequacy of established narrative modes, such as the realist novel, to represent the horrors of the Holocaust, it is felt that conventional representational forms are unsuitable for storying climate change. This is because literary realism is as implicated in the culture and thinking that had led to the Nazi horror as it is historically tied up with 'the period when the accumulation of carbon in the atmosphere was rewriting the destiny of the earth' (Gosh, 2019: 103). This may explain why the climate crisis has often been told through genres such as dystopia, science fiction, speculative fiction, magic realism, and the thriller (becoming the eco-thriller). The impending ecological disaster indeed seems to call for formally innovative representational approaches; while telling engaging stories, which help to spur audiences into mitigating political action, narratives must be capable of documenting, analysing, and interpreting the environmental catastrophe. At the same time, they must not appear overly didactic as, to quote McEwan, 'Novels don't work if you badger people' (Barrett, 2010).

Arguing against Gosh, Erin James defends literary realism by positing that the traditional novel's third-person limited perspective may challenge the readers' own anthropocentrism. She supports her stand with an analysis of McEwan's satirical cli-fi novel *Solar* (2010), which casts an unsympathetic for profoundly immoral scientist involved in combatting climate change for all the wrong reasons. According to Markku Lehtimäki, who shares James's sympathetic position towards realist aesthetics, 'By making readers see the world through the protagonist's self-centred and objectifying vision, *Solar* . . . ask[s] to reflect on our own ecological preconceptions and "unexamined assumptions"' (2020: 93). In contrast, in his by now canonical study *The Environmental Imagination* (1995), Lawrence Buell expresses scepticism regarding the basic ingredients of narrative such as plot, characters, and viewpoint, which he considers to be anthropomorphic concepts (Lehtimäki, 2019: 86). Similarly, Dominic Head wonders whether the novelistic genre, with human action and human consciousness at its core, is capable of generating green ideas (Lehtimäki, 2019: 86). As an alternative to the conventional third-person novel, some critics, including Ursula Heise, advocate multiperspectival narration that 'can become a source of ecological (meta-)knowledge by conveying not any prior environmental "truth" but instead the divergences and contradictions of attitudes to environmental issues' (Zapf, 2016: 59). According to Zapf, such a narrative strategy

> create[s] a textual space for otherness that alone can break up the sermonizing monologues of missionary (eco-)politics and open the text for the active participation of the reader in the continual co-creation of those rational complexities which constitute ecological awareness and existence.
>
> (2016: 60)

If some chapters in this collection manifest their faith in the capacity of the traditional novel to evoke concern about the looming environmental disaster, others turn their attention to post-apocalyptic, postmodern, and post-exotic modes of writing. As we shall see, formally experimental texts have the potential to destabilise anthropocentric thinking predicated on the separation between the human and the non-human, and on the linked belief in humanity's moral and intellectual exceptionalism. They thus highlight the fact that the Anthropocene has called into question human-centred narratives which, regardless of their genre, can be accused of normalising and thus legitimising the destruction that we have wrought upon the natural world, not to mention various forms of interhuman oppression. For example, in contrast to anthropocentric narratives where non-human animals are traditionally used as allegorical figures serving some thematic purpose, the climate crisis has made us realise that its origins lie in our hierarchical understanding of the human/non-human relationship, demanding that animals should be represented as 'physical presences and material agencies' (Lehtimäki, 2022: 50). According to Clark, what is required is 'a demystification of the cultural sphere' which is based in 'modes of food production requiring mass servitude, imprisonment and slaughter,' and which endorses 'pastoral, hunting and religious practices implicated in animal suffering' (2011: 186).

As part of their commitment to such demystification, several of the essays in this volume point an accusatory finger at the Bible as a narrative that sanctions the humans' hubris and their consequent senseless exploitation of the natural world. These essays thus echo the question posed by American historian Lynn Townsend White Jr., who conjectured that the ecological crisis had its roots in the Christian influences of the Middle Ages. White argued that 'what people do about their ecology depends on what they think about themselves in relation to things around them,' and that 'By destroying pagan animism, Christianity made it possible to exploit nature in a mood of indifference to the feelings of natural objects' (1967: 1205). Indeed, some of the contributions to this collection demonstrate that stories about environmental collapse strive to overhaul our thinking not only about animals, but also about the vegetal world. As it is borne out by the texts that these contributions analyse, plants have been not only materially and ideologically exploited by the humans, but also rendered invisible by our culture. Some of these texts go as far as positing a climate change-related extinction of humanity as the only way of liberating the non-human world from anthropogenic dominion. In the post-apocalyptic scenarios that these narratives stage, plants and animals re-join well-balanced ecosystems and the surviving humans engage in non-hierarchical relationships with the non-human world and with each other. In other words, the threat or even the reality of the ecocatastrophe has the potential to enable more-than-human constellations, innovative social and political organisations, or a return to more traditional ways of life.

The discussions about the challenges that the ecological disaster presents to the narrative reveal the profound political engagement of the texts under analysis. This is not unexpected considering that, firstly, 'The environment and how we acquire, disseminate, and legitimate knowledge about it is highly politicized, reflective of relations of power, and contested' (Neuman, 2014: 1). Secondly, many critics strongly believe in the illuminating, inspirational, and transformative power of climate fiction; equating it with environmental activism, Clark, for example, states that the genre shares the eco-activists' 'faith that the public stage is the one of power and possible influence' (2011: 2). According to Lynn Keller, literature has indeed an important role to play; if environmental apocalyptic writing renders vividly the hellish places to which we seem headed and provides motivation to commit to necessary societal and individual change, a literature that goes beyond the apocalyptic 'can keep us on track towards genuine sustainability' (2012: 58). Keller's words find illustration in the influential contribution of literature to the debate about nuclear energy that erupted in West Germany following the Chernobyl disaster. Novels such as Christa Wolf's *Störfall* (1987) [*Accident*] and Gudrun Pausenwang's *Die Wolke* (1987) [*The Cloud*], both published the year following the accident,

> created a climate of public opinion, which, after the Fukushima disaster of 2011, finally led to a broad political consensus that forced even Angela Merkel's conservative party of the Christian Democrats to abandon its long-term belief in atomic energy and helped prepare the way for the opt out decision of 2012.
>
> (Zapf, 2016: 52–53)

Thirdly, 'ecological issues are inextricably intertwined with political and social issues' (Zapf, 2016: 55). Echoing Zapf's words, Clark notes that

> Human abuse of the world is best understood as the corollary of unjust or oppressive systems of government and economics, and forms of social organisation (hierarchy, plutocracy, patriarchy) that both abuse other human beings and which have no hesitation taking a similar stance towards anything else.
>
> (2019: 3)

Indeed, nearly all of the essays in the present collection highlight the inextricability of humanity's abuse of the environment and interhuman violence, and, by extension, of environmentalism and broader political concerns. One of the connections repeatedly made by the following chapters is that between humanity's masterful violence over the natural world and patriarchy. Such a connection is not entirely surprising given that '"nature" has been androcentrically constructed as a domain for males, in contradiction to female-coded domestic space, yet at the same time symbolically coded as female – an area of potential domination analogous to the female body' (Buell, 2005: 109). Feminism can therefore be, in Noël Sturgeon's view, the key to understanding the root of the environmental catastrophe: 'where women are degraded, nature will be degraded, and where women are thought to be eternally giving and nurturing, nature will be thought of as endlessly fertile and exploitable' (1997: 263). Read in an ecofeminist perspective, which has been deemed 'the most sophisticated and intellectually developed branch of environmental criticism' (Clark, 2011: 111), the texts studied in this volume liken humanity's extractive treatment of the vegetal, animal, and material world to sexual violation and cultural silencing of women. They also endow female characters with capacity for care in relation to both humans and non-humans, and with the vulnerability of (prey) animals. Consequently, they cast them as victims of, to borrow Jacques Derrida's term, carnophallogocentrism (Birnbaum et al., 2009).[5] To avoid inscribing the androcentric construction of women as passive, defenceless, and maternal, the analysed representations of women must remain self-conscious and/or subvert the patriarchal stereotypes they appear to endorse. The alternative would be to follow Stacy Alaimo's recommendation and co-opt 'nongendered tropes of nature that emphasize continuity between human and nature while still respecting nature's difference' and treat nature as a space that 'cannot be encompassed by, controlled by, or even entirely known by human culture' (2000: 183).

Apart from drawing attention to the status of both women and the nonhuman world as the marginalised and oppressed Other of man, ecofeminism, as Buell notes, is itself entangled in the fight for democratic and social justice. Just as, inspired by postcolonial thinking, in the 1980s and 1990s feminism abandoned its narrow focus on white Western women, during the first decade of the new millennium ecofeminists moved in a sociocentric direction. By this Buell means their growing engagement with 'the impoverished and socially marginalized: . . . landscapes of urbanization, racism, poverty, and toxification; and . . . the voices of witnesses and victims of environmental injustice' (2005: 112). Several of the contributors to this volume examine narratives that identify a causal relationship between environmental collapse and social inequity and undemocratic rule. As part of the post-apocalyptic scenarios they sketch, some texts project a return of rampant and violent capitalism following an era of social and gender equality. Others foresee a rise of right-wing quasi-fascist regimes which jealously and artificially guard the ethnic and cultural purity of a nation. They consequently show that combatting climate change effectively requires well-functioning democratic and environmentally committed institutions (Hanusch, 2017); a case in point of the correlation between improved climate performance and strong democratic qualities is Germany, whose successive administrations have proven that 'it *is* possible to act in the interest of a larger change of society and the economy toward environmental responsibility and sustainability' (Zapf, 2016: 58, original italics). Other texts studied in this volume illustrate Glenn Albrecht's concept of 'solastalgia,' which defines the feeling of 'loss of an endemic sense of place due to the negative impacts of global warming' (2019: x). They recognise instability, displacement, and degradation as inherent elements of human condition in a world menaced by ecological catastrophe. If some of these narratives are populated with climate refugees, others follow homeless women, men, and children, who endlessly wander through landscapes ravaged by extraction, industrial agriculture, and climate change-induced weather anomalies.

The other of the two major intersections repeatedly foregrounded by this volume is that between the continued human project to assert our will over the natural world, and the drive to dominate, enslave, exploit, and, ultimately, exterminate other humans. Several essays explore the enmeshment of environmentalism with the struggle against racial domination. Terry Gifford's opinion that 'our exploitation of our environment has emerged from the same mind-set as our exploration of each other' (2017: 84) finds embodiment in the discussion of the shared origins of genocide and ecocide, and the linked critique of the persistence of anthropocentric thinking in post-Holocaust reality. It is also reflected in the study of the Black Anthropocenic condition consisting in the (neo)colonial exploitation of the African people and of the natural resources of their homelands. In Clark's view, environmentalism and ecocriticism are inherently postcolonial, since the industrial development of western countries impinges in one way or another on the other parts of the world:

> The distinguishing mark of 'postcolonial environmentalism' has always been its focus on . . . an 'environmentalism of the poor,' on people being deprived of healthy living conditions by the forces of internal and external colonisation and exploitation, of migrants being driven into urban slums by the loss of the forest of wetlands that once sustained them.'
>
> (2019: 138)

Clark supports his position by qualifying the evils of slavery as both colonial/racist and environmental; for instance, the floods that swept through Haiti in 2024 leaving 2,600 people dead can be attributed largely to the 'catastrophic levels of deforestation' of the island, which began in colonial times and which continues today as the destitute descendants of the slaves rely on wood for fuel (Paravisini-Gebert, 2016: 69).

The intersectionality of the environmentalism of the contemporary narratives with feminist and postcolonial concerns engenders an interrogation not only of the concept of nature, whose understanding varies in different parts of the globe, but also of the idea of humanity as a cohesive construct.

While mapping androgenic and colonial oppression onto humanity's destructive domination of the environment, both ecofeminism and postcolonial ecocriticism expose the manifold internal rifts and tensions within mankind, which stem from identity markers such as class, sexuality, gender, and history (Chakrabarty, 2012: 2). Reflecting this situation, several of the twelve essays advocate a resistance to the practice of domination in all forms. To avoid further genocides and prevent ecological collapse, we must, as environmental philosopher Eric Katz urges, 'preserve the free and autonomous development of human individuals, communities, and natural systems. We must understand the moral limits of our power to control nature and our fellow human beings' (1997: 200). The examined narratives thus issue a warning to their audiences by projecting images of a world rendered uninhabitable by a nuclear disaster, of the collapse of democracy and rise of totalitarianism, and even of human extinction. Such apocalyptic, not to say defeatist rhetoric which, on its own, could engender 'inactivity or undemocratic hyper-activity' (Zapf, 2016: 56), is often accompanied by more uplifting messages. For instance, the environmental catastrophe is shown as resulting in a renewed sustainability of habitats, in a liberation of plants and animals from their centuries-old enslavement to humans, and in more harmonious and non-hierarchical relationships, be they interhuman or human/non-human. Some of the studied texts additionally articulate hope for avoiding or least mitigating the seemingly inevitable ecological collapse, whereby they incite their readers to action on both small and large scales. Such narratives, in Zapf's view, have the power of 'production and creativity . . . in the Nietzschean sense of the productive energies of a creative human being'; put differently, they have a transformative potential to 'subvert and enhance human consciousness and existence in rationalized human societies' (2016: 14).

The structure of the volume

The present collection opens with **Markku Lehtimäki**'s study of econarratology with reference to literary realism. While climate change has been the topic of science fiction, graphic novels, feature films, and documentaries, it has been rarely thematised by the traditional novel. Using Ian McEwan's *Solar* (2010) and Jenny Offill's *Weather* (2020) as examples, Lehtimäki investigates their capacity for narrating a crisis that is thought to exceed the narrow confines of human experientiality. The chapter proceeds to ask whether one can make sense of ecocatastrophe, both ethically and emotionally, through anthropocentric stories about ordinary people who grapple simultaneously with everyday concerns and the larger issue of the environmental collapse. That is to say, can daily problems of fictional characters be squared with anxieties provoked by global warming, rising sea levels, and extinction of species? Lehtimäki answers this question by suggesting that the juxtaposition of different levels of experience may in fact be the most effective way of increasing readerly awareness of the environmental crisis and of the consequent and urgent need of political action.

Continuing the inquiry into narrative strategies of climate fiction, **Georgiana Bozîntan** traces the evolution of Norwegian environmental literature from Knut Faldbakken's pioneering *Uår* novels published in the mid-1970s to Maja Lunde's *Bienes historie* (2015) [*The History of Bees*] and its sequels. Bozîntan discusses the shift from the doom-and-gloom scenarios staged by Faldbakken to the more optimistic and solution-oriented, yet at times didactic, contemporary production where, as she posits, didacticism may in fact instil a sense of hope in the readers. The chapter's key aims are, first, to demonstrate the positive role of readerly empathy generated by Lunde's novels, and, second, following scholars such as Lehtimäki, James, and Heise, to argue for the usefulness of econarratology in illuminating the capacity of climate fiction to move audiences to concrete political action aimed at easing the consequences of the climate crisis. Bozîntan concentrates on the potential of processes such as transportation, character identification, and empathy to shape readers' experience of the storyworld and, more pertinently, to foster their understanding of climate change and its large-scale and long-term ramifications. The chapter closes by identifying the recent Norwegian climate fiction, as represented by Lunde's novels, as an example of ethical storytelling and as a more effective tool in the fight against climate change than its earlier examples.

With recourse to econarratology and 'eco-nomics,' defined by Trexler as an entanglement of cultural, geopolitical, ecological, and economic concerns (2015: 170–222), **Xin Liu** addresses the use of the language of economics in discourses surrounding the encroaching ecological collapse. Rather than resolutely dismissing narratives that speak of ecology in economic terms and of the economy with reference to ecology, Liu recommends we subject these narratives to a close reading. Her analysis of three texts – two popular science books and an economic report – demonstrates that putting a value on ecosystems can foster a better understanding of the human-environment entanglement and of the urgent need to change humanity's practices in relation to nature. Liu's discussion focuses specifically on the narrative techniques of spatialisation and defamiliarisation, where the former enables readers to see the value of nature as comparable to their own lived reality, and the latter resensitises them to the true value of the environment which is figured as a plane, a spacecraft, an individual home, or the Moon. Liu's chapter reveals that, as well as making plain the economic consequences of environmental change, the texts she considers ask vital questions about the ecosystemic and economic nature of the human body. They also bring to the fore the ways in which the non-linearity and uncertainty of nature call for an economic valuation that is also ecological, and thus draw attention to the enmeshment of the defining parameters of environment and economy.

Shifting the volume's focus from the aesthetics of environmental narratives to their politics, **Katarina Leppänen**'s chapter addresses a futuristic representation of post-catastrophic Finland in *De hemlösas stad* (2011) [*The City of the Homeless*]. The untranslated YA novel by Finland-Swedish author Annika Luther opens after the country's capital has been moved from Helsinki (renamed in the novel Halsingih) to Jyväskylä, and closes with the new capital's destruction by a nuclear catastrophe. The novel, as Leppänen posits, is structured by the tension between Halsingih, which has been partly flooded and taken over by climate refugees from, among other places, India, and Jyväskylä, where Finns live a highly organised and controlled life. Drawing on Henri Lefebvre's theorisation of the production of space as both a national and global process, Leppänen discusses how the government's repressive measures are intended to halt the influx of immigrants who are, however, transforming the country into a more colourful, cosmopolitan, and bohemian society. Without romanticising this new cultural heterogeneity, Luther satirises the anti-immigrant agenda of the right-wing movements in present-day Finland by staging Jyväskylä as an artificially homogenic and quasi-carceral community. As for Helsinki/Halsingih, by focusing on a city that, familiar to some readers, is recognisable to others through a series of ubiquitous political and religious institutions, Luther hopes, as Leppänen argues, to incite her audience to assume greater environmental responsibility both for their own living space and globally. Without reducing Luther's novel to ecological didacticism, Leppänen believes that its not-all-negative representation of the post-catastrophic world has the power to encourage readers to take environmental action.

The images of rising water levels and flooding are also central to **Faeze Rezaii**'s chapter, which scrutinises the representation of water planets in two films – *Star Wars Episode II – Attack of the Clones* (2002) *and Interstellar* (2014). The chapter mobilises two water-related optical phenomena: 'reflection,' which serves to demonstrate that the films endlessly reproduce images of humans stuck in anthropocentric logic, and 'diffraction,' which Rezaii uses to isolate the rare instances of a departure from anthropocentric agency. Starting with the premise that futuristic climate imaginaries reflect the way we see ourselves as agents in the natural world, Rezaii proposes that, in the two films, the planets Kamino and Miller are projections of Earth affected by global warming. Such a representation of the planets comes across as a critique of the late capitalist drive towards 'limitless expansion and permanent growth without end,' which translates into 'futurological imaginings, to orbital space stations, lunar cities, Martian settlements, asteroid belt mining colonies, sleeper ships to Alpha Centauri, and on and on' (Canavan, 2014: 5). That said, Rezaii argues that, by showing Kamino and Miller to be beset by extreme meteorological events such as floods and tsunamis, the films play down human agency. They suggest consequently that the largely anthropogenic ecocatastrophe has little to do with our actions and that it is beyond our control. Such a figuration of water planets, Rezaii concludes, supports the films' ideological objective of naturalising the environmental collapse, glossing over human responsibility for it and, ultimately, maintaining the status quo.

The theme of the nuclear catastrophe present on the margins of Leppänen's discussion is more fully explored by the following two chapters. Focusing on the award-winning novel *Terminus radieux* (2014) [*Radiant Terminus*] by Antoine Volodine, **Helena Duffy** probes the political investment of the French writer's representation of a post-catastrophic world. To test the novel's apparent condemnation of capitalism and nostalgia for communism, Duffy analyses its critical engagement with the apocalyptic narrative model, as famously theorised by Mark Kermode, and its critique of political teleologies that, drawing on the apocalyptic model, dominated the Cold War era. More specifically, *Radiant Terminus* exposes the fallacy of both Francis Fukuyama's concept of the 'end of history', which predicted the lasting triumph of liberal democracy and market economy, and the propagandist construct of 'radiant future,' which promised Soviet citizens a happy and prosperous tomorrow in a communist idyll. Duffy's discussion of Volodine's scepticism towards ideologies, whether they belong to the Left or the Right, is followed by an examination of his novel's awareness of the destructive impact that these ideologies have had on the environment. A close reading of Volodine's representation of steppes and forests, which, although scarred by the human dream of socio-technological progress, are beginning to regenerate in humanity's near-absence, reveals the French author's efforts to counter plant blindness, to invest the natural world with subjectivity and agency, and to expose the violence underlying our cultural constructions of the environment. Duffy closes her chapter by proposing that *Radiant Terminus* identifies a dramatic decrease in human population caused by ecological disaster as the only hope for a liberation of plants and animals from the human yoke and for a re-establishment of well-balanced ecosystems.

Also centred on post-nuclear scenarios, **Sébastian Thiltges**'s chapter considers two examples of anticipatory fiction recently published in Luxembourg, a small landlocked country with one of the world's highest incomes per capita and with an exceptionally high carbon footprint. Yorick Schmit's Luxembourgish-language short story 'Eng Stëmm an der Stëllt' (2018) [A Voice in the Silence] and Pierre Decock's French-language novel *Luxembourg Zone rouge* (2019) [*Luxembourg Red Zone*] are both set in Luxembourg in the aftermath of a nuclear catastrophe. Due to the well-established connection between the nuclear and ecology, the nuclear catastrophe, which is an ecological disaster in its own right, can be read as a metaphor for a global environmental collapse. Luxembourg, as imagined by Schmit and Decock, therefore becomes a model for a world destroyed by humanity's reckless behaviour. Yet, the firm rooting of the two narratives in the local topography and landscape enables their authors to make Luxembourgish readers better envisage the consequences of environmental disaster with its inevitable result: the decimation of the population and the displacement of the rare survivors. That said, Thiltges's reading of the two texts demonstrates that an environmental catastrophe can potentially engender humanity's more empathetic and less hierarchical relationship with the environment.

While also advocating a connectivity across species, **Elizabeth Tavella**'s analysis of short stories collected in *Multispecies Cities: Solarpunk Urban Futures* (2021) and Tamiko Beyer's volume of poetry *Last Days* (2021) criticises the apocalyptic narrative model which has largely structured environmentalist discourses, including cultural ones. Tavella blames this model not only for upholding the unproductive ontological split between Humanity and Nature, but also for the dominant defeatist response to the ecological disaster. Contrary to dystopian tales, the texts scrutinised by Tavella are capable of arousing hope for social and ecological healing. They also question the very premises of modern western societies – white supremacy, heteropatriarchy, corporate power, and speciesism – and recommend an active investment in the future whose course we are still able to change. Relying on abolitionist frameworks, urban ecology, and multispecies studies, Tavella's examination of narratives within the rubric of ecopoetry and solarpunk (an online subgenre of climate fiction that brings together issues of race, gender, sexuality, class, and ecology) foregrounds these texts' critique of anthropocentric, expansionist, and capitalist logic which is to blame for social inequity and the destruction of the environment. Instead, the analysed short stories, set in and written by authors from the Asia-Pacific region, stage multispecies inclusivity and non-human agency and personhood. They problematise the commodification of the animals that are considered as food and whose enslavement, abuse, and killing are sanctioned within the speciesist logic. Similarly, Beyer's 'queer::eco::poetics' (Beyer, 2010) encourage readers to reflect on their own implication in racism, environmental destruction, and other forms of violence, while advocating trans-species interconnectedness and solidarity. Needless to say, Tavella applauds the forward-looking aesthetics and solution-oriented approaches to socio-ecological adversity embodied by the examined examples of speculative writing.

Chloé Bour--Lang's chapter, which shares with Tavella's contribution its focus on humanity's destructive and controlling impact on the animal and vegetal worlds, offers an ecofeminist reading of Barbara Kingsolver's *Prodigal Summer* (2000). Zooming in on the novel's male characters, Bour-Lang demonstrates an interdependency between their sexism and abusive attitude towards the natural world, and their consequent responsibility for the socio-ecological crisis that plagues the United States. As represented by *Prodigal Summer*, contemporary America is simultaneously patriarchal and ecocidal, which means that both nature and women are the Other of the male subject. Kingsolver conveys her message, as Bour-Lang contends, by embracing the male perspective before identifying it as destructive. Without altogether dismissing American culture, Kingsolver's didactic yet upbeat novel proposes, as Bour-Lang demonstrates with reference to Carol Gilligan's concept of the ethics of care, this culture's transformation through an espousal of behaviours traditionally perceived as feminine. As well as caring, female characters are cast as capable of crossing boundaries between realms largely believed to be unreconcilable and even as sensorially attuned to the supernatural.

Relying also on ecofeminism, in her discussion of the work of two contemporary Polish artists, **Katarzyna Bojarska** relates humanity's dominance and oppression of nature to the Nazi genocide of Europe's Jews and other instances of intrahuman violence. She demonstrates that the creative and critical practice of Agata Siniarska and Karolina Grzywnowicz aims to 'change the story,' that is to unsettle anthropocentric history in the post-Holocaust world. The interhuman and human/non-human violence witnessed over the last eight decades shows that, on the one hand, we have not learnt from our past mistakes and that, on the other, human-focused history is an untenable construct. To counter this trend, Siniarska's ecofeminist and ecological work expands agency to non-human characters and intertwines narratives of the degradation of nature with those of the exploitation of women. Her project *Second Nature* uses the work of Judeo-Polish dancer, choreographer, and Holocaust survivor Pola Nireńska, as a springboard for a meditation on the entanglements of past and present violence, including our abuse of the environment. Alluding to Eric Katz's analogy between genocide and ecocide, Grzywnowicz's installation (which provided the setting for Siniarska's dance performance), consists entirely of plants that the Nazis exploited for their genocidal ends.

The work of another female author, Annie Proulx, is the subject of **Hannah Jocelyn**'s chapter which posits the American writer's novels as representative of 'deep-landscape literature'. As opposed to 'ecolit,' a genre Jocelyn describes as a cultural artefact associated mainly with male writers, deep-landscape literature engages readers, both critically and actively, in the author's ecological agenda, even if, as in Proulx's case, this agenda may be implicit rather than explicit. Proulx's novels stage the disastrous consequences of humanity's exploitative treatment of nature, including uprooted people, decrepit and/or impermanent homes, and landscapes devastated by intensive farming. The images of homelessness and dilapidation are meant, as Jocelyn argues, as a warning against further environmental damage resulting from our rapacious consumption. They can also be connected to 'solastalgia' which, following Albrecht, Jocelyn defines as a sense of loss provoked by the wretched state of one's home and territory. By making her readers experience solastalgia, even if only vicariously, Proulx intends to activate them environmentally.

Even more acutely than in Proulx's America, the effects of colonialism, humanity's reckless exploitation of the environment, and the consequent climate change are felt in Africa, as portrayed by two artworks created in the Democratic Republic of Congo. **Spring Ulmer**'s closing chapter reminds us that the African continent which is responsible for only between 2 and 3 per cent of global energy-related carbon dioxide emissions, has been disproportionately affected by the anthropogenic warming trend: temperatures have risen more than the global average and, due to the exceptional rise in sea levels, the frequency and severity of coastal flooding and erosion have increased. The effects of the Capitalocene, which is 'the geological age during when the maximization of economic profit has made it to the centre stage . . . and dominates whatever other ages have been proclaimed, including the Anthropocene' (Rieger, 2022: 2), can be gleaned from the Congolese works that Ulmer studies through a transcultural materialist lens. Although very different in form, Fiston Mwanza Mujila's award-winning novel *Tram 83* (2014) and Sammy Baloji's photographic series *The Beautiful Time* (2010) both stage the Black Anthropocenic condition which combines genocidal suffering with ecological degradation (Yusoff, 2018). The writer and the photographer thematise the violence wrought by colonial settlers on the human and the non-human through the extraction and theft of natural resources and through the commodification of the bodies of Black slave labourers and prostitutes. But *Tram 83* and *The Beautiful Time* go beyond isolating extractivism as the cause of climate chaos in the African continent; reframing the environmental crisis as a political one, they also show that colonial exploitation continues through neocolonial domination. To step away from these damaging practices, to resist Black Anthropocenes, and to reclaim precolonial memory, Mujila and Baloji recommend, as Ulmer showcases, a revival of African cosmology which would foster harmonious relationships between humans and the non-human world, including the metal and stone that have been coveted by colonisers and global capitalists alike.

Notes

1 The term 'Anthropocene' was first introduced in 2000 by a group of geologists led by Paul Crutzen. According to its proponents, human activity has borne such impact on the Earth that it was necessary to declare a new epoch. This epoch manifests itself through the increase in the anthropogenic emission of greenhouse gasses, the disappearance of tropical rain forests, the increase in human population and human control over Earth's surface, and the transformation of waterways through dams (Trexler, 2015: 6).
2 According to its website, Extinction Rebellion is 'an international movement that uses non-violent civil disobedience in an attempt to halt mass extinction and minimise the risk of social collapse' (Extinction Rebellion). Insulate Britain is an organisation campaigning for a national programme that would ensure homes are insulated to be low energy by 2030.
3 Since 1995, the summit has been held in a different country every year, except in 2020 and 2021 due to Covid.
4 For more information on the 10:10 campaign, see Katz (2009).
5 The term adds carnivorism to aggressive manhood and rationality construed by Derrida as fundamental preconditions of western metaphysics.

Bibliography

Agrawal, Vinita (2020), *Open Your Eyes: An Anthology of Climate Change: Poetry and Prose* (Calcutta: Hawakai).
Alaimo, Stacy (2000), *Undomesticated Ground: Nature as Feminist Space* (Ithaca: Cornell University Press).
Albrecht, Glenn A. (2019), *Earth Emotions: New Words for a New World* (Ithaca: Cornell University Press).
Andersen, Gregers (2020), *Climate Fiction and Cultural Analysis: New Perspectives on Life in the Anthropocene* (New York: Routledge).
Barrett, Brendan F.D. (2010), '*Solar*: A Climate and Energy Novel', *Our World*, 21 June, [Online], [Accessed 10 January 2023], available from: https://ourworld.unu.edu/en/solar-a-climate-energy-novel
Bartov, Omer (2023), 'What Is the Environmental History of the Holocaust?', *Journal of Genocide Research*, 3, 419–428, https://doi.org/10.1080/14623528.2021.1924587
Bell, Matt (2021), 'Climate Fictions: Future-Making Technologies', in Jeffrey Jerome Cohen and Stephanie Foot (eds.), *The Cambridge Companion to Environmental Humanities* (Cambridge: Cambridge University Press), 100–112.
Beyer, Tamiko (2010), 'Notes Towards a Queer::Eco::Poetics', *Doveglion Lit*, 29 November, [Online], [Accessed 28 September 2022], available from: doveglion-lit.wordpress.com/2010/11/29
Birnbaum, Daniel, Anders Olsson and Jacques Derrida (2009), 'An Interview with Jacques Derrida on the Limits of Digestion', *e-Flux Journal*, 2 January, [Online], [Accessed 19 June 2012], available from: www.e-flux.com/journal/02/68495/an-interview-with-jacques-derrida-on-the-limits-of-digestion/
Buell, Lawrence (1995), *The Environmental Imagination* (Cambridge, MA: Harvard University Press).
Buell, Lawrence (2005), *The Future of Environmental Criticism: Environmental Crisis and Literary Imagination* (Oxford: Blackwell).

Canavan, Gerry (2014), 'Introduction: If This Goes On', in Gerry Canavan and Kim Stanley Robinson (eds.), *Green Planets: Ecology and Science Fiction* (Middletown: Wesleyan University Press), 1–21.
Carrington, Damian (2022), 'Climate Endgame: Risk of Human Extinction "Dangerously Underexplored"', *The Guardian*, 1 August, [Online], [Accessed 28 September 2022], available from: www.theguardian.com/environment/2022/aug/01/climate-endgame-risk-human-extinction-scientists-global-heating-catastrophe
Chakrabarty, Dipesh (2009), 'The Climate of History: Four Theses', *Critical Inquiry*, 35:2 (Winter), 197–222, https://doi.org/10.1086/596640
Chakrabarty, Dipesh (2012), 'Postcolonial Studies and the Challenge of Climate Change', *New Literary History*, 43, 1–18, https://doi.org/10.1353/nlh.2012.0007
Chakrabarty, Dipesh (2021), 'The Chronopolitics of the Anthropocene: The Pandemic and Our Sense of Time', *Contribution to Indian Sociology*, 55:3, 324–348, https://doi.org/10.1177/00699667211065081
Chaudhuri, Una (1994), '"There Must Be a Lot of Fish in That Lake": Towards an Ecological Theatre', *Theatre*, 25:1, 23–31, https://doi.org/10.1215/01610775-25-1-23
Clark, Timothy (2011), *The Cambridge Introduction to Literature and the Environment* (Cambridge: Cambridge University Press).
Clark, Timothy (2019), *The Value of Ecocriticism* (Cambridge: Cambridge University Press).
Connolly, Kate (2023), 'Climate Activists "Prepared to Risk Lives" to Stop German Coalmine', *The Guardian*, 10 January, [Online], [Accessed 10 January 2023], available from: www.theguardian.com/world/2023/jan/10/climate-activists-prepared-risk-lives-stop-german-coalmine-lutzerath
de Hoog, Niels and Ashley Kirk (2023), 'Twelve European Countries Broke Temperature Records in 2022', *The Guardian*, 10 January, [Online], [Accessed 10 January 2023], available from: www.theguardian.com/world/2023/jan/10/twelve-european-countries-broke-temperature-records-in-2022
de Steiguer, Edward J. (2006), *The Origins of Modern Environmental Thought* (Tucson: The University of Arizona Press).
Extinction Rebellion, [Online], [Accessed 1 December 2022], available from: https://extinctionrebellion.uk/the-truth/about-us/
Fanjul, Sergio (2022), '"Marxists Are Brainwashing Us": The Conspiracy Theory Taking Hold Among Some on the Right', *El País*, 5 July, [Online], [Accessed 1 January 2023], available from: https://english.elpais.com/society/2022-07-05/marxists-are-brainwashing-us-the-conspiracy-theory-taking-hold-among-some-on-the-right.html
Fay, Jennifer (2018), *Inhospitable World: Cinema in the Time of the Anthropocene* (Oxford: Oxford University Press).
Gifford, Terry (2002), 'Gary Snyder and the Post-Pastoral', in J. Scott Bryson (ed.), *Ecopoetry: A Critical Introduction* (Salt Lake City, UT: University of Utah Press), 77–87.
Glass, Rodge (2013), 'Global Warning: The Rise of "Cli-fi"', *The Guardian*, 11 December, [Online], [Accessed 11 December 2022], available from: www.theguardian.com/books/2013/may/31/global-warning-rise-cli-fi
Gosh, Amitav (2016), *The Great Derangement: Climate Change and the Unthinkable* (Chicago: The Chicago University Press).
Gosh, Amitav (2019), 'The Great Derangement', *Log*, 47 (Fall), 101–105.
Griffiths, Matthew (2017), *New Poetics of Climate Change: Modernist Aesthetics for a Warming World* (London: Bloomsbury).
Hanusch, Frederic (2017), *Democracy and Climate Change* (New York: Routledge).

Holte, James Craig (2022), *Climate Change in Popular Culture: A Warming World in the American Imagination* (Santa Barbara, CA: ABC-CLIO).
James, Erin (2022), *Narrative in the Anthropocene* (Columbus: Ohio State University Press).
James, Erin and Eric Morel (2020), *Environment and Narrative: New Directions in Econarratology* (Columbus: Ohio State University Press).
Johns-Putra, Adeline (2019), *Climate Change and the Contemporary Novel* (Cambridge: Cambridge University Press).
Katz, Eric (1997), *Nature as Subject: Human Obligation and Natural Community* (Lanham: Rowman & Littlefield).
Katz, Ian (2009), '10:10: What's It All about?', *The Guardian*, 1 September, [Online] [Accessed 1 September 2022], available from: www.theguardian.com/environment/2009/sep/01/10-10-campaign-explained
Keller, Lynn (2012), 'Beyond Imagining, Imagining Beyond', *PMLA*, 127, 579–585, https://doi.org/10.1632/pmla.2012.127.3.579
Kormann, Carolyn (2013), 'Scenes from a Melting Planet: On the Climate-Change Novel', *The New Yorker*, 3 July, [Online] [Accessed 1 September 2022], available from: https://www.newyorker.com/books/page-turner/scenes-from-a-melting-planet-on-the-climate-change-novel
Kluwick, Ursula (2014), 'Talking about Climate Change: The Ecological Crisis and the Narrative Form', in Greg Garrard (ed.), *The Oxford Handbook of Ecocriticism* (Oxford: Oxford University Press), 502–516.
Lehtimäki, Markku (2019), 'Narrative Communication in Environmental Fiction: Cognitive and Rhetorical Approaches', in Scott Slovic, Swarnalatha Rangarajan and Vidya Sarveswaran (eds.), *Routledge Handbook of Ecocriticism and Environmental Communication* (New York: Routledge), 84–97.
Lehtimäki, Markku (2020), 'A Comedy of Survival: Narrative Progression and the Rhetoric of Climate Change in Ian McEwan's *Solar*', in Erin James and Eric Morel (eds.), *Environment and Narrative: New Directions in Econarratology* (Columbus: Ohio State University Press), 87–101.
Lehtimäki, Markku (2022), 'Polar Bears and Butterflies: Allegory, Science, and Experientiality in Climate Change Fiction', in Matthias Stephan and Sune Borkfelt (eds.), *Literary Animal Studies and the Climate Crisis* (Basingstoke: Palgrave Macmillan), 49–70.
Mehnert, Antonia (2016), *Climate Change Fiction: Representations of Global Warming in American Literature* (Basingstoke: Palgrave Macmillan).
Meredith, Sam (2021), '"COP Is a Failure": Greta Thunberg Says Climate Summit Has Turned into a PR Event', *CNBC*, 5 November, [Online], [Accessed 5 January 2023], available from: www.cnbc.com/2021/11/05/greta-thunberg-says-cop26-climate-summit-is-a-failure-and-a-pr-event.html
Morens, David, Peter Daszak, Howard Market and Jeffrey K. Taubenberger (2020), 'Pandemic COVID-19 Joins History's Pandemic Legion', *mBio*, 11:3 (May–June), [Online], [Accessed 10 January 2023], available from: www.ncbi.nlm.nih.gov/pmc/articles/PMC7267883/
Neuman, Rod (2014 [2005]), *Making Political Ecology* (New York: Routledge).
Normington, Gregory (2013), *Beacons: Stories for Our Not So Distant Future* (London: Oneworld).
Paravisini-Gebert, Lizabeth (2016), 'Gade nan mizé-a tonbe: Vodou, the 2010 Earthquake, and Haiti's Environmental Catastrophe', in Chris Campbell and Michael Niblett (eds.), *The Caribbean: Aesthetics, World-Ecology, Politics* (Liverpool: Liverpool University Press), 63–80.

Perkowitz, Sydney (2010), *Hollywood Science: Movies, Science, and the End of the World* (New York: Columbia University Press).
Quammen, David (2012), *Spillover: Animal Infections and the Next Human Pandemic* (New York: W.W. Norton).
Rannard, Georgina (2022), 'Cop27: War Causing Huge Release of Climate Warming Gas, Claims Ukraine', *BBC News*, 14 October, [Online], [Accessed 14 October 2022], available from: www.bbc.co.uk/news/science-environment-63625693
Reicher, Stephen (2023), 'More Conflict, More Violence: That's the Future If the UK's New Anti-protest Law Is Passed', *The Guardian*, 19 January, [Online], [Accessed 19 January 2023], available from: www.theguardian.com/commentisfree/2023/jan/19/conflict-violence-anti-protest-law
Rieger, Joerg (2022), *Theology in the Capitalocene: Ecology, Identity, Class and Solidarity* (Minneapolis: Fortress Press).
Rogers, Nicole (2021), *Law, Fiction and Activism in the Time of Climate Change* (New York: Routledge).
Scheckner, Jesse (2023), 'Lauren Boebert to Speak at Heartland Institute Conference on Climate Change in Orlando', *Florida Politics*, 18 January, [Online], [Accessed 19 January 2023], available from: https://floridapolitics.com/archives/582008-lauren-boebert-to-speak-at-heartland-institute-conference-on-climate-change-in-orlando/
Simpson, Kate (2021), *Out of Time: Poetry from the Climate Emergency* (Scarborough: Valley Press).
Snyder, Timothy (2015), *Black Earth: The Holocaust History and Warning* (London: Bodley Books).
Sturgeon, Noël (1997), 'The Nature of Race: Discourses of Racial Difference in Ecofeminism', in Karen J. Warren (ed.), *Ecofeminism: Women, Culture, Nature* (Bloomington: Indiana University Press).
Syal, Rajeev (2022), 'Deportation Dreams and Tofu-eating Threats: Who Is Suella Braverman?', *The Guardian*, 19 October, [Online], [Accessed 19 October 2022], available from: www.theguardian.com/politics/2022/oct/19/who-is-suella-braverman-home-office
Thorpe, David and Dan Bloom (n.d.), 'Interview: Dan Bloom on Clifi and Imagining the Cities of the Future', *Smartcitiesdive*, [Online], [Accessed 1.12.2022], available from: www.smartcitiesdive.com/ex/sustainablecitiescollective/interview-dan-bloom-clifi-and-imagining-cities-future/1037731/
Trexler, Adam (2015), *Anthropocene: The Novel in a Time of Climate Change* (Charlottesville: University of Virginia Press).
Vermeulen, Pieter (2020), *Literature and the Anthropocene* (New York: Routledge).
White, Lynn, Jr. (1967), 'The Historical Roots of Our Ecological Crisis', *Science*, 155:3767, 1203–1207.
Woynarski, Lisa (2020), *Ecodramaturgies: Theatre, Performance, and Climate Change* (Basingstoke: Palgrave Macmillan).
Yusoff, Kathryn (2018), *A Billion Black Anthropocenes or None* (Minneapolis: University of Minnesota Press).
Zapf, Hubert (2016), *Literature as Cultural Ecology: Sustainable Texts* (London: Bloomsbury).

1 Daily life and global crisis

Human experience and narrative fiction in the age of the Anthropocene

Markku Lehtimäki

The dominant modes of storying climate change have been individual survival stories, utilising conventional narrative strategies in the face of what exceeds the human scale. Storification is, however, problematised in the context of the current environmental crisis because narrative is linked to human experientiality and fails to represent broad timescales and non-human agencies. In his book *Anthropocene Fictions* (2015), Adam Trexler suggests that other modes and media 'lack the novel's capacity to interrogate the emotional, aesthetic, and living experience of the Anthropocene' (2015: 6). Following Trexler, we can ask what the large-scale complexities and implications of climate change (which Trexler uses interchangeably with the Anthropocene) do to conventional narrative forms with their focus on human experientiality. Indeed, he adds that 'climate change necessarily transforms generic conventions' and that climate change fiction changes the parameters of storytelling (Trexler, 2015: 14).

In this chapter, I suggest that well-known climate change novels such as Ian McEwan's *Solar* (2010) and Jenny Offill's *Weather* (2020) are more or less anthropocentric narratives which deal with human concerns in the changing global environment. What especially interests me here are the different ways in which *Solar* and *Weather* aim at narrativising the experience of climate change. McEwan's novel does this more through a conventional character- and plot-centred narrativity, while Offill's novel employs a collage-like fragmentary narration. Obviously, climate change and its various environmental consequences are difficult to make sense of in human terms precisely because they radically exceed the human scale. While it may be possible to imagine and represent the effects of climate change on human experience, the phenomenon itself may escape representation because of its size and complexity. Accordingly, it may be argued that stories focusing on human experience are necessarily limited in their grasp of an environmental issue of global proportions. Yet it is mainly through narratives – and perhaps especially fictional ones – that the complex and abstract problem of climate change can be made both ethically and emotionally felt.

DOI: 10.4324/9781032726953-2

In what follows, I discuss the ways in which McEwan employs both narrative comedy and embodied consciousness in order to approach the experience of climate change. *Solar* can be read as a satire with a comic and failed Everyman, physicist Michael Beard, as its anti-hero, trying to find scientific solutions to solar energy and climate change. In Offill's *Weather*, the first-person narrator, Lizzie Benson, a librarian on a university campus, is forced to adjust her daily life to the apocalyptic dimensions of climate change. She gradually understands that in the age of the Anthropocene, we humans simultaneously inhabit multiple levels of experience, from the most mundane daily routines to visions of the planet in peril. In these two novels, the reader is able to sense the simultaneity of daily life and global crisis, whereas the characters of the narratives are only slowly realising the truly global dimensions of climate change. Still, in each narrative the main character gradually grows and transforms to take part in climate action, and this rhetorical move in the narratives may also be interpreted as a call for readers to reflect on their own ecological awareness.

Narrating climate change

While a popular topic in science fiction, fantasy, graphic novels, feature films, and documentaries, climate change has been an issue rarely grappled with in the mode of a conventional novel. McEwan's *Solar* and Offill's *Weather* still fall into the category of the realist novel despite their allegorical dimensions, in the sense that, as Patrick Murphy has it, they are 'set in the here, the now, and the local' (2014: 158). McEwan's and Offill's novels thus aim to narrativise the complex phenomena of climate change from the grassroots level of human experience and daily life.

Therefore, the representation of climate change is not only a matter of imagination and dystopia in the mode of fantasy or young adult fiction, even if those popular genres visibly dominate in the cli-fi market. In Offill's *Weather*, the narrator's brother likes films in which there is 'always some great disaster about to happen and one unlikely person who can stop it' (2020: 11). Indeed, 'In those disaster movies, the hero always says "Trust me"' (2020: 195). In contrast, the novels discussed in this chapter – and there are obviously similar cases in addition to these two – try to dramatise the effect of climate change on the daily lives of ordinary people with whom the reader can more or less identify. Accordingly, *Solar* and *Weather* deal with the very human problem of trying to live on two very different planes: on the one hand, there are the 'internal' affairs of family life, work, and everyday needs, and on the other, there are 'external' issues of the changing global world in the age of the Anthropocene, including climate change and its complex consequences. For instance, Dellarobia Turnbow, the protagonist of Barbara Kingsolver's much-discussed climate change novel *Flight Behavior* (2012), feels that her 'domestic drama' (2012: 124), which includes inglorious daily routines with her unimaginative husband and her two small children, soon gets a new perspective in large-scale global phenomena such as climate change.

The novels discussed in this chapter also show how human characters are embodied consciousnesses and suggest how our mental images are firmly bound up with our bodily experience of the world and with our physical perception of things. In effect, as I argue, human minds are always embedded in a particular environmental context. In what follows, I will focus on the ways in which the textual designs of the two novels try to capture the sense of changing ecologies. I will tentatively ask how a focus on the novelist's exploration of ecological issues can be complemented with scrutiny of the narrative designs by means of which the novelists discussed in the chapter engage with those issues. As James Phelan argues in *Experiencing Fiction*, 'texts are designed by authors in order to affect readers in particular ways' (2007: 4). And while the textual design of *Solar* differs from that of *Weather*, the affective reading experience of these two climate change novels is quite similar.

Whatever their emphasis on natural phenomena and the looming eco-catastrophe, the novels discussed here are stories of human experiences and perceptions of climate change. My approach could be characterised as a mode of humanist ecocriticism or, to be more precise, as an ecologically oriented non-anthropocentric new humanism, with its links to literary theory and moral philosophy (see Iovino, 2010: 32). As Kate Rigby notes,

> an acknowledgement of the centrality of the human actant, however contingent, contextualised, and decentered she might be in herself, is also a necessary condition for there to be such a thing as literature, as commonly understood, along with almost all other kinds of artistic endeavour.
>
> (2004: 427)

It should also be noted that Rigby bases her approach partly on Dominic Head's (1998) model of the ecological text and ecocritical operation, which recentres the human subject (instead of decentring it). Indeed, in his influential essay 'The (Im)possibility of Ecocriticism,' Head suggests a mode of weak anthropomorphism, according to which human action – for good or ill – needs to be at the core of both ecological thought and literary politics (1998: 29). We should note, of course, that both Head's and Rigby's humanist ecocritical projects have been challenged by the subsequent developments in new materialism and posthumanism; yet, in the context of this chapter, a human approach to environmental crisis remains a central one.

Monika Fludernik's project of 'natural' narratology argues that purely scientific accounts cannot give us a sense of experiencing things, since they do not evoke human experientiality in the way that narratives do (1996: 13). Obviously, as another cognitive narratologist, Marco Caracciolo, notes twenty-five years later, in the face of climate change the insights produced by science cannot be ignored: 'Scientific models offer an important – and in some respects the sole – route into the global scale of the climatological processes triggered by human activity' (2021: 12). Consequently, in climate change fiction, a mimetic representation of human action only provides a restricted viewpoint; it is as if a novel cannot be taken seriously if it does not base its approach to climate change on scientific facts and explorations. However, the main question that McEwan's and Offill's novels ask, and that will also be explored in this chapter, concerns the ways in which the complex phenomenon of climate change looks and feels like in the lives of human characters.

It is well known that recent posthumanist approaches have been engaged to problematise the humanist assumptions of literary theory and narrative studies, including the generally held notion that narratives are necessarily anthropomorphic and that they evoke human experientiality. According to Erin James's summary of this large discussion, cultural theorists steeped in deconstructivist philosophy, such as Claire Colebrook, have argued that narrative is a rhetorical mode deeply unsuited to the epoch of the Anthropocene, especially because it is solidly linked to human perspectives and as such 'cannot adequately represent the broader timescales and wider conception of nonhuman lives that our current moment of environmental crisis demands' (2020: 184). Whereas deconstruction has encouraged general scepticism about language, narrative, and narrative theory, I argue that by providing new tools for analysing textual designs, narrative theory can enrich the methods of ecocritical analysis; in other words, a focus on narrative forms may reveal new modes of interconnection between textual practices and natural ecologies (see Lehtimäki, 2013: 119). The recent project of *econarratology*, initiated and developed by James, is as much informed by ecocriticism's referential and political interests as by narratology's textual and structural emphasis. In her study, ecocriticism's interest in the relationship between literature and the physical environment is combined with narratology's focus on literary structures and devices. She suggests, quite rightly to my mind, that 'Focusing on structures – narrative structures in particular – opens up ecocritical discourse to a set of texts that had previously been illegible to ecocritics' (2015: 14). Correspondingly, my analysis pays attention to the narrative structure of *Solar* and *Weather*, even if this kind of structural analysis also includes a larger, contextual question: what are the possibilities of the novel form to address the questions related to climate change? James especially emphasises the reader's experiential immersion in fictional worlds, which are understood in relation to our experience of the natural world outside that fiction.

In what follows, I will read McEwan's *Solar*, which emphasises its protagonist's embodied consciousness against the backdrop of the changing climate and at the same time, beyond the fictional character's restricted viewpoint, attempts to raise the reader's ecological awareness. After my analysis of *Solar*, I will focus on Offill's *Weather*, in which the main character's daily life is embedded in a looming ecocatastrophe. In its fragmentary narrative mode *Weather* may be less mimetic as well as less 'immersive' as a reading experience than *Solar*, but we need to ask whether its intention and design differ significantly from McEwan's more conventional narrative.

Solar: character progression and global warming

Ian McEwan's climate change novel *Solar* dramatises human experience and human concerns in the age of global warming, simultaneously marginalising (at least in the limited consciousness of its protagonist) the importance of non-human nature. The initial idea of *Solar* goes back to 2005, when McEwan took part in an expedition to the North Pole focusing on the crucial topic of climate change. On his part, McEwan's fictional character, Michael Beard, takes a trip to the Arctic Circle with a boatload of artists concerned with global warming. Self-ironically, one of the participants, 'a gangling novelist,' a 'crop-haired fellow with rimless glasses' (2010: 76), resembles McEwan himself. In my reading, *Solar* emerges as a kind of parody of environmental literary studies with their overly pessimistic view of the human impact on nature as well as their overly optimistic notion that you can change things by making literary art.

When reading the complex narrative communication of *Solar*, however, we have to ask whether the ideas about climate change of its protagonist, Nobel Prize-winning scientist Michael Beard, need to be taken seriously:

> Beard was not wholly skeptical about climate change. . . . But he himself had other things to think about. And he was unimpressed by some of the wild commentary that suggested the world was in 'peril,' that humankind was drifting towards calamity, when coastal cities would disappear under the waves, crops fail, and hundreds of millions of refugees surge from one country, one continent, to another, driven by drought, floods, famine, tempests, unceasing wars for diminishing resources.
>
> (McEwan, 2010: 15–16)

As the narrative discloses those threatening visions which are by-products of climate change, we are told that Beard does not believe in these alleged perils. Beard is annoyed by the apocalyptic rhetoric associated with climate-conscious talk about the planet, as well as sick of listening to the 'familiar litany of shrinking glaciers, encroaching deserts, dissolving coral reefs, disrupted ocean currents, rising sea levels, disappearing this and that, on and on' (McEwan, 2010: 36). And yet, behind Beard's back, these global threats are conveyed to the reader in a visually and emotionally evocative way, potentially raising her or his ecological awareness. Therefore, the reader is supposed to interpret the text not only through but also beyond Beard's limited vision and search for the material presence of non-human nature. We, as readers, cannot continue seeing the changing environment through the eyes of this ego-centric scientist.

Beard's attitude strongly reflects humankind's helplessness in the face of something endlessly complex and truly global. The narrative progresses from 2000 through 2005 to 2009, and there are discernible changes in the climate. The Intergovernmental Panel on Climate Change (IPCC) gives its successive reports, and Al Gore challenges George W. Bush in the US presidential election. At the beginning of the new millennium, the British government gets engaged with climate change 'practically rather than merely rhetorically' (McEwan, 2010: 16). Yet Beard himself only 'read about it' and 'expected governments to . . . take action' (McEwan, 2010: 15), and soon 'the century had ended and climate change remained a marginal concern' (McEwan, 2010: 75). The novel consists of three parts, each of which formulates a different viewpoint on the topic of climate change: first, Beard's scepticism about it (Part One, 2000), second, his rhetoric outlining the benefits of solar energy (Part Two, 2005), and third, his action of saving the planet and his own life (Part Three, 2009). Whereas the first part of the novel shows the protagonist's profound scepticism about global warming, the last part shows him convinced of its reality, much like Al Gore in his documentary film, *An Inconvenient Truth* (2006).[1] At the beginning of *Solar*, Beard's egotism strongly reflects humankind's self-centred attitude in the face of a crisis affecting the entire planet, but his subsequent conclusion that in a crisis we understand 'that it is not in other people or in the system or in the nature of things that the problem lies, but in ourselves, our own follies' (McEwan, 2010: 155) signifies his growing ethical vision.

Solar progresses from the 'frozen shores' (McEwan, 2010: 79) of the North Pole to the 'savage heat' (McEwan, 2010: 232) of the New Mexico deserts, so that human thinking and action are enveloped in the harsh and changing realities of the natural environment. The text depicts science, which is Beard's only and true religion, as practised in sterile, lifeless sites, detached from everyday human concerns. But the problem is achingly real; the changing climate, such as the freezing cold or the burning sun, has direct effects on the everyday life of McEwan's protagonist. These effects make Beard seem like a comical figure or an egotist, for his fears about climate change are not really global, but instead very local, concerning his own body (the prospect of his penis freezing and the signs of melanoma on his skin). As the narrative discourse and the story-time progress, climate change is written on Beard's body, as it were, or at least extreme temperatures influence his physical well-being.

As mentioned, Beard's bodily experiences have consequences for his ethical thinking. The world around him gives constant signs of its existence, so that Beard is forced to concede that not all concerns were 'abstract,' for some were 'distinctly embodied' (McEwan, 2010: 184). Here, I think McEwan utilises Beard's bodily experiences for rhetorical purposes, foregrounding human experientiality as a way of dealing with issues as big and complex as climate change. Clearly, human embodiedness should be understood not as a mode of individual experientiality, but, according to Stacy Alaimo, as a kind of 'trans-corporeality, in which the human is always intermeshed with the more-than-human world' (2010: 2). Because of his personal and extreme physical experiences, Beard eventually begins to think that perhaps 'climate change, radical warming above the Arctic Circle, was actually taking place and was not a figment of the activist imagination' (McEwan, 2010: 59). In the world where 'everything is connected to everything else,' because there is only one ecosphere for all living organisms, according to Barry Commoner's First Law of Ecology, it is short-sighted to act only locally (see also Mehnert, 2016: 64). As Timothy Clark, among other scholars, suggests, 'the issue of climate change also undermines the very possibility of acting only locally' (2011: 136; see also Kluwick, 2014: 504).

In *Solar*'s ironic narrative construction, Michael Beard's selfishness is a symptom of the non-sustainable ways in which humans have treated their natural environment, and yet he is the one who takes action in thinking up solutions to save the planet by developing new technologies for utilising solar energy. He is, therefore, a complex character with both negative and positive attributes. In his greedy, selfish, and calculating mind, Beard is finally 'an average type' (McEwan, 2010: 170), a kind of Everyman representing mankind. Thus, in his banality, he is also an allegorical figure. As Ursula Heise has it, however, 'allegorical representations are generally ill suited to reflect dynamic changes in global ecosystems, even as it is difficult to imagine tropes for the planet as a whole that do not in some way invoke allegorical mechanisms' (2008: 64). It even seems that ancient flood narratives which are to be found not only in the Bible but also in other cultural texts and myths are one imaginative resource that contemporary authors utilise in order to make sense of present environmental risks. As Gary Johnson argues, we need to redefine and reconceptualise the use of allegory in contemporary fiction instead of only understanding it as 'a particular kind of didactic-religious-personification composition . . . peopled by upper-case abstractions' (Johnson, 2012: 2–3). For example, in Kingsolver's *Flight Behavior*, the biblical imagery of burning bushes and apocalyptic floods is part of the novelist's rhetorical resource, and we may find the use of allegory also in *Solar* and *Weather*, although in a somewhat ironic mode.

As Heise further argues, perceptions of climate change and other environmental risks are shaped by narrative modes and rhetorical tropes, which serve as a means of 'organizing information about risks into intelligible and meaningful stories' (2008: 138). Consequently, classical figures, tropes, and allegorical story models, such as pastoral, apocalypse, irony, tragedy, and comedy, retain their vitality when novelists and other artists try to come to terms with climate change and communicate its occurrences and consequences to their audiences. *Solar* also serves to flesh out the argument made by Joseph Meeker in his pioneering work on environmental imagination, *The Comedy of Survival* (1972). Meeker regards literature as an important resource of the human species or even as a means of its survival, since it offers insights 'into human relationships with other species and with the world around us' (Meeker, 1972: 4). For Meeker, *comedy* is a mode that locates human beings in their natural environment, while *tragedy* is about humanity's doomed attempt to put ourselves above the world. Meeker seeks to determine whether literature contributes to the survival of the human species or to its extinction (1972: 25), and his emphasis on the comic mode as a model for ecological behaviour may help us appreciate McEwan's use of comedy in *Solar* or, for that matter, Offill's employment of a tragicomic mode in *Weather*.

Beard's comical ventures around the globe are firmly rooted in his physical experiences. For example, it is part of the black comedy of McEwan's novel that whereas Beard is saved from the jaws of a *living* polar bear during his extreme North Pole expedition (2010: 70–72), his younger colleague Tom Aldous is killed by a *dead* polar bear in the seeming safety of Beard's London townhouse, or, to be exact, by a 'polar-bear rug on the polished floor, with its open mouth and yellow teeth' and its 'frozen laugh' (McEwan, 2010: 89, 92). After Aldous's accidental death, Beard continues his colleague's pioneering work on artificial photosynthesis, the use of carbon dioxide, water, and sunlight as an endless resource for the solar energy industry. It appears, however, that Beard is in it for commercial gain rather than the saving of the planet, so that, indeed, climate change is something he *profits* from. Compared to Aldous and his 'deep interest in global warming, ecology, [and] sustainable development' (McEwan, 2010: 270), Beard comes across as an opportunistic and selfish villain. This makes it difficult for readers to view him and his actions with sympathy.

One of the most revealing conversations occurs in Part Three between the scheming Beard and his not-so-bright assistant Toby Hammer, who is worried about the possibility, mentioned on television, that the planet is *not* heating:

> 'Here's the good news. The UN estimates that already a third of a million people a year are dying from climate change. Bangladesh is going down because the oceans are warming and expanding and rising. There's drought in the Amazonian rainforest. Methane is pouring out of the Siberian permafrost. There's a meltdown under the Greenland ice sheet that no one really wants to talk about. . . . Two years ago we lost forty per cent of the Arctic summer ice. Now the eastern Antarctic is going. The future has arrived, Toby.' . . . Beard laid a hand on his friend's arm, a sure sign that he was well over his limit. 'Toby, listen. It's a catastrophe. Relax!'
>
> (McEwan, 2010: 216–217)

There is clear irony here, for Beard needs the sun to get scorching hot in order to make his solar panels function more effectively. The very contradictoriness of Beard's practice is at the core of *Solar*'s narrative rhetoric, for the novel dramatises the potential dead ends to which even well-meaning individual or governmental practices may lead. Therefore, his telling Toby to 'Relax' – because *it will be* a catastrophe – is not very reassuring. We can view this white male physicist as an embodiment of a general refusal to think about the future of the planet. Beard's vices, which include gluttony, greediness, sexual promiscuity, opportunism, calculation, and profit-seeking, hold up a mirror to humankind. His 'grotesque body' is the very centre of the novel's 'satirical allegory' (Garrard, 2013: 125, 130), as is the symbolic figure of the polar bear.[2] This has also been suggested by Astrid Bracke, who reads *Solar* as an allegory of 'the kind of greed that brought humankind to the brink of extinction' (2019: 54). Beard consumes, steals, and exploits everything around him, whether it is food, women, or research plans of other scientists. As Beard's body starts to collapse and his ethics gradually deteriorates, he becomes a symbol of humankind's grossly exploitative way of treating the planet, the polar bear being a sorrowful mirror-image of his actions.

As we have seen, McEwan gives his main character pronouncedly physical attributes as well as allegorical dimensions which, however, Beard is unable to live up to precisely because of his physical shortcomings. At the beginning of the narrative we are told that 'ethereal Beard of planetary renown' (2010: 11) has 'given up hopes of being the mortal chosen to find [the] grail' (2010: 66), but near the end he believes he is in 'his quest to rescue humankind from self-destruction' (2010: 223). He thus aspires to be an allegorical figure and a bigger man than he is capable of being; indeed, like Milton's god of fire, Hephaestus, he is a falling star in the setting sun, aspiring in vain to the role of his namesake, the Archangel Michael, who leads the heavenly forces against Satan in *Paradise Lost*. Despite his allegorical dimensions, Beard is totally mundane; his 'messy' life, filled with junk mail and empty bottles, unmatching socks and uncleared attics, never reaches 'Eden, purged of clutter and distraction' (2010: 222, 226).

As a fictional figure, Michael Beard fulfils the three character functions or dimensions suggested by Phelan in his rhetorical theory of narrative: the *mimetic* component involves readers' interest in the narrative world like their own and in characters like themselves; the *thematic* component involves readers' interest in the cultural, ideological, philosophical, or ethical issues that the narrative addresses; and the *synthetic* component involves readers' close attention to the narrative as an artificial construct (2007: 5–6). Beard's figure is, for better or worse, very human and thus mimetic. His allegorical quality obviously serves thematic functions. In addition, he is a useful narrative device for McEwan, enabling metafictional and intertextual commentary, and is therefore synthetic. Beard's developing skin cancer is an indexical sign of global warming, which is partially caused by short-sighted human actions, but Beard also symbolises the planet's looming peril, even becoming an iconic image resembling the Earth: 'an oily nausea at something monstrous and rotten from the sea stranded on the tidal mudflats of a stagnant estuary, decaying gaseously in his gut and welling up, contaminating his breath, his words, and suddenly his thoughts' (2010: 148). Indeed, the descriptions of his bodily features sometimes draw on geographic metaphors: 'the archipelago of his disrupted selfhood' (2010: 142); 'the northern hemisphere of his eyeball' (2010: 244). It may be argued, then, that Beard's body reads as a planetary metaphor for exploitation and that his character fails to be Milton's glorious Archangel; indeed, the novel's allegorical thematics are undercut by McEwan's parodic characterisation of Beard's banal figure.

Following Phelan's definition, we may see how the abstract dimensions of Beard's character are converted into concrete functions of his character in the narrative progression (1989: 9), and how the character's mimetic component comes to the foreground. In the next section, I will discuss a different kind of character development and narrative progression in Offill's *Weather*, which in its collage-like fragmentariness may also be a less immersive reading experience than McEwan's more conventional narrative.

Weather: domestic fiction and narrative fragmentation

Jenny Offill's short novel *Weather* focuses on the daily life of its autodiegetic narrator Lizzie Benson, a university librarian living with her husband Ben and son Eli in Brooklyn, New York. The daily life with its mundane details ('There's a note on the fridge saying we are out of milk, cheese, bread, and toilet paper' [2020: 34]) is, however, embedded in the rapidly shifting conditions in the global climate, 'climate' here meaning not only natural phenomena such as weather, but also the radical changes in US politics on the brink of Donald Trump's election. Interestingly, while McEwan's *Solar* charts the earlier political atmosphere in which Al Gore with his 'inconvenient truth' loses the presidency to George W. Bush, *Weather* envisions a dystopic future in a country – and in a larger world – led by a new president whose name is not even mentioned in the narrative. Apparently, emptying the narrative of any references to the historical and political reality involves a risk that the text remains abstract.

Like Michael Beard, Lizzie Benson is simultaneously preoccupied by the large-scale phenomenon of climate change and by the small-scale demands of everyday life. Whatever large and complex issues the text is addressing, at the centre of the story is constantly Lizzie herself, worrying about her son Eli's future and her brother Henry's drug addiction, her own bad knee, and her mother's teeth that need to be fixed. It is as if the painfulness of the global situation was felt on one's body, just as Beard feels global warming on his skin. Lizzie is, then, often frustrated and anxious when facing the demands of daily life. The narrative also suggests, through mimetic action rather than didactic lecturing, that the age of the Anthropocene requires that people live on multiple planes simultaneously. Walking away from her son's elementary school, Lizzie ruminates: 'The problem with Eli's school is it's not on a human scale' (2020: 27). This can also be read as an allegory of global climate change: it is not on a human scale.

Climate change becomes a part of Lizzie's everyday life also in a more concrete sense. She is a college 'dropout' with a half-finished dissertation titled 'The Domestication of Death: Cross-Cultural Mythologies' (2020: 59). Lizzie accepts an unlikely job answering letters that respond to her former supervisor Sylvia's climate change podcast, almost apocalyptically titled *Hell and High Water*. Having worked at a help desk in a university library, she is not altogether unfamiliar with the nature of the new job. Indeed, when asked how she is able to know things, Lizzie answers: 'I'm a fucking librarian' (2020: 170). Still, it is obvious that information is now everywhere and that a single librarian cannot be a source of all knowledge anymore. In the world of *Weather*, everything happens rapidly and it happens 'now,' without clear history or context: 'Everything is happening much faster than expected' (2020: 76). There are only disconnected senses of the present and the quick questions and answers in social media.

Lizzie's husband Ben has a 'complicated relationship to modern things' (2020: 36). He holds a PhD in classics and antiquity, and likes to read the Stoics with his morning coffee, but has been forced to find a new job as a video game designer. Ben refers to ancient concepts such as *anachoresis*, which means 'retreat into the desert,' or to Greek values, such as *xenia*, meaning 'hospitality,' but these ideas are losing their meaning in the world in which there is no possibility of retreat to any safe place or to show hospitality to strangers. As Lizzie herself reflects, this is 'twenty-first-century everything' (2020: 159). One of the distinguishing aspects of this contemporary 'everything' is that human beings are attacked by information from every direction and consequently it becomes difficult to concentrate on anything particular.

Accordingly, as Lizzie deals with various emails about ecological disaster and global warming, she comes across questions veering towards evangelical and apocalyptical thinking: 'Did Noah's flood cover the whole earth or just the places where people lived?' (2020: 35), 'What is the philosophy of late capitalism?' (2020: 44), 'What are the best ways to prepare my children for the coming chaos?' (2020: 93), 'What would it mean to bioengineer humans to be more efficient?' (2020: 167), 'Do we live in the Anthropocene?' (2020: 170). One of the themes in Sylvia's podcast is 'The Center Cannot Hold,' which refers to a line in W. B. Yeats's famous poem 'The Second Coming' (1919). The podcast includes discussions about climate change and employs apocalyptic rhetoric ('the invisible horsemen galloping toward us' [2020: 10]). In the same context, Lizzie looks out the window and sees '[s]omething in the distance, limping toward the trees' (2020: 10).[3]

Ursula Heise notes that climate change has begun to make its way into the cultural imagination and that 'it poses a challenge for narrative and lyrical forms that have conventionally focused above all on individuals, families, or nations, since it requires the articulation of connections between events at vastly different scales' (2008: 205). Consequently, Heise argues that in their portrayal of climate change, some popular films rely on the conventions of 'apocalyptic narrative' with its 'clichés of urban disaster and alien invasion' (2008: 206). These popular notions are also addressed by Michael Beard in *Solar* with increasing cynicism: for McEwan's protagonist, 'the end of the world' is a 'fantasy,' and when it comes to the special case of climate change, 'the apocalyptic tendency had conjured yet another beast' (2010: 16). According to Sylvia's last words in *Weather*, 'the world continues to end' (Offill, 2020: 198), and yet she herself goes on 'to water her garden' (Offill, 2020: 198). Here, the allusion is to the closing words of Voltaire's classic philosophical satire *Candide* (1759), according to which all is well, but we must cultivate our garden.[4]

Therefore, *Weather* deals with the simultaneity of daily life and global crisis, two realities on vastly different scales. Lizzie's 'internal' world, which is filled with seemingly small everyday concerns, is finally a part of an 'external' world which is changing in an alarming sense and which cannot be ignored. This is a world of simultaneity, and it is part of the design of the narrative that Lizzie tells her story in the present tense. As Greg

Garrard notes, 'climate fiction *reflects on changes as they are in the process of occurring*' (2016: 302; original italics). In the rather hopeless mood of Lizzie's narrative, there is no outside position anymore from where to evaluate the past happenings, as in the classical novel. People who cannot sleep at night and the melting glaciers in the Arctic are part of the same reality. But in order for the contemporary novel to deal with the issue as large as climate change, it needs to focus on concrete particulars instead of universal abstractions.

In *Weather*, the different scales of daily life and global crisis are brought together in one short paragraph which juxtaposes two different levels of experience:

> Eli is at the kitchen table, trying all his markers one by one to see which still work. Ben brings him a bowl of water so he can dip them in to test. According to the current trajectory, New York City will begin to experience dramatic, life-altering temperatures by 2047.
>
> (Offill, 2020: 106)[5]

Here, a safe and peaceful family life is constantly threatened by the looming future, which Eli, who still plays with toys, will have to face in more severe ways than his parents. Eli's father, Ben, is Jewish, and in Hebrew 'Eli' means 'high' or 'elevated.' There is, then, an allegorical dimension in *Weather* but it is possible to interpret that biblical names and stories have lost their power.

As in McEwan's novel which focuses on the thoughts and actions of its egoistical protagonist, there are characters in Offill's novel who can only think of the global crisis on a very local and personal level. One of the motifs connecting the two narratives is Sylvia's lecturing about the reasons and consequences of climate change, even though Beard's motives may be somewhat different from those of Offill's character. In *Weather*, 'people are really sick of being lectured to about the glaciers' (2010: 73). For instance, a 'red-faced man,' while listening to a lecture about Arctic ice melting, is concerned only with his own living environment: 'But what's going to happen to the American weather?' (2020: 73). There are also young academics in the story, going on about 'transhumanism' (2020: 39), but not patient enough to wait when it comes to satisfying their immediate needs. Or, as Sylvia puts it: 'These people long for immortality, but can't wait ten minutes for a cup of coffee' (2020: 39).

When answering to the multitude of letters responding to Sylvia's podcast, Lizzie also learns through this experience and familiarises herself with the complex facts and statistics about climate change. In another context, when discussing with her mentor in meditation, Margot, Lizzie learns a new word, 'enmeshment' (2020: 58). Later, she is listening to a podcast episode 'about something called the "mesh"' (2020: 67). Timothy Morton's metaphor of the 'mesh' is an alternative to linear and hierarchical ways of thinking about the relationship of human beings with non-human animals and about agencies (2010: 28–33). In Marco Caracciolo's reading of Morton's work, '[t]he image of the mesh suggests fragile interdependence, but also formidable complexity of pattern and connection' (2021: 12). What we see in both *Solar* and *Weather* is the enmeshment of humans with the natural, non-human world, also in cases where human characters are not conscious of this reality or where they try to maintain their human 'exceptionality.'

Morton also writes about *hyperobjects*, meaning objects that 'exist on almost unthinkable timescales [and] confound our limited, fixated, self-oriented frameworks' (2010: 19). Caracciolo, referring to Morton's concept, adds that 'climate change confronts us with planetary rhythms and cycles that . . . we struggle to know experientially,' because '[o]ur sensory apparatus is attuned to our immediate surroundings, not to atmospheric carbon dioxide or global sea levels' (2021: 10). Because of its large scale, climate change easily remains an abstract and remote problem instead of being a concrete and graspable thing in the everyday life of human beings. Novels such as *Solar* and *Weather*, however, try to envision the ways in which the global crisis manifests itself in the daily lives of its characters. This is an example of the power of fiction: it enables us to see how people experience and feel things. In his book *Narrating the Mesh: Form and Story in the Anthropocene*, Caracciolo studies narrative form in the context of climate change, but instead of relying on age-old structuralist conceptions of narrative, he argues that 'the form of narrative is fundamentally embodied and affective' (2021: 5).

Instead of the mimetic narrative mode which emphasises plot and character development and which is represented by *Solar*, *Weather* is a more experimental text, consisting of brief and often seemingly disconnected fragments. The novel's disconnected structure is even reflected upon in Lizzie's narration that 'regular life becomes more fragmented' (Offill, 2020: 44). The fragmentariness is also shown at the level of literary self-consciousness. Ben studies surviving fragments of ancient classics, while Lizzie 'always had an obsession with lost books, all the ones half written or recovered in pieces' (50), including her own unfinished dissertation. In a self-reflexive way, Offill's chosen technique implies that the human experience of the Anthropocene cannot be storied as a clear continuity; rather, it emerges as a chaotic collage of more or less meaningful moments. There are nostalgic references to age-old conventions of human storytelling from times when '[w]e built fires and told stories to make the dark nights pass by' (2020: 84), but it seems that the days of comfort and joy provided by stories are over. In addition, *Weather* is not informed by the conventions of realist fiction as much as it is by quick and sparse communication of social media. In the novel, the discussion about climate change and other larger-than-life issues, such as religion, is mainly conducted in social media platforms which do not allow space for deep and closely argued philosophical thinking. As compared to the rather conventional narrative mode applied in McEwan's novel, *Weather*, published ten years later, reflects on the possibilities of the novel in the age of the Anthropocene. In other words, novelists such as Offill are deeply aware of the limitations of their slow medium when it comes to the heated discussion about the rapidly changing environment. As a reading experience, *Weather* does not resemble a novelistic narrative anymore; rather, its verbal and typographical means simulate a Twitter feed and social media platform.

Indeed, it is possible to argue that digital technologies and social media have had a profound influence on the ways in which people write and read printed books (see, e.g., Johnson, 2022: 9–19). However, instead of being a "Twitterfiction" (Johnson, 2022: 6) with a strict word limit, Offill's novel rather self-consciously employs the means of social media in its separate paragraphs, simultaneously reflecting upon the problems and possibilities of a novel form in the age of digital information and climate change.[6] In *Weather*, Offill uses short paragraphs and blank space between them for dramatic and thematic purposes. The story told by Lizzie takes place in urban and often enclosed settings such as the library, corridors, hallways, and lifts. This spatiality produces a kind of claustrophobic sense in the narrative. There are only rare lyrical and melancholic glimpses of the natural environment: 'It's the first day of spring, weird clouds, hazy sun' (2020: 26); 'The sky is dull, a soft feathery gray, streaked here and there with clouds' (2020: 120). In the relatively nondramatic succession of events, Lizzie interacts with her family members, colleagues, and friends, and even has a brief encounter with a war journalist, but there is no time for an adulterous romance. Whereas classical realistic narrative succeeded at forming relationships between people in their shared social space with the help of linear narrativity, *Weather* is an example of a contemporary novel that stresses social distancing through its fragmentary narration, in which society has been replaced by social media. Lizzie is forced to listen to a young 'techno-optimist guy' chanting that 'soon everything in our lives will be hooked up to the internet of things' and that 'we will be connected through social media to every other person in the world' (2020: 38–39).

The reader needs to do active work to 'piece together' (2020: 164) separate events into a coherent narrative, only finding it to be a futile attempt; the typography and short paragraphs of the novel cut off narrative progression and break links between events. In my reading, Offill's technique is designed to reflect on our general inability to perform a continuous, long-term action in the changing world of ours. Bleakness, blankness, and despair are the governing senses of the narrative. Sylvia, for instance, has previously included an 'obligatory note of hope' (2020: 67) in her podcast.[7] But, in the age of a growing environmental despair, she now reflects that 'there's no hope any more, only witness' (2020: 133).

One might differentiate *Weather*'s fragmentary and non-linear narrativity from *Solar*'s more plot-oriented narrative about character development. In addition, while *Weather* is ultimately about a collective, despite Lizzie's being the narrator-protagonist, *Solar* focuses almost solely on Michael Beard's thoughts and actions. At least metaphorically, *Weather* could be constructed as a feminine 'domestic' fiction (cf. Armstrong, 1987), differing from *Solar*'s masculine, picaresque style. What especially interests me here, however, is how both these novels bring the often mundane demands of daily life and larger-than-human agencies of climate change into a dialogic clash, forcing their characters to live simultaneously on two different planes of reality.

At the same time, we should note that Lizzie is an aspiring novelist herself, having taken courses in creative writing at some earlier point of her life. *Weather* can be read as Lizzie's 'autofictional' attempt at shaping her fragmented experiences of daily life into a literary form.[8] As the narrative is filled with anecdotes, jokes, quotations, and email conversations, it comes to resemble a collage of sorts, and the reader may recognise Lizzie's literary ambitions inside Offill's novel. Indeed, Wojciech Drag, in his analysis of twenty-first-century collage novels, including Offill's earlier *Dept. of Speculation* (2014), comments on the general ideas about 'art in crisis' as print publications being replaced by electronic literature and audiobooks (Drag, 2020: 5). Drag, however, notes that contemporary collage novels, which utilise narrative fragmentation, white space on the page, and juxtaposition of various materials, typically also deal with crises, either of a personal nature or on a global scale. In *Weather*, we grasp the idea of 'the coupling of two realities,' which Drag mentions as one of the meanings of collage (2020: 10), as both personal grief and the despair about the changing climate occupy the space of the same page and even the same paragraph.

Conclusion

The mental and physical atmosphere of Jenny Offill's *Weather* is even more grim and hopeless than that of Ian McEwan's *Solar*, published a decade earlier. Both novels are partly written in a tragicomic mode, and there has been discussion whether comedy or tragedy are appropriate ways of approaching environmental issues and especially climate change. In *Weather*, however, comedy or satire are not principle means of dealing with ecocatastrophe, as they may be in *Solar*; rather, Offill's novel, with its occasional and sometimes desperate jokes and parody, is not able to find any particular generic mode to narrativise and conceptualise climate change. Instead, *Weather* self-consciously reflects upon the problems and possibilities of a conventional novel form in the age of digital media and environmental crisis.

What the two novels discussed in this chapter seem to suggest is that global climate change is something too large-scale and abstract to be grasped in any concrete sense by individual human beings, even if *Solar*'s Michael Beard may feel the warming of the planet on his own skin. It may be argued that narratives typically emphasise human experience and concrete particulars; or, in other words, that the mimetic dimension is the springboard for the reader's ethical, emotional, and affective experience of fictional narratives. The mimetic level of experience is foregrounded in both novels discussed in this chapter; indeed, whatever their allegorical, intertextual, and metafictional devices, *Solar* and *Weather* deal with the apocalyptic threats posed by climate change from a grassroots level. Therefore, the world of these narratives is necessarily the world that the actual readers of the novels also share. Put differently, there must be something very recognisable (working, eating, human relationships, family issues, childcare, etc.) in order to make the depiction of daily life in the midst of global crisis effective and graspable. Accordingly, 'internal' and experiential issues provide the narratives with a richer ground than 'external' and abstract ones, even though the informed reader of these fictions is not supposed to forget what is going on in the extratextual reality.

Notes

1 In fact, some elements of Beard's talk in front of his audience, such as his calculated expression in a staged theatre performance or his statement that 'the planet . . . is sick' (2010: 148) are parodies of Gore's populist phrasings. Gore's celebrated documentary is, ultimately, a conventional masculine narrative employing familiar rhetorical strategies. At the centre of the film is not only climate change but, even more so, the autobiographer Gore himself, reflecting on his life, lecturing to sponsors, and using his laptop to conduct scientific research (see Heise, 2008: 209; cf. Slovic, 2008: 132–133).

2 Janice Cavell argues that traditionally the visual representation of the Arctic has been very limited in its means of representation and imagination, so that 'a few standard images of the Arctic were repeated over and over' and 'they eventually became a kind of shorthand that represented the far north in the popular imagination mainly through clichés and stereotypes' (Cavell, 2008: 40). Edwin Landseer's famous painting *Man Proposes, God Disposes* (1864) with its grisly polar bears is one of those classical representations, constructing the polar bear as a 'standard image' of the Arctic.

3 This may evoke in the potential reader's mind the closing lines of Yeats's poem: 'And what rough beast, its hour come round at last,/slouches towards Bethlehem to be born?'

4 The topic of cultivating the garden as the most important action in the world recurs in Antoine de Saint-Exupéry's *Le Petit Prince* (*The Little Prince*, 1943) and in J.M. Coetzee's *Life & Times of Michael K* (1983).
5 This vision is therefore even more dystopic than the one presented in Kim Stanley Robinson's novel *New York 2140* (2017). See also Ameel (2021: 63).
6 An interesting example of a novel in the age of digital media is Ben Lerner's *10:04* (2014), a combination of autofiction and environmental narrative, and in that sense resembling Offill's *Weather*; according to an analysis of Lerner's novel, being a 'self-reflexive and reactive "new novel" in the age of digital media [it] continuously breaks down its own fictions and short-circuits the novelistic form' (Hansen, 2017: 12). Indeed, novels such as *10:04* and *Weather* are examples of 'new' novels in the age of both digital media and environmental crisis, reflecting upon the shortcomings of story, plot, and characterisation and instead corresponding to the fragmentariness and open-endedness of twenty-first-century experience.
7 Interestingly, obligatorynoteofhope.com, which is mentioned in a paratext of the novel, is the author Jenny Offill's own website.
8 Obviously, in the strict sense *Weather* is not the actual author Offill's autofiction – in which case the narrator-protagonist should be called Jenny according to the traditional structuralist definition of the genre – but it rather partakes in the contemporary 'autofictional writing' as a larger cultural phenomenon (see Effe and Lawlor, 2022).

Bibliography

Alaimo, Stacy (2010), *Bodily Natures: Science, Environment, and the Material Self* (Indianapolis: Indiana University Press).

Ameel, Lieven (2021), 'Balloon Explorers, the Panorama, and the Making of an Arctic *Nomos* in Contemporary Fiction', in Markku Lehtimäki, Arja Rosenholm and Vlad Strukov (eds.), *Visual Representations of the Arctic: Imagining Shimmering Worlds in Culture, Literature and Politics* (New York: Routledge), 61–77.

Armstrong, Nancy (1987), *Desire and Domestic Fiction: A Political History of the Novel* (Oxford: Oxford University Press).

Bartosch, Roman (2019), *Literature, Pedagogy, and Climate Change: Text Models for a Transcultural Ecology* (Basingstoke: Palgrave Macmillan).

Bracke, Astrid (2019), 'Science and Climate Crisis', in Dominic Head (ed.), *The Cambridge Companion to Ian McEwan* (Cambridge: Cambridge University Press), 45–59.

Caracciolo, Marco (2021), *Narrating the Mesh: Form and Story in the Anthropocene* (Charlottesville: University of Virginia Press).

Cavell, Janice (2008), *Tracing the Connected Narrative: Arctic Exploration in British Print Culture, 1818–1860* (Toronto: University of Toronto Press).

Clark, Timothy (2011), *The Cambridge Introduction to Literature and the Environment* (Cambridge: Cambridge University Press).

Cook, Jon, Sebastian Groes and Victor Sage (2009), 'Journeys without Maps: An Interview with Ian McEwan', in Sebastian Groes (ed.), *Ian McEwan: Contemporary Critical Perspectives* (New York: Continuum), 123–134.

Drag, Wojciech (2020), *Collage in Twenty-First-Century Literature in English: Art of Crisis* (London: Routledge).

Effe, Alexandra and Hannie Lawlor (eds.) (2022), *The Autofictional: Approaches, Affordances, Forms* (Basingstoke: Palgrave Macmillan).

Fludernik, Monika (1996), *Towards a 'Natural' Narratology* (London: Routledge).

Garrard, Greg (2013), '*Solar*: Apocalypse Not', in Sebastian Groes (ed.), *Ian McEwan: Contemporary Critical Perspectives* (New York: Bloomsbury), 123–135.

Garrard, Greg (2016), 'Conciliation and Consilience: Climate Change in Barbara Kingsolver's *Flight Behaviour*', in Hubert Zapf (ed.), *De Gruyter Handbook of Ecocriticism and Cultural Ecology* (Berlin: De Gruyter), 295–312.

Hansen, Denise Rose (2017), 'Ben Lerner and the Novel in the Age of Digital Media', *The International Journal of the Book*, 15:3, 11–21, https://doi.org/10.18848/1447-9516/CGP/v15i03/11-21

Head, Dominic (1998), 'The (Im)possibility of Ecocriticism', in Richard Kerridge and Neil Sammels (eds.), *Writing the Environment: Ecocriticism and Literature* (London: Zed Books), 27–39.

Heise, Ursula K. (2008), *Sense of Place and Sense of Planet: The Environmental Imagination of the Global* (Oxford: Oxford University Press).

Iovino, Serenella (2010), 'Ecocriticism and a Non-Anthropocentric Humanism: Reflections on Local Natures and Global Responsibilities', in Laurenz Volkmann, Nancy Grimm, Ines Detmers and Katrin Thomson (eds.), *Local Natures, Global Responsibilities: Ecocritical Perspectives on the New English Literatures* (Leiden: Brill), 29–53.

James, Erin (2015), *The Storyworld Accord: Econarratology and Postcolonial Narratives* (Lincoln: University of Nebraska Press).

James, Erin (2020), 'Narrative in the Anthropocene', in Erin James and Eric Morel (eds.), *Environment and Narrative: New Directions in Econarratology* (Columbus: Ohio State University Press), 183–202.

Johnson, Gary (2012), *The Vitality of Allegory: Figural Narrative in Modern and Contemporary Fiction* (Columbus: Ohio State University Press).

Johnson, Miriam J. (2022), *Books and Social Media: How the Digital Age Is Shaping the Printed Word* (London: Routledge).

Kingsolver, Barbara (2012), *Flight Behavior* (New York: HarperCollins).

Kluwick, Ursula (2014), 'Talking about Climate Change: The Ecological Crises and Narrative Form', in Greg Garrard (ed.), *The Oxford Handbook of Ecocriticism* (Oxford: Oxford University Press), 502–516.

Lehtimäki, Markku (2013), 'Natural Environments in Narrative Contexts: Cross-Pollinating Ecocriticism and Narrative Theory', *Storyworlds: A Journal of Narrative Theory*, 5, 119–141, https://doi.org/10.5250/storyworlds.5.2013.0119

McEwan, Ian (2010), *Solar* (London: Jonathan Cape).

Meeker, Joseph (1972), *The Comedy of Survival: Studies in Literary Ecology* (New York: Scribner and Sons).

Mehnert, Antonia (2016), *Climate Change Fictions: Representations of Global Warming in American Literature* (Basingstoke: Palgrave Macmillan).

Morton, Timothy (2010), *The Ecological Thought* (Cambridge, MA: Harvard University Press).

Murphy, Patrick D. (2014), 'Pessimism, Optimism, Human Inertia, and Anthropogenic Climate Change', *ISLE: Interdisciplinary Studies in Literature and Environment*, 21:1, 149–163, https://doi.org/10.1093/isle/isu027

Offill, Jenny (2020), *Weather* (London: Granta).

Phelan, James (1989), *Reading People, Reading Plots: Character, Progression, and the Interpretation of Narrative* (Chicago: The University of Chicago Press).

Phelan, James (2007), *Experiencing Fiction: Judgments, Progressions, and the Rhetorical Theory of Narrative* (Columbus: The Ohio State University Press).
Rigby, Kate (2004), 'Earth, World, Text: On the (Im)possibility of Ecopoiesis', *New Literary History*, 35:3, 427–442.
Slovic, Scott (2008), *Going Away to Think: Engagement, Retreat, and Ecocritical Responsibility* (Reno: University of Nevada Press).
Trexler, Adam (2015), *Anthropocene Fictions: The Novel in a Time of Climate Change* (Charlottesville: University of Virginia Press).

2 Feelings of hope and helplessness in Knut Faldbakken's and Maja Lunde's climate change novels

An econarratological reading

Georgiana Bozîntan

Introduction: econarratology and transportation into storyworlds

Literary renderings of climate change have a strong tradition of environmental apocalypticism, but the capacity of doomsday scenarios to elicit social and political action is still debatable (see Buell, 2010). While Lawrence Buell states that 'Apocalypse is the single most powerful master metaphor that the contemporary environmental imagination has at its disposal' (1995: 285), Ursula Heise contrarily underlines that 'a steady drumbeat of gloom-and-doom rhetoric is liable to discourage and alienate individuals more than incite them to action' (2008: 142). In his empirical study of the impact climate fiction has on readers, Matthew Schneider-Mayerson indicates that cli-fi readers typically show 'dramatic emotional response to these works of fiction' (2018: 489) and points out that disaster scenarios usually engender feelings of helplessness and passivity that can potentially 'present an obstacle to successful persuasion and mobilization' (2018: 489–490).

Drawing insight from recent research in econarratology, this chapter aims to compare two Norwegian cli-fi works and shed light on what narrative strategies and techniques might influence readers' feelings and attitudes towards the climate crisis. Norwegian cli-fi started receiving international acclaim with Maja Lunde's *Bienes historie* (2015) [*The History of Bees*], followed by its sequels *Blå* (2017) [*The End of The Ocean*], and *Przewalskis hest* (2019) [*The Last Wild Horses*].[1] However, Norwegian literature's tradition of the genre dates back to the 1970s when Knut Faldbakken's *Uår* [*Un-year*] series – *Aftenlandet* (1974) [*Twilight Country*] and *Sweetwater* (1976) – became one of the earliest literary texts to engage with anthropogenic climate change (Hennig, 2021: 20).[2] There is, nonetheless, a clear demarcation line one can draw between these series in terms of the outlook conveyed by the two authors. Reinhard Hennig analyses Faldbakken's series as a fictional account of a future world

DOI: 10.4324/9781032726953-3

irreversibly affected by climatic instability and environmental disruptions, in which adaptation measures are required (2013: 46). He further emphasises that these adaptation scenarios might generate a sense of gloom and helplessness. On the other hand, Sissel Furuseth (2019) illustrates a rather optimistic trend in contemporary Norwegian climate fiction, with Maja Lunde being one of its most representative voices. While these scholars mostly draw on thematic considerations, this chapter aims at directly comparing Faldbakken's *Uår* series with Lunde's climate novels, by casting light upon formal and narrative strategies that engender the different outlooks conveyed by each cli-fi series. Grounding my approach in econarratology, I begin by laying out the theoretical foundations of my analysis, which will then be divided into two parts. Firstly, I discuss how the spatial and temporal structures of the novels can be interpreted with the help of concepts such as storyworld and transportation. Secondly, I shed light on character identification to discuss how empathy potentially fosters positive feelings of hope, while the lack of sympathetic attitudes can lead to a negative perception of the storyworld. Given the temporal distance that separates the publication of these novels – over 40 years – the comparative analysis pinpoints a possible evolution in Norwegian cli-fi that has recently incorporated more ethical narrative strategies, such as giving voice to vulnerable characters and fostering empathy through the construction of sympathetic characters, which often pave the way for a certain didacticism. Finally, one overarching aim of this discussion is to show how ecocriticism's use of narratological concepts can illuminate the potential of climate fiction to change attitudes and beliefs.

The first to propose a reading method based on the conjunction between ecocriticism and narratology was Markku Lehtimäki in his article 'Natural Environments in Narrative Contexts: Cross-Pollinating Ecocriticism and Narrative Theory,' where he sheds light on 'the world-shaping functions of rhetorical designs that unites ecocritics and narratologists' (2013: 125). With her seminal work *The Storyworld Accord: Econarratology and Postcolonial Narratives*, Erin James has established econarratology as an approach that pairs 'ecocriticism's interest in the relationship between literature and the physical environment with narratology's focus on the literary structures and devices by which writers compose narratives' (2015: xv). She explains that

> Econarratology studies the storyworlds that readers simulate and transport themselves to when reading narratives, the correlations between such textual, imaginative worlds and the physical, extratextual world, and the potential of the reading process to foster awareness and understanding for different environmental imaginations and experiences.
> (xv)

Econarratology thus represents the outcome of the need 'to shift from content to form, and to explore the ways in which literary structures might offer up environmental insight' (James, 2015: xiv), which has been a common preoccupation in ecocriticism. Heise, for instance, was one of the major ecocritical scholars to call attention to the importance of narrative forms in texts that portray environmental crises, in her *Sense of Place and Sense of the Planet: The Environmental Imagination of the Global* (2008). While James's work focuses on postcolonial literature, the use of econarratological approaches is not limited to a particular type of texts and is open to multiple perspectives. James and Eric Morel (2020) have identified three major directions in econarratology: the representation of non-humans, the turn towards cognitive science, and narrative ethics. Mostly interested in studying how narrative strategies used in cli-fi storytelling potentially elicit an emotional response from readers, my analysis employs a cognitive perspective and discusses elements such as transportation, character identification, and empathy as underlying components that influence the readerly experience of the storyworld.

Drawing on cognitive approaches to narratology, James's theoretical framework relies on the concept of storyworld, defined by David Herman as 'the world evoked by a narrative text or discourse; a global mental model of the situations and events being recounted' (2009: 193). This perspective thus entails recognition of the means and strategies by which texts invite readers to inhabit mentally the fictional world. As Morel and James put it, the 'concept of "storyworld" calls attention to the worldmaking power of narrative, or its potential to immerse or transport readers into virtual environments that differ from the physical environments in which they read' (2020: 11). The two scholars refer here to Richard Gerrig's *Experiencing Narrative Worlds* (1993), which brings forth the metaphor of being transported to explain the process of immersion in the textual world. Gerrig describes this cognitive process as a way in which readers 'travel' to the fictional world by performing certain actions which distance them from the world of origin and then finally return to the actual world 'somewhat changed by the journey' (1993: 10–11). According to Gerrig, some consequences of being transported include what he calls anomalous suspense, anomalous replotting, and the reader's emotional responses (1993: 157–191). Melanie Green and Timothy Brock posit that 'transportation is a distinctive process in which affect, imagery, and cognition play integral roles' (2000: 718). In their study which investigates 'the role of transportation in the persuasiveness of public narratives,' they highlight that a compelling consequence of being transported is a change in the beliefs and attitudes of individuals.

Green and Brock also emphasise that 'attachment to a protagonist may be an important determinant of the persuasiveness of a story' (2000: 702).

This assumption hints at an essential element in transportation, namely character identification, or the 'imaginative process through which an audience member assumes the identity, goals, and perspective of a character' (Cohen, 2001: 261). Considering James's observation that 'in addition to providing readers access to new, possibly unfamiliar environments, storyworlds also provide readers access to highly subjective understandings of what it is like to live in, conceptualise, and experience a given space and time' (2015: xii), I seek to demonstrate that narrative structures, use of narrative voice, and focalisation are elements of paramount importance in guiding readers' perception of the storyworld. As I will try to show in what follows, narrative strategies of representing spatial and temporal dimensions of climate change play a fundamental role in conveying an understanding of the phenomenon's large-scale implications and shaping readers' beliefs and general worldview.

Knut Faldbakken's *Twilight Country* and the gloomy climate change

Drawing inspiration from Club of Rome's *The Limits to Growth* (1972), but also from Oswald Spengler's *Der Untergang des Abendlandes* (1918) [*The Decline of the West*], Faldbakken imagines a climate-changed world in a novel where 'efforts to introduce counter-measures against the process of decline' are presented as 'futile' (Hennig, 2013: 47).[3] Climate change narratives are often anchored in 'a futuristic dystopian and/or postapocalyptic setting' (Johns-Putra, 2016: 269), and they excel in cognitive estrangement as a literary strategy (Flinker, 2018: 50).[4] As this section aims to demonstrate, in Faldbakken's *Twilight Country*, the process of defamiliarisation is not only generated by the portrayal of an apocalyptic world, but stems mostly from the main character's perception of this dystopian world, which is often in dissonance with the narrator's voice.

Twilight Country opens *in medias res*: 'Allan Ung og Lisa flyttet ut på Fyllingen en ettermiddag i begynnelsen av mars' [Allan Ung and Lisa moved out to the Dump one afternoon at the beginning of March] (1974: 7, 1993: 5). The Dump is, literally, a giant landfill site situated on the outskirts of the megalopolis Sweetwater, to where the protagonists of the novel withdraw in the face of a looming collapse, a catastrophe generated by overpopulation and overconsumption. Together with their son, simply named Boy, Allan and his eighteen-year-old wife, Lisa, set out in a camping van to the Dump, renouncing their 'trangt, tungvint og forfallent' [cramped, inconvenient, and decayed] (1974: 17, 1993: 13) existence in the apartment on 'den larmende April Avenue' [the noisy April Avenue] (1974: 16, 1993: 12). On the Dump, they try to recreate normal living conditions and, while Allan goes to work at a petrol station in the city, Lisa remains at home, taking care of their son. Together with other inhabitants of the Dump – Doc and his sick wife Marta, the immigrants Felix and Run-Run, and Mary Diamond and her partner Smiley – they work out solutions to cope with the worsening living conditions and to protect themselves. Hennig thus remarks that '*Uår* can be read as a literary study par excellence of what kinds of adaptation might be required in a future characterised by ecological collapse, resource depletion and climate change' (2013: 46). Glimpses of hope also appear with the birth of Allan and Lisa's daughter at the end of *Twilight Country*. In the second volume, *Sweetwater*, the apocalyptic scenario is completely set in place. Accepting that the Dump no longer offers them protection and necessary resources, they return to the ravaged city where they struggle to survive the harsh winter.

As the action unfolds in a fictional city set in an unnamed country, a detailed description of Sweetwater's geography at the beginning of the novel facilitates the reader's 'transportation' to the storyworld, allowing her or him to become acquainted with the space that the characters inhabit. Located in the east, between Paradise Bay and the Motorway, the Dump was first meant to become 'forlengelsen av havneanleggene i Sweetwater City' [an extension of the harbour in Sweetwater City] (1974: 11, 1993: 8), but ended up as 'byens søppelplass, både privat og offentlig' [the city's garbage tip, public as well as private] (1974: 12, 1993: 9). Nonetheless, the third-person narrator does not merely deliver a detached depiction of the storyworld. Using indirect and free indirect discourse in the opening paragraph, the narrator adopts Allan's perspective to guide the reader in the fictional world:

> selv om det jo ikke var forbudt å gå på Fyllingen og lete etter ting en trengte, ville nok koffertene og den store skipssekken ha vakt en vaktmanns mistanke. Skjønt gikk det vaktmenn rundt omkring her ute lenger, selv i arbeidstiden? Neppe. . . . Selv om en kunne merke krisetidene, var det på en forunderlig, ulogisk måte samtidig blitt lettere å eksistere, tenkte Allan der han gikk og slepte på den tunge kofferten og skipssekken.
>
> [Although it was not prohibited to go to the Dump, their suitcases and the big seaman's bag would have aroused any guard's suspicion. Were there any still around there, even in working hours? Unlikely. . . . As he humped along the case and heavy bag, Allan thought that although the crisis was obvious, in a strange illogical way life had become easier.]
> (1974: 7, 1993: 5)

Thus, from the outset, the reader's experience of the storyworld seems to be mediated by Allan. However, the narrator – or what I would rather call the implied author – is not completely absent from the narrative but occasionally intervenes, through free indirect speech, with subtle commentaries that are at times difficult to distinguish from Allan's own thoughts, but which make the reader question the character's way of understanding his own world. For instance, Allan sees the Dump as the ideal place for a withdrawal away from the urban agglomeration, which might evoke at first a pastoral retreat, since the Dump bears the last traces of nature as it used to be:

> De to-tre husene som var blitt stående på denne strekningen, var snart blitt fraflyttet . . . En kunne her og der skimte ruinene av dem dypt inne mellom kratt og villnis som var alt som var igjen av de *idylliske hageanleggene*, en naken gavl, en røys av sten nedgrodd av elde, av subtropiske hagevekster som trivdes i vill tilstand.
>
> [The two or three houses left had soon been abandoned . . . Here and there, their ruins could just be seen in the undergrowth and thickets, all that was left of *idyllic gardens* – a bare gable, a heap of stones overgrown by the years, a subtropical garden enjoying its wild state.]
>
> (1974: 14, my emphasis; 1993: 10, my emphasis)

However, the reader soon finds out that it is not the nostalgia of these 'idyllic gardens' that draws Allan to the Dump. In fact, the wistful voice that speaks in this paragraph of the reminiscence of bygone times is more likely to express the implied author's stance, rather than Allan's reflections. The protagonist's understanding of nature later appears to be extremely alienated from the implied author's perspective. When he visits the countryside, Allan realises that living close to nature is not as he had imagined; he is rather bored by what rural life has to offer and frightened by the wildness of animals (1974: 30–33). But, inspired by a documentary about the South Sea Islands, Allan still dreams of escaping to a place where he could find harmony and peacefulness. However, lush nature is not what Allan is looking for. Instead, the idyll is precisely what the Dump represents:

> Det var slik han siden hadde drømt om at det ville være, borte fra byen, ute på landet. Iallfall lignende, idyllisk og fredelig . . . Han følte den som et konkret nærvær her ute, enda dette var en 'natur' de kjente, det var en 'natur' bygd opp av de samme grunnstoffer som deres egen tilværelse var bygd opp av, det var *deres* 'natur'.
>
> [He had dreamt it would be like that away from the city, out in the country. Or something like it, idyllic and peaceful, happy and harmonious . . . Out there on the Dump he felt a concrete presence, 'a countryside' they knew, a 'countryside' based on the same basic materials as their own existence – it was their 'countryside'.]
>
> (1974: 41–42, 1993: 33)

If we take a closer look at the translation of this fragment, it is interesting to notice how the word 'nature' is replaced in the English version with 'countryside,' which is probably meant to emphasise how nature is understood in opposition to the urban space, as a setting untouched by industrialisation and technological development. In this way, it would not be wrong to assume that, to Allan, countryside is the equivalent of nature. Ironically enough, although he sees it as an alternative to his life in the megalopolis, the Dump is not the opposite of the city, but a symptom of its own collapse. Besides, the distance between the implied author and the character also produces subtle irony, for, while the implied author acknowledges that there were 'dårlige tider' [the times were bad] (1974: 15, 1993:11), to Allan, 'var dette gode tider, lovende tider' [the times were good, even promising] (1974: 15, 1993: 12). This gap is also echoed by a distance between the reader's outlook on the storyworld and the character's own thoughts. The world to which the reader is transported is undoubtedly a doomed version of the future, which Allan, however, perceives as an opportunity. This seems a rather erroneous outlook, given the impending collapse described in the first pages of the novel. Therefore, the immersion in the storyworld through identification with the character becomes ambiguous and challenging. Caught between the implied author's gloomy, pessimistic stance, on the one hand, and Allan's optimistic and hopeful attitude, on the other, the reader might understand this dystopian pastoral as a sort of 'best-case scenario' to expect in a future where failure is taken for granted. In the end, Allan's wishful thinking generates a pessimistic perspective, as the only possible solution this scenario proposes for dealing with climate change remains to see the good in the bad.

Besides, a possible restoration of the lost world seems to be impossible for the simple reason that the novel's protagonists are alienated from the traditional idea of nature. Although challenging our understanding of nature is something environmental literature is indeed supposed to do, *Twilight Country* presents a gloomy future in which humans do not take into account responsibility for other-than-human beings. For instance, when Allan's father nostalgically talks about 'noe han kalte "Svaler"' [something he called 'swallows'], Allan simply cannot relate to his father's experiences:

> Slike ting, og andre som Allan hadde forstått like lite av – da – hadde han sittet der på balkongen i sommerheten og snakket om, mens asken dalte ned over byen og la sitt klebrige, grå og etsende forfallsbelegg over alt og alle.
>
> [He had talked like that, and about other things of which Allan had then understood little, on the balcony in the heat of summer, as ash floated down over the city, leaving a sticky grey layer of decay on everything and everyone.]
>
> (1974: 48, 1993: 38)

The irony contained here engenders an even wider gap between reader and character. As Allan is completely ignorant of the importance of other species for the ecosystem, the world around him slowly falls into ruin, precisely because his society has failed to acknowledge the entanglement between human and non-human as the basic principle of ecological balance. The futility of any adaptation scenario, without a critical stance towards the state of the world, fully emerges then in all its irony.

Zooming out, it can be argued that lack of concrete references to the actual world also impedes immersion in the textual world, which remains too abstract, as Lehtimäki similarly indicates in his discussion of McEwan's *Solar* and Offill's *Weather* (Chapter 1, this volume). Sweetwater epitomises the Western capitalist world working towards economic profit until resources are completely depleted. Textual cues, such as English toponyms, names, and nicknames – Sweetwater, Abott Hill Road, Mary Diamond, Smiley, Doc, Boy, Run-Run – suggest the anchoring of the storyworld in the anglophone world. However, by the lack of concrete geographical landmarks, the setting remains too loosely attached to a specific reality, preventing readers from bridging the real and the fictional worlds. The temporal setting, besides the internal chronology of the novel spanning approximately three years, is also vague. In this respect, the title casts a gloomy light over the novel's scenario from the very beginning. *Uår*, which in English could be translated as 'un-year,' is a word 'used in Old Norse literature to designate years of natural disasters, failed harvests, famine and epidemics' (Hennig, 2013: 46). Avoiding more coherent and explicit references to the actual world, the narrative rather builds upon an ominous rhetoric of the humanity's doomsday. Besides, the absence of any ties between the actual world and the textual world engenders a rupture that can lead to a passive attitude in the reader, as, in the climate change scenario depicted, humanity's failure to address ecological issues is envisaged as the only possible outcome. As Richard Kerridge notes:

> In part, the problem is that conventional plot structures require forms of solution and closure that seem absurdly evasive when applied to ecological questions with their extremes of timescale and complexities of interdependency. And that is the challenge for novelists and ecocritical theorists of narrative, if we are to have environmentalist novels that do not take failure for granted.
>
> (2002: 99)

Indeed, it can be argued that the lack of a more complex narrative structure in the *Uår* novels entails a faulty understanding of the large-scale implications of the ecological collapse. As Axel Goodbody and Adeline Johns-Putra explain, 'Particular difficulties in representing climate change in literary or filmic narrative result from the complexity of its causes and manifestations, and the discrepancy between its enormous spatial and temporal scale and that of individual human experience' (2019: 10).[5] In this respect, compared to Faldbakken's account of the impending ecocatastrophe, Maja Lunde's recent climate series can represent a more satisfactory literary response to this challenge, as the next section will seek to demonstrate.

Spatial and temporal entanglements in Maja Lunde's novels

In stark contrast with Faldbakken's *Uår* novels, Lunde's *The History of Bees*, *The End of the Ocean*, and *The Last Wild Horses* are all structured upon broad narrative frameworks, both taken separately and as a series. The three novels thus position human and non-human beings within a wider web of entanglements with their environments, exploring the roots and the fallouts of the unfolding ecological crisis. Formally, the fragmentary plot structure intermingles three or two narrative planes which alternately move between past, present, and future scenarios.

Opening the series, *The History of Bees* narrates stories of three characters. Tao lives in China in 2098 when, after the complete extinction of bees, trees and plants must be hand pollinated to ensure balance in the ecosystem. George, a beekeeper living in Ohio, USA, in 2007, faces the sudden disappearance of his bees, caused by a phenomenon known as Colony Collapse Disorder. Finally, in Hertfordshire, England, we meet William living in the year 1852, a naturalist working to construct a new beehive model. *The End of the Ocean* tackles the issue of global warming and water scarcity, and is built on two narrative threads. The first one is set in 2017, where we meet Signe, a former journalist and environmental activist from Norway, while the second one takes place in 2041, in France, during a drought. In this dystopian scenario of the near future, David and his daughter Lou, separated from the rest of their family, arrive at a refugee camp and strive to survive in the hope of a positive outcome. *The Last Wild Horses* finally depicts three scenarios set in Russia in 1880, Mongolia in 1992, and Norway in 2062. Eva, living at her family farm in Viken, Norway, struggles to provide for her daughter and the animals that remain in her care, while the whole of Europe faces drought, famine, and lack of resources. Mikhail is a zoologist living in Russia in the nineteenth century, who initiates the journey that will bring to Europe the first wild horses discovered in Mongolia. Meanwhile, at the end of the twentieth century, Karin is a veterinarian who travels to the Mongolian steppes in the attempt to bring the species back to its natural habitat.[6]

While these novels, especially *The History of Bees* and *The Last Wild Horses*, engage with more-than-human issues, paying heed to the lives of other species, their first-person narration privileges an anthropocentric perspective that focuses on human experiences. Nonetheless, relocating to a different storyworld with each chapter, readers alternately identify with characters living in contrasting environments and actively participate in the unravelling of the global meaning of the text. The first and the third novels of the series begin by directly transporting the reader to a dystopian world. In *The End of the Ocean*, the opening chapter first introduces Signe, and David appears in the second chapter. However, it is noteworthy that all these characters briefly expose the historical context describing how climate change irrevocably transformed the places they inhabit, thus facilitating the understanding of the future world the novels depict. Tao relates how 'Biene her hadde forsvunnet allerede i 1980-årene, lenge før Kollapsen, sprøytemidler tok livet av dem' [The bees here had disappeared back in the 1980s, long before The Collapse; pesticides had done away with them] (2015: 8, 2018: 1). David recounts how drought has slowly made living conditions impossible, forcing him and his family to abandon finally their hometown in 2041: 'først het det toårstørke, så tre år, så fire. Dette var femte året. Sommeren virka uten ende. . . . Men strømmen kom og gikk, butikkene tømtes for matvarer, og byen ble stadig tommere, stillere. Og varmere' [First it was called the two-year drought, then the four-year. This was the fifth year. The summer seemed to be without end . . . the power came and went, the stores were emptied of food staples and the city became emptier, quieter. And hotter] (2017: 27, 2019b: 25). In the third volume of the series, set another twenty years further into the future, Eva portrays an even worsening situation, with reference to both the drought in *The End of the Ocean* and the disappearance of bees in *The History of Bees*: 'Hele Europa *gikk*, uten retning, uten mål. I mange år hadde folk vandret, tørke drev dem på flukt. Nord stengte grensene allerede da jeg var ungdom. Så kom Kollapsen. Og krigen' [All of Europe was walking, in a state of disorientation, without a destination. For many years people had been displaced; the drought drove them into exile. The borders were closed in the north when I was still a child. Then The Collapse came. And the war] (2019a: 14, 2022: 7).

However, these dystopian scenarios take up only a third of each novel, and a realist strand remains paramount in Lunde's climate fiction.[7] Their narrative structure explores past and present events in order to inspire reflection on the future. To quote Caitlin Anderson, 'Rather than operating progressively without reference to the past, Lunde emphasises the importance of reaching back in time in order to holistically situate and scale human responsibility in an extensive, deep time framework' (2022: 24). Besides, by alternating between a familiar and a dystopian environment with each storyline, Lunde engenders feelings of regret, nostalgia, and concern for the places we usually take for granted. In this way, readers are invited to critically reflect upon the causes and effects of the impending catastrophe, which can further inspire them to take concrete measures aimed at mitigating the consequences of climate change. In this respect, Hennig points out how *The History of Bees*, which was the best-selling novel of 2017 in Germany, inspired voters 'to sign a petition supporting the protection of biodiversity under the motto "save the bees," forcing the state parliament to make the petition's demands into law' (2021: 127).

When discussing innovative narrative strategies in Norwegian contemporary novels addressing climate change, Hennig views Lunde's series as lacking textual complexity and asserts that 'While there are interconnections between these stories, these are mainly restricted to the motif level and usually easy to discern, since the respective chapters tend to follow on from each other' (2021: 126). To this, I would add that, without over-simplifying the discussion on global warming, Lunde successfully rises to the challenge of depicting climate change as a phenomenon unfolding on a large temporal and spatial scale, while raising awareness of the ecological crisis in a comprehensible manner, accessible to a wide range of readers. For this reason, her series can successfully be used as a didactic tool, fulfilling an important function of the cli-fi novel (Loock, 2016: 7).

Even if the use of multiple temporal axes is not necessarily a formal innovation, it is interesting to take a closer look at how temporality functions internally in Lunde's novels by analysing the use of verbal tenses. All the stories in the first volume are narrated in the past tense. However, in *The End of the Ocean* and *The Last Wild Horses*, the narrators Signe and Karin, living respectively in 2017 and 1992, tell their stories in the present tense, while David and Eva employ the past tense, despite their narratives being set in 2041 and 2064. This curious use of different verbal tenses in the different storylines opens the possibilities of the novel. Because recounting the story in the past involves a retrospective look on the events, I would argue that the use of the past tense in the dystopian storylines conveys the idea of continuity and generates a feeling of hopefulness. The prospect of first-person narrators looking back and recounting their past suggests a positive outcome for the characters. Paying attention to the use of verbal tense can thus become an important clue for interpreting the text, as the open denouements of each novel, despite their optimistic tone, do not explicitly elucidate the fate of their characters. As Anna Kvam (2020) in her reading of *The End of the Ocean* similarly suggests, the possibility of surviving the ecocatastrophe and finding solutions that enable the characters to tell their stories in its aftermath conveys a dose of hope for the reader.

Character identification and empathy

In 'Transported into a Story World: The Role of the Protagonist,' Anneke de Graaf and Lettica Hustinx link the process of transportation with a central element of narratives: the characteristics of the protagonist. They seek to demonstrate that transportation can be facilitated by character identification, as 'readers of a story who identify with a character also experience the story from within and feel emotions in reaction to the narrative events' (2015: 117). Relying on Dolf Zillmann's Affective Disposition Theory, de Graaf and Hustinx explain that the emotions and attitudes of the reader are influenced, throughout the reading process, by her or his dispositions towards characters, or, in other words, by 'the extent that [she or he] cares for the character' (2015: 119).[8] Their study demonstrates that 'The protagonist's portrayal as sympathetic, neutral, or unsympathetic influenced the extent to which readers were transported' (2015: 130). Although the hypothesis that 'The effect of the protagonist's portrayal on beliefs will be mediated by transportation into the story world' (2015: 122) was not confirmed by their study, I propose it as a possible explanation for the contrastive worldviews conveyed by Faldbakken's and Lunde's novels.

The previous section of this chapter also largely touched upon the question of character identification to explain the difficulty of becoming immersed in Faldbakken's story from Allan's perspective. However, based on the idea that 'the dispositions that readers have toward characters are dependent on the character's moral conduct' (de Graaf and Hustinx, 2015: 120), I wish to further suggest that the ethical dimension of the storyworld influences the reader's attitudes in a positive or negative direction. Compassion for and attachment to characters are dictated by the attitudes of these characters in the storyworld: 'sympathetic behaviour and personality should lead to a positive disposition, which evokes empathy with the character, whereas unsympathetic behaviour and personality should lead to a negative disposition, which evokes counter-empathy' (2015: 120). Referring to strategies that engender positive feelings in climate fiction, Schneider-Mayerson has also shown that 'most of the positive responses have been elicited through identification with sympathetic and resilient characters' (2018: 489). However, it is important to bear in mind that characters are not always evenly represented in the narrative, which can also influence the reader's ability to identify with them. As Wayne C. Booth notes, 'in centering our interest, sympathy, or affection on one character, [the novelist] inevitably excludes from our interest, sympathy, or affection some other character' (1983: 78–79). Similarly, Martha Nussbaum argues that frequently 'the text will cultivate sympathy unevenly, directing our attention to some types of human beings and not to others' (1997: 100). On this account, I will now touch upon the question of character representation in the two climate fictions by Faldbakken and Lunde, and I will discuss how the representation of vulnerable characters becomes a key issue in depictions of climate change.

Cultivating hope through care and compassion

In her reading of *The History of Bees*, Ylva Frøjd brings to light the multiple levels on which vulnerability is displayed in the novel, drawing mainly on Judith Butler's understanding of vulnerability as the necessary premise for creating a global community (Frøjd, 2021: 212). Despite being associated with negative experiences, vulnerability can become an important ethical resource once it leads to compassion and empathy, and thus a great value can be attached to it. On this premise, Frøjd analyses the interactions between the characters in the novel focusing on the way different expressions of vulnerability invite readers to reflect on the ethical responsibilities in the context of climate change.

Because the future generations are the ones most vulnerable to the effects of the ecological crisis, children often become a central motif in climate fiction, and 'the discourse of environmentalist crisis in the Anthropocene is peppered with . . . references to parental obligations to posterity' (Johns-Putra, 2019: 5). In *Uår*, children are also highly exposed to risk and therefore need to be secured. Preoccupied by their son's health and well-being, Allan and Lisa discuss the advantages of moving to the Dump:

> Tenk hvor mye han kan finne å leke med her, hvor mye friere han vil få det . . . Dette hadde han også tenkt på. Den trange leiligheten i April Avenue, bare med adgang til en helsefarlig bakgård og til gaten, hadde vært det rene fengslet for gutten.
>
> [What a lot of things he'll be able to play with here. How much freer he'll be! . . . He had thought so, too. Their cramped apartment with access only to the street and an unhealthy backyard had been prison for the boy.]
>
> (1974: 21, 1993: 16–17)

In Allan's eyes, although children mean more responsibilities, they also bring hope, and Lisa's pregnancy is for him a sign of health and vitality. However, representation of children is ambiguous in Faldbakken's novel, because their voice is almost entirely absent from the narrative. Boy – whose name deprives him of an actual identity – is not considered by Allan as an individual worth listening to. When he returns home after spending a day with his new friend Run-Run and enthusiastically tells his mother that his new friend taught him how to fish, Lisa makes an effort to show support and talks to him, although she feels rather insecure about the maternal role she has to play. However, when Allan makes his appearance, he gets angry at this news, because he perceives Run-Run as a dangerous person. Moreover, because he does not understand how Boy can communicate with Run-Run, who does not speak, Allan concludes that 'Det hadde ingen hensikt å diskutere med en femårings fantasi' [There was no point arguing with what went on in a five-year-old's imagination] (1974: 136, 1993: 109). Both Allan and Lisa lack the ability to show compassion towards their five-year-old boy, who is rather neglected and even brutalised by his father in an effort to persuade him to talk.

Children are essential characters in Lunde's cli-fi novels, and their voice is used to transmit important messages about the effects of climate change. Dialogues are one strategy that allows children to express themselves in the narrative. In *The History of Bees*, for instance, Tom, George's son who comes home from college to visit his parents and helps his father build new beehives, is the one who makes George contemplate upon human responsibility towards nature: 'Biene dør, pappa. Det er bare vi som kan gjøre noe med det' [The bees are dying, Dad. We're the only ones who can do anything about it] (2015: 389, 2018: 288). Another strategy is building the narrative scheme upon a plurality of perspectives. *The Last Wild Horses* is built on a more fragmentary structure than the two earlier volumes, which grants the reader access to multiple viewpoints on the storyworld. Eva's narration is succeeded at times by letters her teenage daughter, Isa, writes to her friend Lars. However, because Lars and his family have left and Isa never gets the chance to actually send the letters to her friend, they function more like a diary, where she can freely give voice to her thoughts. In the middle of the novel, Louise, a character that first appears in *The End of the Ocean* as David's daughter, also gets to recount her life story in a letter she writes to Eva.[9] Moreover, the final chapter of the book is narrated by Karin's son, Mathias, who is a secondary character in the main storyline, accompanying her mother in the journey to return the wild horses to the Mongolian steppes in 1992. At the end, Mathias speaks from the vantage point of 2019, when he revisits Mongolia together with his own daughter.

It is interesting to notice how in Karin's and Eva's narratives, intergenerational care is paired with interspecies care, as both women wish to create a better future for their children and for the endangered animals they are responsible for. When she travels to Mongolia to reintroduce the wild horses back to their natural habitat, Karin takes her son Mathias with her. Mathias is a father himself, but he was not able to raise his daughter due to his history with substance abuse. The journey to Mongolia is therefore a chance for him to complete the recovery from drug addiction and to get closer to his mother. In the post-apocalyptic world depicted in Eva's story,

Isa lives alone with her mother at their farmhouse and animal park in Heiane, a place that has been owned by the family for three generations. When almost everyone has left the village, including her sister and her close neighbours, Eva refuses to do the same. Instead, affectively attached to her homelands and to the farm animals, she lives with the hope that she might be able to save them. She is mostly dedicated to the wild horses. Aware that these vulnerable animals are menaced with extinction, Eva dreams of the day everything would come back to normal, and the species could get a chance to survive. Despite Lunde's use of an anthropocentric perspective that privileges human-centred narration, her novels call for human responsibility for other species and, as Călina-Maria Moldovan shows in her reading of the climate series in a materialist and post-speciesist perspective, the urgency to acknowledge the intrinsic value of the non-human other (2021).

If animals are silent, children are given a voice to express their own understanding of the world they inhabit. In her letters to Lars, Isa writes about how much she misses having friends and contact with other people (Lunde, 2019a: 29). Although she appreciates the efforts made by her mother, she often feels suffocated by her presence and despises the exaggerated attachment and empathy towards animals, that, in Isa's opinion, make Eva blind to the needs of other humans: 'Til å være så opptatt av dyr er mamma lite oppdatert på menneskedyret. Hvordan vi bør leve, og hva vi trenger' [For being so concerned with animals, mom is not so updated on the human animal. How we should live, and what we need] (2019a: 29; my own translation). The reader understands, however, that Eva's devotion derives from an ecocentric ethical responsibility that she assumes: as Eva puts it, animals 'var der jo ikke for oss, vi var der like mye for dem' [weren't here for us, after all; it was more that we were here for them] (2019a: 15, 2022: 8). Her care is, however, not limited to non-humans, as it becomes evident when Eva, seeing Louise walking the streets alone in the rain, invites the woman to stay on the farm. The first-person narration contributes here prominently to fostering character identification, which Suzan Keen considers to be "The most commonly nominated feature of narrative fiction to be associated with empathy" (2007: 93). Besides identification through the subjective perspective, it is also her compassionate attitude, the disposition to always help those in need, the attachment to her homeplace, as well as her regrets and concerns that make Eva a sympathetic character. In the meantime, the reader is also granted access to Isa's intimate thoughts and can relate to her experiences.

Education is central in Eva's relationship to Isa because it represents a way of transmitting human values to her daughter, such as respect, kindness, empathy, and responsibility. But studying also serves as an opportunity to escape the harsh reality in which they live, as one particular scene suggests. After many days of incessant, heavy rain, the two women sit inside 'i stillhet foran ovnen' [in silence before the stove] (2019a: 397, 2022: 312), and Eva tells Lisa to focus on her lessons and not worry about the wind outside. But the girl cannot take her mind off the weather and gives voice to her worries: 'En dag kommer ikke vindusrutene til å klare mer. De kommer til å gi etter' [One day the windowpanes won't be able to take any more. They will give in' (2019a: 397, 2022: 313). Eva tries to distract her daughter by asking her about probability calculus and, when Isa replies that there is no point in doing mathematics if they are not safe, her mother suggests she should try to calculate what the probabilities for their windows to break actually are. Watching them, 'Louise smilte over stoppetøyet hun hadde i fanget' [Louise smiled above the mending she had in her lap] (2019a: 397, 2022: 313). This conversation, illustrating a typical familial scene in which parents try to come up with reasons for studying, probably evokes childhood memories both to Louise and the reader. Importantly, the reader's emotional response is mediated here by Louise's perspective. Bearing in mind that she had lost her family due to the extreme drought in France, seeing this scene through her eyes makes the reader consider how the environmental crisis can affect parent-child relationships. The struggle to protect her daughter from the harsh reality becomes evident when Eva admits to herself: 'Jeg var glad jeg klarte å tulle, at jeg klarte å skjule hvordan jeg egentlig følte meg' [I was glad I was able to joke, that I managed to conceal my true feelings] (2019a: 397, 2022: 313), showing her concern for making Isa feel safe, although this requires hiding her own worries and fears from her daughter.

In *Uår*, Allan also attempts to get involved in Boy's education, once he loses his job at the petrol station and finds the necessary time to dedicate to this purpose. Despite his initial enthusiasm, the reading lessons with his son do not go as hoped, as he often loses his temper over Boy's difficulty to learn, but also over his own inability to understand and explain things. And so, 'Som regel endte undervisningen med at han fløy opp i sinne, skjelte gutten huden full og jaget sin fullstendig uanfektede og sannsynligvis takknemlige elev tilbake til hans hemmelighetsfulle sysler annetsteds' [The lessons usually ended with him losing his temper, shouting at the boy and sending his totally unaffected and probably grateful pupil back to his secret pursuits elsewhere] (1974: 218, 1993: 171). In fact, because he uses old magazines as textbooks, Allan's attention is easily diverted by the pictures of 'avkledde kvinner' [scantily dressed women] (1974: 218, 1993: 171), and the episode directs the reader's attention to Allan's sexuality and masculinity.

One of the reasons why it is rather difficult to identify a form of genuine compassion and care in Faldbakken's cli-fi is that almost all interaction between characters becomes a pretext for sexual fantasies for the men in the novel, who remain the main focus of the series. It is important to place the *Uår* novels in the context of Faldbakken's authorship, as he belongs to a generation of writers interested in exploring how the modern man was affected by the feminist movement of the 1960s and 1970s. The majority of his novels are characterised by an 'erotic psychology' (Rottem, 1997: 396–398), and they often depict 'menns problematiske tilpasning til en ny samfunnsmessig situasjon' [men's problematic adaptation to a new social situation] (Rottem, 1985: 96, my own translation). Considering the socio-historical framework in which the novels were written, the representation of gender roles and masculine sexuality in *Uår* remains a complex subject that deserves a separate discussion. Here, I would like to pinpoint the importance of character representation when one reads the novel as climate fiction. It is necessary to bear in mind that today's readers most certainly experience the storyworld in *Uår* through a different set of values from that of 1970s readers, which makes identification with Faldbakken's characters less likely nowadays. In contrast, it seems that contemporary cli-fi shifts the focus from the depiction of crisis situations and problematic identities to actually offering examples of what kind of attitudes and values are necessary for coming to grips with climate change, as illustrated in Lunde's novels.

The friendship between Doc and Lisa is one of the few relationships portrayed in the *Uår* novels that are based on a desire to help the other. Doc offers Lisa medical assistance during her pregnancy and later offers to care for her when Allan sells her in exchange for a gun (Faldbakken, 1976: 110–112). But Doc's patriarchal position and the way he infantilises Lisa do not inspire an authentic form of compassion: 'Helt siden første gang han så henne, hadde han følt et faderlig instinkt, en trang til å vise omsorg for denne piken, dette inntagende barnet' [Ever since he had seen her, his paternal instinct had been aroused and he had felt a need to care for this young creature, this appealing child] (1974: 153, 1993: 120). Seen through the male gaze, Lisa is sexually objectified and portrayed as a weak character throughout the novel. Allan describes her as 'hans barnebrud, hans tenåringshustru' [his child bride, his teenage wife] (1974: 10, 1993: 7), but also as 'utrolig ubehjelpelig . . . upraktisk' [helpless and impractical] (1974: 24, 1993: 18), barely managing to fulfil her maternal role. In the brief episodes where the narration is focalised through Lisa, her thoughts only seem to confirm the male perspective:

> Men iallfall, tenkte hun der hun strevde, er han nær nok til at jeg kan rope på ham om jeg virkelig trenger hjelp. Det var også en trøst, for hun var ikke sterk, ikke utholdende, hennes oppmerksomhet løp hit og dit, lett å avlede, bringe på avveier.
>
> [As he struggled on, she thought, he is near enough so that I can call him if I really need help. That was also a consolation, for she was neither strong nor persevering, her attention easily diverted.]
>
> (1974: 59, 1993: 47)

Marta, Doc's wife, is similarly portrayed as a woman dependent on her husband's help. She does not even play a significant role in the plot development, as she almost always lays sick in bed, 'plaget av gikt' [plagued with arthritis] (1974: 75, 1993: 59), until she finally dies at the end of the first novel.

Mary Diamond, the most powerful female character of the series, also shows compassion towards Lisa when she meets the teenage mother. Mary Diamond consoles and encourages her when Lisa expresses her fears about giving birth, and then assists her during labour. Mary Diamond's willingness to help Lisa finds an echo in *The Last Wild Horses*, where Eva welcomes Louise into her house and offers her help to bring her baby into the world and raise the child. Nonetheless, the solidarity between the two women in *Uår* is scantly portrayed and, in their relationship, they are never equal. While Lisa is always defenceless and weak, Mary Diamond appears a mystified Mother Earth figure (Rottem, 1997: 401). When Lisa gets too sick to care for her newborn daughter, Rain, Mary Diamond gladly takes on the maternal role but, once Allan decides to leave the Dump, shows no more interest in the teenage mother.

The novel's final chapters depict the emergence of a new society, based on egalitarian values and collective participation to the reconstruction of the city. Separated once they arrive in the military basis of Sweetwater, Mary Diamond and Rain finally reunite with Allan and Boy after approximately five to six years. *Uår. Sweetwater* ends with Mary Diamond changing Rain's name to Alviva (1976: 240), which symbolically sets an optimistic tone to the story's denouement. Thus, the ending also shifts the focus from men to the female characters, suggesting that women occupy an important place in the new society. That said, the novel is largely focalised through masculine characters, leaving women and children in a marginalised position. Because they fail to show sympathy with other characters, male protagonists rather provoke counter-empathy, and make readers' identification with them difficult. Therefore, it could be argued that representing vulnerable characters, such as children, and sick or pregnant women, through a masculine, patriarchal gaze hinders a sympathetic response from readers.

In what ways, then, could climate fiction inspire hope, despite the dystopian scenarios it sets in place? In *The Last Wild Horses*, Eva's obstinate efforts to remain optimistic paradoxically engenders a feeling of helplessness; much like Allan's naïve optimism in *Twilight Country*. However, I argue that a genuine feeling of optimism derives from a shared feeling of care and compassion, despite all misunderstandings between the characters. Since the multitude of subjective narrative perspectives provides insight into multiple experiences of the storyworld, I propose reading Lunde's novel in light of Johns-Putra's model of 'eudaemonistic reading'. As she explains, this reading method highlights the capacity of character identification and reader's immersion in the storyworld to lead 'to the development of sympathetic acknowledgment of shared vulnerability with others and of a eudaemonistic desire to address that common vulnerability and promote a common flourishing' (2019: 45).

When one of the horses that had been set free returns to the farm, Eva sets the animal off when she tries to chase it away, and the horse attacks her, leaving her unconscious. Eva and Isa have an argument just before the accident happens. When Isa asks why she keeps feeding the horses, although they are running out of hay, her mother replies: 'Du kommer uansett ikke til å forstå hvordan jeg tenker' [You won't understand my thinking, no matter what I say] (Lunde, 2019a: 428, 2022: 340). Indeed, Isa often mentions in her letters to Lars how difficult it is to understand her mother's decisions, but, after Eva gets hurt, Isa writes about the love and gratitude she feels towards her mother: 'Jeg har aldri fortalt deg hvor utrolig glad jeg er i mamma. . . . Jeg vet ikke hva jeg skal gjøre hvis hun forsvinner. . . . Dra eller bli, det spiller ingen rolle hvis hun ikke er her. Kjære, kjære mamma' [I have never told you how incredibly much I love mum. . . . I don't know what I will do if she goes away. . . . Leave or stay, it doesn't mean anything if she's not here. Dear, dear mum] (2019a: 433; my own translation). The care Isa shows for her mother puts forward the importance of compassion for Eva as well:

> Hun satte seg på en stol ved sengen min, løftet hånden og strøk meg over pannen. Ansiktet var en revne av engstelse, men det var noe annet der også, noe jeg tidligere bare hadde sett små blaff av i møtet med dyrene, en altoppslukende omsorg, en slik omsorg der man glemmer seg selv.
>
> [She sat down on a chair beside my bed, lifted her hand and strokes my forehead. Her face was riddled with anxiety, but there was something else there, as well, something I had only seen in brief fleshes with the animals – an all-consuming compassion, the kind of caring that makes one forget about oneself]
>
> (2019a: 453, 2022: 359)

Always putting her daughter's health first, Eva forgets that she is just as exposed to risk. As a woman, Eva also remains vulnerable to men who try to take advantage of her, although she manages to stand up for herself in most of the cases. When a fisherman makes advances towards her, Eva confronts the man, making him feel ashamed. When her ex-partner Einar and his friend Peter break into Eva and Isa's home one night, Louise is not afraid to threaten the men with a toy gun to get them out. However, later, after Einar and his friends steal their food, Eva is ready to let Peter take advantage of her, because she knows this to be the only way to convince him to give back the stolen food.[10] Eva's story is one of the novel's examples that illustrate how those who care for others are themselves vulnerable.

In the relationship between Eva, Louise, and Isa, compassion is what leads to a strong friendship between the three women, suggesting that an empathetic response to vulnerability can create the basis for a community capable to face the challenges posed by the effects of climate change. Providing insight into all the characters' points of view and allowing access to their subjective understanding of the storyworld give an account of what Johns-Putra calls a 'sympathetic acknowledgment of shared vulnerability with others' (2019: 45). I thus argue that this shared feeling of sympathy endorses the possibility of hope in fictional depictions of ecocatastrophes.

The ability of characters to share their experiences and vulnerabilities with others, but also to respond with empathy and compassion, might thus influence the reader's understanding of the storyworld in a positive way. In contrast, in *Uår*, Allan 'har vokst opp i et samfunn, i en tid, som gjorde alt for å få kamuflert de ytre tegn på menneskelig lidelse og lidelsens kulminasjon, livets naturlige kulminasjon: døden' [had grown up in a society which had done everything to camouflage outward signs of human suffering and the culmination of suffering, the natural culmination of life – death] (1974: 250–251, 1993: 196). Failing to acknowledge the existence of pain and suffering, it becomes rather difficult for the characters in the *Uår* novels to display empathy as a response to others' vulnerability.

Nussbaum understands compassion as 'the recognition that another person, in some ways similar to oneself, has suffered some significant pain or misfortune in a way for which that person is not, or not fully, to blame' (1997: 90–91). Compassion, she adds, requires 'a sense of one's own vulnerability to misfortune' and 'promotes an accurate awareness of our common vulnerability' (1997: 91). For Nussbaum, literature has the ability to inspire prosocial action and form democratic citizens when sympathetic reading is paired with critical reading (1997: 101). While the failure to convey compassion towards vulnerable characters in the *Uår* novels might elicit counter-empathy and negative attitudes in the reader, Lunde proposes alternative scenarios in which relationships between characters are essentially based on sympathetic attachments. If cli-fi novels have the potential to influence readerly attitudes, strategies of identification with characters and compassion towards them can become essential in eliciting positive responses to the imagined storyworlds.

With vulnerability being understood as a great ethical resource, Frøjd (2021) places emphasis on how Lunde's cli-fi can be explored in an educational context. The same didactic impulse is investigated by Furuseth (2021), who argues that the novels in the climate series are 'påfallende didaktiske' [strikingly didactic] (2021: 86, my own translation) and shows how the dialogues between characters can lead to fruitful discussions on sustainable development in the classroom. Although didacticism often raises suspicion among literary scholars, a certain didactic impulse might become an essential component for creating a sense of hope in climate fiction, in a way that could, in the end, change the reader's attitudes towards the unfolding climate crisis. Ecocritical readings of the novels, such as those performed by Frøjd and Furuseth, suggest that this type of narrative is much needed for discussing environmental issues and nature-culture relationship in the humanities. Of course, novels such as *Uår* can also serve didactic purposes and as a way of helping readers to become aware of certain phenomena that are at the root of the environmental crisis. However, as I have tried to show throughout my comparative analysis of the two novel series, the narrative strategies of character identification in Lunde's books are based on the reader's emotional involvement, which may facilitate her or his understanding of the storyworld.

Conclusions

In the 1960s and 1970s, when apocalyptic rhetoric dominated public discourses on the ecological crisis, fiction writers aimed to 'reveal' what the climate apocalypse would look like (Buell, 2010: 18) and to 'purvey doom' (20) as a way of raising awareness. Faldbakken's *Uår* series is pervaded by the same ominous imagery. In contrast, Lunde's climate series is rather focused on depicting human and interspecies relationships against the backdrop of the climate crisis in a way that assumes responsibility for the reader's experience of the storyworlds and for her or his understanding of climate change. When studying the impact cli-fi literature has upon readers, central questions that emerge are: what kind of emotions do literary characters elicit, and how are readers changed by their experience after being transported to the storyworld? Econarratology ultimately makes an important contribution to understanding the mechanisms cli-fi narratives employ to depict climate change and its implications for the world we live in.

We can further link the econarratological reading of Faldbakken's and Lunde's novels with what James Phelan names the ethics of the telling, which draws attention to the 'ethical responsibility of storytellers to their audience,' 'the ethical dimensions of the narrative's techniques' and how 'the use of these techniques [implies] and [conveys] the values underlying the relations of the storytellers . . . to their materials . . . and their audiences' (2014: 531). My analysis of the two cli-fi series has tried to underline that those positive feelings related to hope and compassion have the potential to elicit action aimed at mitigating the effects of climate change, while apocalyptic scenarios and lack of empathy in characters convey a sense of helplessness and passivity. Therefore, I suggest that the idea of hope in recent Norwegian cli-fi literature is strongly related to a mode of ethical storytelling that, paired with a didactic impulse, takes account of the vast ramifications of climate change, while inciting to debate and challenging the reader's beliefs and worldviews.

Notes

1 In 2022, the last volume in Lunde's climate series was published in Norwegian with the title *Drømmen om et tre* ("The Dream of a Tree"). The novel was not yet translated into English. Other Norwegian writers who have contributed to the rise of cli-fi are Jostein Gaarder, Gert Nygårdshaug, and Aksel Selmer. However, Norwegian cli-fi can be considered as part of a wider Scandinavian tradition; see Flinker (2018).

2 The translation of the *Uår* novels was published in the UNESCO Collection of Representative Works. The project was initiated in 1948, and its aim was 'to make available in translation in two international languages, English and French, the masterpieces of world literature' (Maunick, 1986: 6). The essential criteria for the works proposed were to be considered 'representative of the values of their cultures' (8).

3 Club of Rome is 'a group of leading economists and scientists which studies the effects of economic growth from 1968 onwards' (Rutheford, 2013: 92). In 1972, the group founded the publication of *The Limits to Growth: A Report for the Club of Rome's Project on the Predicament of Mankind*, written by Donella H. and Dennis L. Meadows, Jørgen Randers, and William W. Behrens III. In '*The Limits to Growth* Revisited,' Tim O'Riordan shows how the book introduced the idea of limits to 'unending growth' and had an important contribution to the environmental debate, mostly due to its use of computer-based complex modelling to forecast the future of humankind (2012: 93–96).

4 Flinker borrows the term from Darko Suvin's *Metamorphoses of Science Fiction* (1979).

5 The challenge of scale framing climate change in fictional texts has been discussed by Heise (2008) and Trexler (2015).

6 The Norwegian title of this volume, *Przewalskis hest* [Przewalski's Horse], makes reference to the geographer Nikolay Przhevalsky, who discovered the existence of the Takhi horse in the Mongolian steppes. The English version of the title, *The Last Wild Horses*, emphasises the fact that these horses, considered to be the only remaining purely wild horses, are an endangered species.

7 In the Norwegian editions, at the end of each novel, readers can find a list of resources used by the author. Scientific references can thus endorse the arguments conveyed by the novels.

8 See, for instance, Zillmann (1994).

9 The letters Isa writes to Lars have not been translated in the English version of the novel.

10 Just like in the case of Faldbakken's novels, a discussion on gender representation in Lunde's series is worth an in-depth discussion. Analysing *The History of Bees* from an ecofeminist perspective (2019), Hennig explains that although it is generally women who are stereotypically portrayed as altruistic and caring, traditional gender roles are questioned throughout the novel.

Bibliography

Anderson, Caitlin (2022), 'Bee and Tree Temporality in *The History of Bees* and *The Overstory*', *Journal of Ecohumanism*, 1:1, 19–30, https://doi.org/10.33182/joe.v1i1.1844

Booth, Wayne C. (1983 [1961]), *The Rhetoric of Fiction*. Second Edition (Chicago: The University of Chicago Press).

Buell, Frederick (2010), 'A Short History of Environmental Apocalypse', in Stefan Skrimshire (ed.), *Future Ethics: Climate Change and Apocalyptic Imagination* (London: Continuum), 13–36.

Buell, Lawrence (1995), *The Environmental Imagination: Thoreau, Nature Writing, and the Formation of American Culture* (Cambridge, MA: Harvard University Press).

Cohen, Jonathan (2001), 'Defining Identification: A Theoretical Look at the Identification of Audiences with Media Characters', *Mass Communication and Society*, 4:3, 245–264, https://doi.org/10.1207/S15327825MCS0403_01

De Graaf, Anneke and Lettica Hustinx (2015), 'Transported into a Story World: The Role of the Protagonist', in André Lardinois, et al. (eds.), *Texts, Transmissions, Receptions* (Leiden: Brill), 114–132.

Faldbakken, Knut (1974), *Uår: Aftenlandet* (Oslo: Gyldendal).

Faldbakken, Knut (1976), *Uår. Sweetwater* (Oslo: Gyldendal).

Faldbakken, Knut (1993), *Twilight Country*, trans. Joan Tate (London: Peter Owen).

Faldbakken, Knut (1994), *Sweetwater*, trans. Joan Tate (London: Peter Owen).

Flinker, Jens Kramshøj (2018), 'Den skandinaviske cli-fi: En ny genre i Antropocæn?' ['Scandinavian Cli-fi: A New Genre in the Anthropocene?'], *Spring*, 42, 41–66.

Frøjd, Ylva (2021), 'Sårbarhet i Maja Lundes *Bienes historie*: En økokritisk lesning' [Vulnerability in Maja Lunde's *Bienes Historie*: An Ecocritical Reading], in Adriana Margareta Dancus, Silje Hernæs Linhart and Annika Bøstein Myhr (eds.), *Litteratur og sårbarhet* (Oslo: Universitetsforlaget), 211–229.

Furuseth, Sissel (2019), 'Håpets betydning i norsk klimafiksjon' [The Meaning of Hope in Norwegian Climate Fiction], *Forfatternes Klimaaksjon*, [Online], [Accessed 2 September 2021], available from: https://forfatterne sklimaaksjon.no/2019/08/12/sissel-furuseth-hapets-betydning-i-norsk-klimafiksjon/

Furuseth, Sissel (2021), 'Bærekraftdilemmaer i Maja Lundes klimaromaner' ['Sustainability Dilemmas in Maja Lunde's Climate Novels'], in Marcus Axelsson and Barbro Bredesen Opset (eds.), *Fortellinger om bærekraftig utvikling: Perspektiver for norskfaget* [Narratives of Sustainable Development. Perspectives for the Norwegian Subject] (Oslo: Universitetsforlaget), 85–102.

Gerrig, Richard (1999 [1993]), *Experiencing Narrative Worlds: On the Psychological Activities of Reading* (New Haven: Westview Press).

Goodbody, Axel and Adeline Johns-Putra (2019), 'Introduction', in Axel Goodbody and Adeline Johns-Putra (eds.), *Cli-Fi: A Companion* (Oxford: Peter Lang), 1–17.

Green, Melanie C. and Timothy C. Brock (2000), 'The Role of Transportation in the Persuasiveness of Public Narratives', *Journal of Personality and Social Psychology*, 79:5, 701–721, https://doi.org/10.1037/0022–3514.79.5.701

Heise, Ursula K. (2008), *Sense of Place and Sense of Planet: The Environmental Imagination of the Global* (Oxford: Oxford University Press).

Hennig, Reinhard (2013), '"It Is Immoral to Be a Pessimist" Climate Change Adaptation and Mitigation in Norwegian Literary Fiction', *Tvergastein. Interdisciplinary Journal of the Environment*, 3, 44–50.

Hennig, Reinhard (2019), 'Å trosse naturen i seg selv: Kjønn og kjønnsroller i Maja Lundes roman *Bienes historie*' [Defying the Nature in Oneself: Gender and Gender Roles in Maja Lunde's *Bienes Historie*], in Nora Simonhjell and Bendikt Jager (eds.), *Norsk litterær årbok* (Oslo: Det norske samlaget), 272–290.
Hennig, Reinhard (2021), 'Anthropocene Aesthetics: Norwegian Literature in a New Geological Epoch', *NORDEUROPAforum: Journal for the Study of Culture. Special Issue: Changing Concepts of Nature in Contemporary Scandinavian Literature and Photography*, 105–130.
Herman, David (2009), *Basic Elements of Narrative* (Oxford: Wiley-Blackwell Publishing).
James, Erin (2015), *The Storyworld Accord: Econarratology and Postcolonial Narratives* (Lincoln: University of Nebraska Press).
James, Erin and Eric Morel (2020), 'Introduction: Notes Toward New Econarratologies', in Erin James and Eric Morel (eds.), *Environment and Narrative: New Directions in Econarratology* (Columbus: The Ohio State University Press), 1–24.
Johns-Putra, Adeline (2016), 'Climate Change in Literature and Literary Studies: From Cli-fi, Climate Change Theatre and Ecopoetry to Ecocriticism and Climate Change Criticism', *WIREs Climate Change*, 7:2, 266–282, https://doi.org/10.1002/wcc.385
Johns-Putra, Adeline (2019), *Climate Change and the Contemporary Novel* (Cambridge: Cambridge University Press).
Keen, Suzanne (2007), *Empathy and the Novel* (New York: Oxford University Press).
Kerridge, Richard (2002), 'Narratives of Resignation: Environmentalism in Recent Fiction', in John Parham (ed.), *The Environmental Tradition in English Literature* (Farnham: Ashgate), 87–99.
Kvam, Anna (2020), 'Skandinavisk klimalitteratur: "Skrøbelige handlingsmuligheder" eller "dystopisk lysning"?' ['Scandinavian Climate Literature: "Fragile Possibilities of Action" or "Dystopian Lightning"?'], *Salongen – Nettidsskrift for filosofi og idéhistorie*, [Online], [Accessed 29 April 2022], available from: www.salongen.no/skandinavisk-klimalitteratur/
Lehtimäki, Markku (2013), 'Natural Environments in Narrative Contexts: Cross-Pollinating Ecocriticism and Narrative Theory', *Storyworlds: A Journal of Narrative Studies*, 5, 119–141, https://doi.org/10.5250/storyworlds.5.2013.0119
Loock, Kathleen (2016), 'Cli-fi and the Dystopian Tradition', in *The Dystopian Impulse of Contemporary Cli-Fi: Lessons and Questions from a Joint Workshop of the IASS and the JFKI (FU Berlin)* (Potsdam: Institute for Advanced Sustainability Studies), 6–8.
Lunde, Maja (2015), *Bienes Historie* (Oslo: Aschehoug).
Lunde, Maja (2017), *Blå* (Oslo: Aschehoug).
Lunde, Maja (2018), *The History of Bees*, trans. Diane Oatley (London: Simon and Schuster).
Lunde, Maja (2019a), *Przewalskis hest* (Oslo: Aschehoug).
Lunde, Maja (2019b), *The End of the Ocean*, trans. Diane Oatley (London: Simon and Schuster).
Lunde, Maja (2022), *The Last Wild Horses*, trans. Diane Oatley (New York: Harper Collins).
Maunick, Edouard J. (1986), 'A Library of World Classics', *The UNESCO Courier*, 39, 5–8.

Moldovan, Călina-Maria (2021), 'Human and Non-human Representations in Maja Lunde's Fiction: A New Materialist and Post-Speciesist Approach', *Studia Scandinavica*, 5:25, 51–62, https://doi.org/10.26881/ss.2021.25.03

Nussbaum, Martha (1997), 'The Narrative Imagination', in *Cultivating Humanity. A Classical Defense of Reform in Liberal Education* (Cambridge, MA: Harvard University Press), 85–112.

O'Riordan, Tom (2012), '*The Limits to Growth* Revisited', in Sofia Vaz (ed.), *Environment. Why Read the Classics?* (Sheffield: Greenleaf Publishing), 93–113.

Phelan, James (2014), 'Narrative Ethics', in Peter Hühn et al. (eds.), *Handbook of Narratology*. Volume 2 (Berlin and Boston: De Gruyter), 531–546.

Rottem, Øystein (1985), 'Sprekker i mannsbildet. Om Knut Faldbakkens "Adams dagbok" og Erling Gjelsviks "Dødt løp"' ['Cracks in Men's Identity. About Knut Faldbakken's "Adams Dagbok" and Erling Gjelsvik's "Dødt Løp"'], *Vardøger*, 16:85, 86–112.

Rottem, Øystein (1997), 'Knut Faldbakken – Kulturradikal 68'er', in Edvard Beyer (ed.), *Norges Litteraturhistorie. Inn i medietidsalderen*. Bind 7 (Oslo: J.W. Cappelens Forlag A/S), 396–407.

Rutherford, Donald (2013 [1992]), 'Club of Rome', in *Cambridge Dictionary of Economics*. Third Edition (London: Routledge), 92.

Schneider-Mayerson, Matthew (2018), 'The Influence of Climate Fiction: An Empirical Survey of Readers', *Environmental Humanities*, 10:2, 473–500, https://doi.org/10.1215/22011919-7156848

Trexler, Adam (2015), *Anthropocene Fictions: The Novel in a Time of Climate Change* (Charlottesville: University of Virginia Press).

Zillmann, Dolf (1994), 'Mechanisms of Emotional Involvement with Drama', *Poetics*, 23:1–2, 33–51, https://doi.org/10.1016/0304-422X(94)00020-7

3 Narrating the economic value of nature in the Anthropocene

Xin Liu

Introduction

What do economising narratives do to imaginations of and sensibilities about the environment? In recent years, the language of economic value provides the dominant grammar and vocabulary that frame public and political debates about environmental sustainability. The economising narratives have been the object of numerous critiques. However, there is much less analysis of *how* nature is configured in economic terms. This chapter addresses that question. In so doing, it aims to challenge assumptions about nature and economy, and to enable alternative imaginations of the relation between the global and the local, deep time and embodied experiences, as well as human society and the more-than-human. The chapter argues that close reading, rather than quick rejection, of the economising narratives is especially crucial in the face of a crisis in imagination (see, for example, Graeber and Wengrow, 2021; Mulgan, 2020). As Claire Colebrook explains, the crisis in imagination has resulted in the coupling of '[a] literal globalism' – 'the stark reality of there being no escape, no outside, nowhere else to flee now,' with an incompatible global configuration – '[t] hings will cycle back to recovery' (2014: 61–62).

This chapter explores how the economic value of nature is narrated, and what forms of relation between people and the environment are imagined in narratives such as economic reports and popular science texts that configure the environment in economic terms and economy as a form of ecology. It asks how the translation of nature into economic terms might be reconceptualised as a narrative form. And, relatedly, what kind of narrative strategies are mobilised in enabling certain kinds of economisation? How might these narratives contribute to complicate the narrow definition of the environment 'as wilderness or the pastoral' (James, 2015: 33) as well as reformulate the notion of economy? Before proceeding, a clarification of what constitutes an economising narrative is in order. A good example is Gretchen C. Daily's description of ecosystem services, to which I now turn.

DOI: 10.4324/9781032726953-4

Ecosystem services

Imagine that you are going to move to the moon and set up a good day-to-day life. The ecologist Gretchen C. Daily opens the introduction to the edited volume titled *Nature's Services: Societal Dependence on Natural Ecosystems* by inviting the readers to consider the essential items for starting a life on the moon. As Daily depicts this scenario,

> Tackling the problem systematically, you could first choose from among all the species exploited directly for food, drink, spice, fiber and timber, pharmaceuticals, industrial products (such as waxes, lac, rubber, and oils), and so on. Even being selective, this list could amount to hundreds or even several thousand species. The space ship would be filling up before you'd even begun adding the species crucial to supporting those at the top of your list. Which are these unsung heroes? No one knows which – nor even approximately how many – species are required to sustain human life. This means that rather than listing species directly, you would have to list instead the life-support functions required by your lunar colony; then you could guess at the types and numbers of species required to perform each. At a bare minimum, the spaceship would have to carry species capable of supplying a whole suite of ecosystem services that earthlings take for granted.
>
> (1997: 3)

This passage captures two important aspects concerning ecosystem services. First, there is a lack of knowledge about the relation between the type and the number of species that constitute an ecosystem and their functions in it. For example, as Daily explains, 'A single gram (a pinch) of soil has yielded an estimated 30,000 protozoa, 50,000 algae, 400,000 fungi, and billions of individual bacteria', and '[m]ost of these species have never been subjected to even cursory inspection' (1997: 4). Second, there is a disparity between our inescapable dependence on, and our lack of knowledge of, and in many cases outright ignorance about, the operation of the ecosystems.

Through this thought exercise, Daily hopes to make visible such a disparity – how irreplaceable and critical ecosystems are to our life and how little we know and value them – and its dire consequences for sustaining functioning ecosystems. This is also the main aim of the edited volume *Nature's Services: Societal Dependence on Natural Ecosystems*. It is significant not simply because it provides insights into many aspects of ecosystem functions, such as the ecosystem services of marine, freshwater, forest, and grasslands, but also, and especially relevant for this chapter, because the book played a key role in ushering the necessity of economic valuation of nature into academic and public debates and policymaking practices. As Daily explains, 'economic markets play a dominant role in patterns of human behavior, and the expression of value – even if imperfect – in a common currency helps to inform the decision-making process' (1997: 10). The pricing of the functions of ecosystems makes intelligible and felt the limits of the biosphere and the urgency of changing the overreaching practices of humanity. For Daily and the other contributors to the edited volume, the issue is not that the problem of biodiversity loss and species extinction is a new finding that policymakers and the general public just did not know about. The problem seems to be that the framing of the issue in scientific terms did not evoke or make felt a sense of urgency that could be used to mobilise action for concrete changes by both policymakers and the public. Daily's description of ecosystem services in terms of what it would take to move to the moon should be read against this backdrop.

In recent years, the term 'ecosystem service' has been widely used and has often been interchangeable with terms such as natural capital, natural solution, and ecological asset. For example, the concept of 'ecosystem service' was adopted in the United Nations publication *Millennium Ecosystem Assessment* in 2005, and the inaugural World Forum on Natural Capital was held in Edinburgh in 2013. And yet, outside the still relatively marginal but growing scholarly field of ecological economics, these different notions are often seen as simply different ways of talking about economising nature. This vagueness informs the increasingly fashionable green discourses as, for example, a marketing strategy, on the one hand, and a critique of the lip-service status of these terms and their underlying economic rationale, on the other. As Donna Haraway's comparison between the terms of Anthropocene and ecosystem services illustrates,

> 'ecosystem services' became an indispensable term for monetarizing all matters. It, too, promised to break down nature and culture, but at the cost of turning everything into circuits of monetarization and accounting. I think Anthropocene is similar . . . [T]he scale is global because the models are global . . . It is a particular model: a view from space of how the chemical cycles of the Earth are influenced in really profound ways by something called . . . Homo sapiens or Anthropos . . . ignoring other species . . . in the same manner that ecosystem services represent the Earth [a]s [*sic*] if it were an accounting system and thereby became a tool for the capitalization of the planet.
>
> (Haraway et al., 2016: 539)

Haraway takes issue with the way in which terms such as 'Anthropocene' and 'ecosystem services' adopt and perpetuate the anthropocentric perspectives that value and measure nature according to economic interests at the global scale. While I find Haraway's observation important, I argue that it leaves unaddressed questions such as who and what, and what kind of relations, are metaphorised, and what such metaphorical moves do. Moreover, Haraway's starting point is the assumption that economy is bad for nature, which is problematic. It reproduces even as it seeks to challenge certain understandings of the environment and economy that are ahistorical and decontextualised. For example, as scientific research on biodiversity has shown, due to uncertainties about the various components and functions of biotic and abiotic environments, what counts as a diverse biome is itself changing and subject to debate (see, for example, Dasgupta, 2021; Lähde, 2022). Moreover, understanding and categorisation of what constitutes an ecosystem or biodiversity are shaped by and inform economic practices such as the disclosure-based approach to measuring financial risks (see for example Buller, 2022). Instead of starting out from this assumption, this chapter makes visible and analyses what constitutes economy and nature, and their relations, in specific economising narratives.

This chapter proceeds as follows. First, I will provide a brief elaboration of the theorisation of 'econarratology' (James, 2015) and 'eco-nomic criticism' (Trexler, 2015), which this chapter draws on. Second, I will closely read three narratives that economise nature and capture the complexities of the human environment entanglement. They are Gretchen Daily's discussion of ecosystem services in the edited volume *Nature's Services: Societal Dependence on Natural Ecosystems*, Paul and Anne Ehrlich's description of biodiversity in *Extinction*, and Partha Dasgupta's report on the economics of biodiversity. Finally, and by way of conclusion, I will underscore how these narratives may allow for reconsideration of the question of the local and the anthropomorphic perspective.

Econarratology and eco-nomics

Given that the examples analysed in this chapter include for example the Dasgupta report on the economic value of biodiversity, which may seem quite different from the literary genres analysed in ecocriticism, I will first clarify how narrative is understood here. I draw on Monika Fludernik's theorisation of natural narratology that posits experientiality as the defining element of narrative. As Fludernik explains, 'The specific narrative needs not rely on the teleology of plot . . . but can be produced also by mimetically motivated evocation of human consciousness and of its (sometimes chaotic) experience of being in the world' (1996: 22). I also take inspiration from Ursula Heise's approach to global biodiversity databases in terms of narratives, which, as I see it, sits well with Fludernik's assertion of the centrality of experientiality. As Heise writes, biodiversity databases 'include lengthy narrative entries on some species, and their metadata and classification schemata are in some cases clearly shaped by story templates that also appear in more properly aesthetic ways of engaging with biodiversity loss' (2016: 15). That said, I now turn to the framework of econarratology and eco-nomic criticism. My aim here is not to provide a thorough discussion of these two frameworks or debates within ecocriticism more generally, but to introduce and underscore the key insights and approaches that I adopt in my reading of the economisation of nature narrative.

The question of how experientiality is evoked is central to Erin James's theorisation of econarratology, which 'studies the storyworlds that readers immerse themselves in when they read narratives, the relationship between these worlds and the physical/actual world, and the potential of the reading processes to raise awareness of different environmental imaginations and environmental experiences' (2015: 242). Following James, I understand environmental imagination in terms of 'a conception of experience of a given space and time based upon a subjective understanding of a particular environmental site' (2015: 243). Importantly, the foregrounding of subjectivity and experientiality means that these imaginations are always embedded in a specific context, even when what is imagined is supposedly global in scale. James's theorisation of econarratology draws attention to three aspects of environmental imagination that are especially relevant to the discussion at hand. First, part of the world-creating power lies in the narrative strategies that enable the readers to transport themselves by experientially imagining and immersing themselves in the environment represented by the narrative, and in the narrative as an environment (which begs the question of where environment begins and ends, and troubles the contours and defining parameters of environment). Second, econarratology foregrounds the practice of comparison between the familiar and the unfamiliar, the physical and the textual environments, as the condition for and expression of readers' experience of the environments created by the narratives. Importantly, as James makes clear, comparative reading practices are embodied and embedded in specific socio-historical and cultural contexts. Moreover, it allows for the imagination and experiences of environments of varying scales and temporalities, which might help bridge gaps between different imaginaries, for example, the ecological and economic figurations of environments. Last but not least, econarratology embraces the anthropomorphic perspectives without resorting to the anthropocentric logic of human domination.

The question of the human perspective is a point of contention within the field of ecocriticism. The difference in how nature is approached from deep ecology and social ecology perspectives is a good example here. Whereas deep ecology foregrounds a pastoral depiction of wilderness and the value of local knowledge of nature and the more-than-human, social ecology puts emphasis on the value of nature for human use (see Heise, 2006). At stake here is how nature is valued, and further, how such valuation defines not only how nature is understood, but also how the relation between humanity and the environment is configured.

The question of the value of nature is central to considerations of the broader social, political, and economic implications of environmental change. In *Anthropocene Fictions: The Novel in a Time of Climate Change*, Adam Trexler observes an emerging literary form that describes and 'sensitizes the reader to what is before us, rather than demanding we turn away in revulsion or "solution" of a future that "may be heightened, stylized, even horrifying"' (2015: 170). Trexler calls this literary form 'eco-nomics'. As Trexler explains, eco-nomics is different from 'economics' as a disciplinary field and a realm of practice, as well as the Marxian economics in literary studies, which 'has been insufficiently concerned with the operation of real economies, the production and distribution of real things' (2015: 171). Instead, eco-nomics writing expands the scope of the environment by making visible the intertwinement of social, cultural, geopolitical, ecological, and economic dimensions.

Surveying eco-nomics climate novels, Trexler outlines the following three levels of 'eco' involved in eco-nomics. The first concerns the changing configuration of home and quotidian domestic practices such as 'the fitness of twentieth-century homes, food, family arrangements, work, and shopping' (2015: 171). The second level concerns the economic, which according to Trexler, is about 'the material circumstances of human beings' such as 'scarcity, security, technology, enterprise, energy, trade, production, markets, and consumption' (2015: 171). Last but not least, the third level of 'eco' refers to the embeddedness of human domestic and economic practices within the more-than-human context. This aspect seeks to move beyond the anthropocentric point of view to account for 'how geology, geography, and species radically shape human experience' (Trexler, 2015: 171).

Importantly, and especially relevant to the present discussion, the eco-nomics texts articulate environmental concerns not simply as compartmentalisable or isolated events, but as the intertwined and complex interplays of 'species, weather, social groups, and financial interests' (Trexler, 2015: 200). This allows for analysis of environmental imaginations in contextualised and local economic and ecological practices. In so doing, the analysis moves beyond the debate about whether economy is good or bad for the environment, which reduces both the economy and the environment to abstract and unitary systems. For example, economy is often unproblematically reduced to a certain understanding of 'free-market,' a supposedly isolated realm whose logic and operation is radically different and separate from the environment. As Trexler asserts,

> Markets should not be understood as a framework for purified, rational humans, but rather as a terrain to test the uneven, heterogeneous forces

> of corporations, self-interested individuals, national trade objectives, and environmental restrictions. Plants, animals, bacteria, machines, financial capital, and weather all materially shape markets as well, in ways that are not precisely predictable, even if they are often decisive.
> (2015: 220)

Trexler's formulation of eco-nomics creates license for considering two important aspects of narrating the economic value of nature. First, the problems of climate change and environmental degradation are unavoidably economic concerns, in the sense that they are materialised in and through economic practices that are local but also global, mundane but also abstract, socio-political as well as environmental. As I see it, the eco-nomics narrative expresses economic valuation of nature in ecological terms, understood not as harmonious systems that only become interrupted by external forces, but as a more open and dynamic systematicity shaped by heterogeneous and conflicting relationalities that are internal to it. Second, and closely related to the first aspect, eco-nomics narrative captures, foregrounds, and evokes a crucial question for environmental imaginations, namely, how the environment is lived. This echoes and supports the suggestion to distinguish between anthropomorphic narratives and anthropocentrism. That is, the humanisation of the more-than-human is 'a narrative expedient' that articulates 'the comingling of humans and nonhuman matter in the construction of meaning and the ways in which the human imagination can provide access to the stories told by such matter' (James, 2015: 32; see also Iovino and Oppermann, 2012).

The econarratological and eco-nomic approaches briefly outlined in this section provide tools for analysing the environmental imagination in narratives of the economic value of nature. As I will show, these narratives do not simply express the economic consequences of environmental change and the ways in which the changing environmental configurations affect how life is lived. More importantly, they also further confound the defining parameters of environment, economy, and humanity by asking, for example, how the human body is itself ecosystemic and economic, and how the non-linearity and uncertainty of nature define and demand economic valuation that is at the same time ecological.

Narrating the economic value of nature

In this section, I provide analysis of three examples of narrating the economic value of nature. Before proceeding, a brief clarification of the debate between economic and environmental approaches to nature is in order. A good example is the debate between economist Julian Simon and ecologist Paul Ehrlich, which took place in 1980. Whereas Ehrlich anticipated rising commodity prices based on increased resource scarcity and population growth, Simon argued that prices would decline because the capacity to substitute these commodities would increase as a result of the innovation and development of technologies. However, Ehrlich expressed distrust in economic thinking, seen as guided by and perpetuating 'notions of limitless growth and substitution on a bounded planet' (Dempsey, 2016: 60). In many ways, it could be said that such an understanding of economic valuation practices informs, explicitly or implicitly, the ongoing critique of the economisation of nature.

The edited volume *Nature's Services*, mentioned earlier in this chapter, presents one of the first attempts at reconciling the dispute between economists and ecologists, and to create conversations about and vocabularies of the economic value of nature. Those involved in the production of the volume participated in a five-year project that had begun in 1991 at the Beijer Institute, which focused on the economics and ecology of biodiversity loss. The project brought together experts from fields such as ecology, biology, ecological economics, and environmental economics, as well as policymakers. It aimed to promote 'a deeper understanding of the interplay between ecological systems and social and economic development' (Dempsey, 2016: 61).[1] In effect, the collaboration produced important common grounds that bridged these disciplines, which can be observed in, for example, the usage of mathematical equations as well as the mobilisation of the environmental economist theorem of externalities to frame the loss of biodiversity as an ecological *and* economic problem. In many ways, the texts that are analysed here should be seen as born out of this project. Although they were published at different times and through different outlets, the authors were key participants in the Beijer Institute project.

My aim here is to contextualise and historicise the narratives of economising nature, which provides insight into the epistemic, political, and economic specificities that shaped the environmental imaginations presented in these narratives. The analysis that follows will focus on two narrative strategies – spatialisation and defamiliarisation – that express these ecological economic imaginations. The reason for this focus is that these narrative strategies are often used together in narratives that economise nature. I will closely read three examples to demonstrate how these narrative strategies are used and discuss their implications.

Spatialisation and defamiliarisation

Spatialisation is an important narrative cue that encourages the reader to relocate to that imagined environment. Different forms of spatialisation can be utilised to make visible and felt what ecosystem services are, and their importance and value for humanity. It is important to note that spatialisation is not neutral, but is informed by relations of power and different valuation practices. For example, different spatialisations of what constitutes biodiversity inform different political economic investments in what is valuable and for whom.

Daily defines ecosystem services in terms of 'the conditions and processes through which natural ecosystems, and the species that make them up, sustain and fulfil human life' (1997: 3). This short sentence encapsulates a few key features of ecosystem services. First, these systems are natural processes performing important functions that make possible human life. The functions include cleaning water and air, recycling nutrients to keep soil fertile, and decomposing waste. Second, they are constituted by species. Third, they provide resources and 'ecosystem goods' such as 'seafood, forage, timber, biomass fuels, natural fiber, and many pharmaceuticals, industrial products, and their precursors' that are vital for economic activities (Daily, 1997: 3). Last but not least, they are essential for sustaining the social and cultural aspects of human life by supporting the diversity of human cultures and providing 'aesthetic beauty and intellectual stimulation that lift the human spirit' (Daily, 1997: 4).

Daily's description of ecosystem services re-spatialises and defamiliarises the environment by making visible the interconnectedness of domains that are usually thought to be separate from each other – species diversity, ecosystem functions, economic practices, culture, and well-being. In so doing, it underscores the radical dependency of human life on nature in two different ways. First, the figure of the moon is used in Daily's introduction to represent a familiar yet different, universal yet specific place, and in that way frames the discussion about the importance of ecosystems. The moon is familiar in the sense that just like the earth, it is habitable by humans as it provides 'the basic conditions for supporting human life, such as atmosphere and climate similar to those on earth' (Daily, 1997: 3). However, this basic condition does not seem to support what counts as good life that is experienced and evaluated subjectively. As Daily asks, 'After inviting your best friends and packing your prized possessions, a BBQ grill, and some do-it-yourself books, the big question would be, which of earth's millions of species do you need to take with you?' (1997: 3).

Changing from the third-person neutral and scientific description to a second-person narrative that directly addresses the reader, this question evokes a subjective and affective environmental imagination by asking the reader to compare three different environments. The first environment is the home as a place of happiness ('inviting your best friends') and of ownership and properties whose values are specific to each individual ('your prized possessions'). The second environment is the earth, which in this depiction is characterised by the millions of species that inhabit it and are specific to it. The third environment is the moon as an empty space that lacks and awaits the construction of individual homes, perhaps with the help of 'some do-it-yourself books' (Daily, 1997: 3).

When the reader is asked to imagine these three spaces and environments, the comparability of them lies in the defamiliarisation of each environment, which also allows for an immersive and transportive reading experience. That is, only through the defamiliarisation of home – moving from the one on the earth to the one that is yet to be built on the moon – is the reader able to experience and imagine what it is like to live on the moon *simultaneously* and in relation to what it is like to live on the earth. Moreover, the readers are invited to make comparisons in affective and embodied ways. For example, the comment that you have invited your best friends and the mentioning of the BBQ grill conjure the image of celebrating the move to the new home that involves embodied practices such as eating good food and having nice company.

The spatialisation and defamiliarisation of the narrative thus provide means to compare, that is, to make intelligible and felt, the value of nature and the ecosystem not just to human life as such, but to the lived reality that is specific to each individual. Comparison is thus central to comprehending the narrative of the value of nature. As James explains, the process of comparison

> involves determining how the actions and events depicted in a narrative relate to other possible past events, alternative presents, or potential happenings in the future, and is often aided by familiar representations, particularly those involving stereotypical sequences of actions and events. Such a comparison thus calls on readers to directly engage the defamiliarizing aspects of a text as they compose and inhabit their mental models of that text's world.
>
> (2015: xi)

However, it is important to note the coloniality and anthropocentric dimensions of both strategies. For example, the moon is represented as a *terra nullis* – an empty space that awaits the inhabitation and cultivation of humans. Moreover, the species and the ecosystems are depicted as objects that are simply up for grabs depending on individual needs and desires. There is also the colonial fantasy of transcending the limits of material conditions and local specificities by dislocating, transporting, and relocating earthly species to a different environment. This risks perpetuating the assumption that there is an unlimited number of species and resources to be taken and used, *without cost*. Yet, importantly, these colonial and anthropocentric underpinnings are challenged and reconfigured through the depiction of a fourth space – the spaceship.

The spatiality of the spaceship adds an interesting dimension to the story. The spaceship is a bounded space that creates a sense of isolation – a lone vehicle floating in the universe. It puts a limit on the list of items to bring ('The spaceship would be filling up before you'd even begun adding the species crucial to supporting those at the top of your list'). On top of this, there is an immense gap in knowledge about what kind of species and how many of them are essential for a specific ecosystem function. The rational, thinking human subject is faced with an inevitable conundrum: we are radically dependent upon the transporting capacity of the spaceship, as well as the species and the ecosystem functions, but the limited space of the spaceship and the lack of knowledge about the right kind of species to prioritise make the creation of an earth-like happy home on the moon extremely challenging, if not completely impossible. This simultaneous radical dependence upon and lack of knowledge about the species and the ecosystem services thus throw into doubt the exceptionality of the human subject and their spatialising and colonising capacities.

Similar spatialising and defamiliarising strategies are also utilised in Paul and Anne Ehrlich's description of the problem of species loss and extinction. In the book *Extinction*, published in 1981, the Ehrlichs invite the reader to imagine an airplane on which she or he is going to fly:

> As you walk from the terminal toward your airline, you notice a man on a ladder busily prying rivets out of its wing. Somewhat concerned, you saunter over to the rivet popper and ask him just what the hell he's doing.
>
> 'I work for the airline – Growthmania Intercontinental,' the man informs you, 'and the airline has discovered that it can sell these rivets for fifty cents apiece.'
>
> 'But how do you know you won't fatally weaken the wing doing that?' You inquire.
>
> 'Don't worry,' he assures you. 'I'm certain the manufacturer made this plane much stronger than it needs to be, so no harm's done. Besides, I've taken lots of rivets from this wing and it hasn't fallen off yet. Growthmania Airlines needs the money; if we didn't pop the rivets, Growthmania wouldn't be able to continue expanding. And I need the commission they pay me – fifty cents a rivet!'
>
> (1981: xi–xii)

In this narrative, the plane represents not simply the vehicle that carries the human body, but the ecosystem itself. As the Ehrlichs explain, the rivets stand for the species that constitute the ecosystems. The ecosystem is collapsing because of the extraction of the rivets/species. The rivets seem substitutable, both in terms of their appearance and price ('fifty cents a piece').

It is thus difficult to determine which rivet/species might be critical to the ecosystem. As Dempsey explains,

> we cannot know which rivet is critical – which rivet is 'the one' . . . [T]he result is an irresolvable uncertainty about how individual species relate to the function of the ecosystem more broadly; we cannot distinguish species that are necessary from those that are ostensibly surplus.
>
> (2016: 29)

Here, the importance of species diversity is made intelligible and felt in a way that challenges the economic rationale of the substitutability and unlimitedness of species.

In a similar vein as Daily's scenario about moving to the moon, the depiction of the ecosystem as a plane functions to spatialise and defamiliarise the environment. Like the spaceship, the plane is a confined space that is isolated and floating in the air.[2] However, it is different from the spaceship in the sense that it is not simply a vehicle that is used to transport humans and species from the earth to the moon, but it represents the ecosystem that is destroyed by the extractive economic practices that threaten human survival. The 'confidence' that the plane is built to be stronger than necessary is put in stark contrast to the uncertainty of when, not just if, the wing would fall off.

In terms of the eco-nomics configurations, the first narrative by Daily changes the economic configuration of nature from one of resources that often follow a 'reductionist, single-use commodity model' adopted by fisheries and forest management, that 'sidelined the complexities of ecological systems,' to one that foregrounds the processes of ecological functions that are characterised by 'dynamism and uncertainty' (Dempsey, 2016: 63).[3] In other words, nature is seen not only as constituted by raw materials which are passively awaiting extraction and usage, but also as a provider of various essential and interlinked services. In the second example, Paul and Anne Ehrlich make the relation between economic practices and environmental consequences more explicit. It casts into doubt the cost-and-profit calculation of the supposedly rational economic subject – it is unfathomable to risk losing life in order to gain fifty cents from each rivet. The name of the company, 'Growthmania Intercontinental,' provides a satirical commentary on the market-oriented approaches that treat species as resources that are substitutable and limitless, as well as the discourse of economic growth and global expansion that aims at ever greater speed and an ever wider reach.

In the aforementioned two examples, spatiality is organised according to the logic of externality. That is, the human, the ecosystem, and the economy are considered distinct realms, and each functions according to its own laws and mechanisms. For example, in the first narrative about moving to the moon, the ecosystem that could be simply transported to another location is understood as a system that is different and separate from the human body. In the second narrative, the ecosystem is represented by the plane. It carries the human body and is affected by human economic decisions (whether or not to extract the rivets). However, even in the second case, the ecosystem and the human are considered two separate entities. For example, the passenger can decide whether to board the plane. In the narrative that will be analysed next, the boundaries that separate the human body from the environment and from the economy are further displaced through spatialisation and defamiliarisation.

Before proceeding, it should be noted that the third example is quite different from the two narratives previously analysed. Whereas the first two narratives are excerpts from edited volumes and monographs that popularise scientific knowledge, the third one is part of an economic report. It was commissioned in 2019 by Her Majesty's Treasury and supported by an advisory panel drawn from areas such as public policy, science, economics, finance, and business. In analysing this example, I wish to show how the genre of economic report can and should be read as a form of eco-nomics narrative. In *The Economics of Biodiversity: The Dasgupta Review*, Partha Dasgupta describes the human body as an ecosystem:

> The human body is an ecosystem. Someone experiencing a breakdown in health would at a formal level be exhibiting a regime shift. Non-linearities in ecosystem processes also lead to segmentations among ecosystem populations . . . Likewise, non-linearities in human metabolic processes lead, most prominently in poor countries, to a partitioning of human population into groups that are healthy (for example, people who enjoy recommended body mass indices) and groups that are malnourished.
>
> (2021: 92)

Dasgupta uses this example to illustrate the spatiality of non-linearity, which governs the biosphere. Despite its wide usage, the function and implication of non-linearity for biological, ecological, and socio-economic systems often remain vague. The term is often used to connote the temporality of transitions such as the accumulation of extinction of species that leads to a sudden collapse of an entire ecosystem (the temporality of non-linearity is made visible and felt in the story of prying rivets from a plane, discussed earlier). In this narrative, the correlation between nutrition status and working capacity provides spatial specification of the abstract notion of non-linearity.

Two levels of spatiality are at work in Dasgupta's report. First, non-linearity is measured in corporeal and individual terms. Dasgupta gives the example of the relation between nutritional status – measured through BMI, that is, 'weight (in kilograms) divided [by] the square of height (in metres)' (Dasgupta, 2021: 92) – and work capacity. As Dasgupta writes,

> The relationship between nutritional status and work capacity has been found to be sigmoid: its gradient, although positive beyond a BMI equal to, say, 17, is very small but then increases rapidly until tapering off from values of 24 or so, and then is negative for people whose BMI exceeds 30.
>
> (Dasgupta, 2021: 92)

The non-linearity of the relationship between nutritional status and work capacity means that it is important to ensure access to high nutrition, that is, to be in good health, which is measured as having a BMI above 16 (a lower score means undernourishment) and below 30 (which signifies obesity). This correlation demonstrates the non-linearity of the metabolic process and working capacity at the individual corporeal level, conceptualised as an ecosystem.

Second, the dynamic interplay between stability regimes and the regime shift of an ecosystem takes the shape of the segmentation of two socio-economic groups, divided according to nutritional status. Stability regimes of ecosystems are typically discussed in terms of a narrow understanding of the environment; for example, large-scale landscapes such as forest and grassland.[4] Given this, the narrative that economises ecosystems on the one hand and naturalises economy on the other requires the reader to defamiliarise the conventional spatialisation of economy, environment, and the human body, that is, to rethink where each begins and ends. After having introduced the correlation between nutrition status and working capacity, Dasgupta asks the reader to imagine the following scenario:

> Imagine a country that is so poor that no economic system could ensure that everyone has a high nutritional status. At the same time, no efficient economic system in the country would countenance on grounds of equity that everyone should have a low nutritional status, for that would reduce aggregate labour productivity to very low levels and keep the country in poverty. The choice is cruel, but an efficient economic system would bifurcate the population into one whose nutritional status is low and another whose nutritional status is high. The former group is trapped in poverty, which is formally a stability regime.
>
> (2021: 93)

This example sheds light on two aspects of the non-linearity of the human body and the realm of the socio-economic as ecosystem. First, the two socio-economic groups, understood as two stability regimes marked by two correlated nutritional and income levels, are sustained by feedback loops that are metabolic/biological/environmental and socio-economic. That is, for example, the poorer a person is, the less access to nutrition and the lower the working capacity they have, and the poorer they become. Second, seen in the light of the ecosystem functions, the bifurcation of the population into the two stability regimes is considered efficient because it makes the socio-economic ecosystem resilient. As Dasgupta clarifies, 'the *resilience* of an ecosystem in a stability regime is the magnitude of disturbance it can tolerate before it tips into a different stability regime' (2021: 87, emphasis in original). Importantly, in reconfiguring socio-economic processes in terms of ecosystems governed by the law of non-linearity and metabolic feedback loops, notions such as resilience and efficiency of the environment, often associated with a certain romantic and anthropocentric conception of nature, are called into question. Moreover, the spatial and temporal configuration of non-linearity and feedback loops throws into question the exceptional status of the human condition characterised by the ideal of the management and control of the environment that is seen as linear and passive.

Here, the spatial relation between the human body, the environment, and the economic system is reorganised. Taking inspiration from the approach of econarratology to the body and its environment, the body in this example can be read as shaping the narrative about inequality in a way that is not recognised otherwise.[5] For example, in this narrative, issues such as the health of the human body measured by body mass indices, access to nutrition, capacity to work, segmentation of the population in terms of income, and economic policy such as taxation, are not seen as located in separate realms, in which each functions according to its own logic and laws. Not only is the body the location where the environmental and economic conditions become intelligible and felt, it is also eco-systemic and eco-nomic through and through. Instead of the logic of exteriority that structures the two aforementioned narratives, here the mechanism of non-linearity becomes the defining feature of the eco-nomic, expressed through the corporeal as its form and content.

Conclusion

If you are going to set up a happy life on the moon, which species are you going to bring with you? You might have felt that it was impossible to answer this question when you encountered it at the beginning of this chapter, because there are simply too many species to list, and too many species whose role in an ecosystem remains unclear, and too many species that we just simply do not know anything about. But by the time you have reached the conclusion, you are probably wondering if this question can still be posed, as the defining parameters of the relation between the environment and human life are no longer self-evident.

In this chapter, I have closely read three narratives of the economisation of nature. I have taken inspiration from James's theorisation of econarratology as well as from Trexler's formulation of eco-nomics to tease out the different ways of imagining the environment and its relationship with economic practices and human life. More specifically, I have paid close attention to spatialisation and defamiliarisation as narrative strategies and made visible how they evoke affective responses to and comparisons between various imagined and lived environments. As I have sought to show, there are different modes of spatialisation in these narratives, which contribute to the changing configuration of the economic value of nature. For example, Daily's narrative shifts the configuration of nature from stocks and resources to processes and functions, understood as ecosystem services that are essential to human biological, economic, and cultural life. The earth, the uncanny and empty moon, the individual home, as well as the spaceship are figures that spatialise and defamiliarise the relation between human life and the environment.

In the Ehrlichs' text, the configuration of the environment shifts from a detachable and relocatable system whose species are too many to be fully known or extracted, to foreground the problem of species extinction and unsustainable economic practices that are extractive and exploitative. Both the first and second narratives evoke emotional responses to environmental change by defamiliarising experiences such as moving home and travelling on a plane. They mobilise spaces such as the spaceship and the plane that represent a sense of limitations and uncertainties, which calls into question the anthropocentric notion of control and management. In the first narrative, human agency, understood in terms of rational subjects who are free to choose, is restricted and destabilised by the limited space of the spaceship and the vast number of species that would be essential but are outside the reach of the human capacity to know. The second narrative decentres the human rational subject by making visible the lack of knowledge of the species and the ignorance of the economic calculus (as the name 'Growthmania Intercontinental' so excellently captures) through the depiction of the visually identical rivets and how they are wrongfully considered substitutable and accorded low value. It also expresses the temporality and the implication of the law of non-linearity through evoking the visual imaginary of the imminent but sudden crash of the plane as the accumulated effect of extracting rivets.

Dasgupta's description of the human body as an ecosystem spatialises the operation of non-linearity by detailing the correlation between nutritional status and working capacity and their relation to socio-economic divisions and policies. Whereas the first two narratives economise nature by either shifting the narrative of nature from a simplistic notion of resource and goods to complex systems whose function human life depends on, in the third narrative the boundaries between the human body, the environment, and the economy undergo further destabilisation. Instead of the spatial arrangement of exteriority where the human body, the environment, and the economy, although interacting, are located outside each other, and each operates according to its own laws and mechanisms, the law of non-linearity is here seen to organise the entire biosphere. That is, what is considered natural and biological, and its other, economic practices in this case, are not separate realms, but are deeply intertwined and shaping each other. The human body, at once biological, ecosystemic, and economic, provides the eco-nomic form and content.

Importantly, the reconfiguration of the human body in terms of an ecosystem functions to renaturalise the human, all the while raising the question of what 'naturalising' might actually mean and bridging the gap between different modes of imagination. In a similar vein as the imagination of moving to and setting up a new life on the moon and travelling by plane, the spatialising of the human body as an ecosystem and vice versa performs the narrative strategy of defamiliarisation which in turn gives a sense of experientiality that is local and subjective. As I have tried to show, these narratives embrace the anthropomorphic perspective *and* call into question the exceptional status of the human condition. In analysing the narratives of the economisation of nature, I hope to have shown that the economisation of nature does not simply reduce the materiality of nature to numerical measurements, nor is it simply a question of whether nature could or should be marked by price tags. Rather than beginning with the question of whether economy is good or bad for nature – a framing that reproduces certain imaginations of the environment and the economy, paying attention to narrative strategies such as spatialisation and defamiliarisation allows for insights into and alternative imaginations of the relation between the human body, the environment, and the economy. The narratives analysed here stretch our imagination in various ways and offer opportunities to ask one of the most important questions in the Anthropocene, namely, how the environment is lived.

Notes

1 As Dempsey writes, 'The first project proposed by the Beijier participants focused on the economics and ecology of biodiversity loss . . . Participants included the now "who's who" of this area: ecologists and biologists like Paul Ehrlich, Gretchen Daily, Brian Walker, Joan (née Jonathan) Roughgarden, and Crawford

Stanley "Buzz" Holling; members of the discipline of ecological-economics like Robert Costanza and Carl Folke; and more "mainstream" environmental economists like [Charles] Perrings, Partha Dasgupta, Timothy Swanson, Edward Barbier, David Pearce, and Kenneth Arrow, many of whom were prominent not only as academics but also as advisors to governments and international institutions. Other participants, important because of links to policy and practice, included Jeffrey McNeely, the chief scientist from IUCN, as well as representatives from the World Bank' (2016: 62).

2 Both the spaceship and the plane echo the imagery and the metaphor of the earth as an island. As Kelly Oliver writes, 'If our conception of earth determines our conception of ourselves and vice versa, then we must rethink the earth as island metaphor. The island metaphor suggests that the earth is isolated, insular, and self-contained. The initial reactions to the photographs of earth were illusions in part because they perpetuated this view of earth. If we conceive of the earth as a network of dynamic living relationships, however, then we also change the way we see ourselves. No longer maverick space cowboys desperately holding on to the fantasy of the autonomous self-contained individual, species, nation, we can embrace our earthbound existence, sharing our planetary home with diverse worlds beyond our comprehension' (2018: 422).

3 However, the changing understanding of ecological systems through the notion of ecosystem service still draws on economic theory that emphasises utility. This is especially the case, as Dempsey notes, 'in the drive to identify points of equilibrium between economic growth and some "level of biodiversity"' (2016: 63).

4 The coexistence of forest and grassland as two different stability regimes for the ecosystem of tropical land is an excellent example of eco-system efficiency. As Dasgupta explains, 'Forest and Savannah are alternative stability regimes for tropical land, separated by biotic and abiotic drivers, and maintained by feedback loops of fire (in the case of Savannah) and fire suppression (in the case of rainforest). When rainfall levels are low, vegetation is dominated by grasses with very limited wood cover to provide shade. The intensity of the sun can thus start fires easily, creating a 'trap' for any developing woody cover, and making the return shift to forest extremely unlikely. In contrast, high levels of rainfall enable wood cover to grow and develop into canopies, which then maintain the necessary water cycle . . . and provide shade which limits the possibility of fire breaking out. Such dynamics give rise to a tropical world that admits two broad stability regimes, one dominated by grass, the other dominated by trees, with the potential for sudden flips as conditions change' (2022: 90).

5 James's formulation of econarratology draws on, for example, the theorisation of corporeal narratology, which attends on one level 'how the body is used as a component of stories via traditional narrative elements such as plot, setting, and character as a means of "see[ing] the body at work in elements of the story where we may not have recognized it otherwise"' (2015: 240).

Bibliography

Buller, Adrienne (2022), *The Value of a Whale: On the Illusions of Green Capitalism* (Manchester: Manchester University Press).

Colebrook, Claire (2014), *Death of the Posthuman: Essays on Extinction* (Ann Arbor: Open Humanities Press).

Daily, Gretchen C. (1997), 'Introduction: What Are Ecosystem Services?', in Gretchen C. Daily (ed.), *Nature's Services: Societal Dependence on Natural Ecosystems* (Washington, DC: Island Press).
Dasgupta, Partha (2021), *The Economics of Biodiversity: The Dasgupta Review* (London: HM Treasury).
Dempsey, Jessica (2016), *Enterprising Nature: Economics, Markets, and Finance in Global Biodiversity Politics* (West Sussex: Wiley Blackwell).
Ehrlich, Paul and Anne Ehrlich (1981), *Extinction: The Causes and Consequences of the Disappearance of Species* (New York: Ballantine Books).
Fludernik, Monika (1996), *Towards a 'Natural' Narratology* (London and New York: Routledge).
Graeber, David and David Wengrow (2021), *The Dawn of Everything: A New History of Humanity* (Oxford: Signal).
Haraway, Donna, et al. (2016), 'Anthropologists Are Talking – About the Anthropocene', *Ethnos*, 81:3, 535–564, https://doi.org/10.1080/00141844.2015.1105838
Heise, Ursula K. (2006), 'The Hitchhiker's Guide to Ecocriticism', *PMLA*, 121:2, 503–516, https://doi.org/10.1632/003081206X129684
Heise, Ursula K. (2016), *Imagining Extinction: The Cultural Meanings of Endangered Species* (Chicago: University of Chicago Press).
Iovino, Serenella and Serpil Oppermann (2012), 'Material Ecocriticism: Materiality, Agency, and Models of Narrativity', *Ecozon@*, 3:1, 75–91, https://doi.org/10.37536/ECOZONA.2012.3.1.452
James, Erin (2015), *The Storyworld Accord: Econarratology and Postcolonial Narratives* (Lincoln: University of Nebraska Press).
Lähde, Ville (2022), 'Life Matters Everywhere. The Notion of Biodiversity in the Dasgupta Review', [Online], [Accessed 10 August 2022], available from: https://bios.fi/en/life-matters-everywhere-the-notion-of-biodiversity-in-the-dasgupta-review_eng/
MEA (2005), *Millennium Ecosystem Assessment: Ecosystems and Human Well-Being* (Washington, DC: Island Press).
Mulgan, Geoff (2020), 'The Imaginary Crisis (and How We Might Quicken Social and Public Imagination)', [Online], [Accessed 6 February 2024], available from: https://demoshelsinki.fi/wp-content/uploads/2020/04/the-imaginary-crisis-web.pdf
Oliver, Kelly (2018), 'Earth: Love It or Leave It?', in Matthias Fritsch, Philippe Lynes and David Wood (eds.), *Eco-deconstruction: Derrida and Environmental Philosophy* (New York: Fordham University Press).
Sullivan, Sian (2017), 'Making Nature Investable: From Legibility to Leverageability in Fabricating "Nature" as "Natural Capital"', *Science and Technology Studies*, 31:3, 47–76, https://doi.org/10.23987/sts.58040
Trexler, Adam (2015), *Anthropocene Fictions: The Novel in a Time of Climate Change* (Charlottesville: University of Virginia Press).

4 Building a new world on the ruins of Helsinki

Critical utopia in Annika Luther's *The City of the Homeless*

Katarina Leppänen

Introduction

The Finnish-Swedish author Annika Luther's untranslated novel *De hemlösas stad* (*The City of the Homeless*) (2011) takes place in Finland in 2050. Water covers large parts of the country, and the capital city has been moved from Helsinki on the south coast to Jyväskylä in central Finland.[1] The author creates a tension between the two cities, the new and the old capital, which is depicted through a young woman's eyes. In *The City of the Homeless*, the environmental disaster is caused by an earthquake in Greenland in 2035, which led to the ice calving off the glacier, which in turn created monster waves and a lasting rise in sea levels. The natural disaster is combined with anthropogenic climate change in the form of global warming, all of which becomes apparent in the description of the weather as wet and warm, and therefore as uncharacteristic of Finland. The story takes place in the near future, and life in Jyväskylä invokes contemporary Finnish society. Both everyday situations in family life and social institutions such as schools, workplaces, police, and the European Union, remain intact. Meanwhile, the coastal parts of the country in the south and the west, including the current capital Helsinki, have been flooded or destroyed, and are now populated by climate refugees.

The author Annika Luther is a biologist and a teacher, and she has published more than ten books, many addressing environmental concerns. *The City of the Homeless* belongs to the rapidly growing genre of climate fiction, and it further belongs in the genre of young adult fiction.[2] Although Luther's novel is set in the future, the verisimilitude of Jyväskylä sets the book in a recognisable society. Luther's narrative of the near future addresses the critique put forth by Amitav Ghosh, that the modern novel tends to situate climate catastrophe in the future when it should instead emphasise the present: 'The future is only one aspect of the Anthropocene: this era also includes the recent past, and, most significantly, the present' (Ghosh, 2016: 72). Many climate change novels display an overlap

DOI: 10.4324/9781032726953-5

between the realistic and the futuristic, and Axel Goodbody and Adeline Johns-Putra note that

> what is crucial about both future worlds marked by climate catastrophe and everyday milieus torched by climate concerns is that they provide drama, and thereby engage readers' attention to the way which non-fiction cannot replicate without recourse to elements of fictionalisation and personification.
>
> (Goodbody and Johns-Putra, 2019: 234)

The depiction of climate change in *The City of the Homeless* has been termed an 'Orwellian dystopia' (2019: 202). However, regardless of its dystopian starting point in environmental catastrophe, the book conveys an optimistic curiosity towards a changing world. I therefore read Luther's novel more along the lines of Tom Moylan's 'critical utopia' that can contain the ideology of a dominant class or society yet has room for exploring oppositional ideologies (1986: 1–2, 10). A critical utopia combined with critical sustainability disrupts the sharp distinction between utopia and dystopia and instead approaches environmental change from a humbler literary position. Such a position is directed towards the future and 'rather than aspiring towards a utopian "stationary state" . . . might be seen as something the human species has continually to strive towards, while knowing it will never reach it' (Johns-Putra et al., 2017: 5). *The City of the Homeless* includes constructive solutions for the future, and its near-future setting gives it an air of realism, rather than utopia (Ahlbäck, 2019). And this strange familiarity is one of the advantages of near-future climate fiction. Science fiction that deals with distant times and planets or new life-forms does not have the same possibility to connect its narratives to the reader's everyday world.

On cities

Since 2007, some 55 per cent of the world's population have lived in cities, according to the World Bank.[3] Popular culture is swamped with stories of disaster-hit cities that have spiralled into poverty and uncontrolled violence. From countless films such as *Escape from New York* (John Carpenter, 1981) and *Bladerunner* (Ridley Scott, 1982) to books like Kim Stanley Robinson's *New York 2140* (2017) and Octavia Butler's *Parable of the Sower* (1993), we know that cities are especially vulnerable to environmental and social catastrophe, while at the same time attracting ever-increasing populations. Urban planning in the form of waste management, transport systems, energy supplies, and food pose great challenges in real life, and figure as problems in urban fiction (Choen and Dong, 2021). A city can be defined as 'a place where people live that is larger or more important than a town: an area where many people live and work' or 'networks of social interaction embedded in physical space,' or even as a complex super-organism (*The Britannica Dictionary*; Ortman et al., 2020: 1–2). But urban environment is not nature's other or outside; rather, a city is 'a biological environment, a social environment, a built environment, a market environment, a business environment, and a political environment'; or, to borrow Lewis Mumford's terms, 'an institutional process, a theatre of social action, and an aesthetic symbol of collective unity' (Ortman et al., 2020: 2, 4). World cities or global cities have the additional characteristic of holding economic world power by being centres of trade and are sometimes mistakenly viewed as existing outside and beyond the reach of the nation state. Yet, as Delphine Ancien (2011: 2477) convincingly argues, nation states and global cities co-act in creating these specific conditions by, for example, tax legislation or other concessions.

For the purpose of this chapter, the city's symbolic and political power is investigated during a process of change; a city nearly wiped out by disaster opens to new global possibilities. My reading of *The City of the Homeless* as a story about the conflict between state control and the potential re-imagining of global cities is inspired by Henri Lefebvre's theory of the social and political production of space. Lefebvre may not be the immediate choice when it comes to understanding the workings of fictitious future cities. His writing is firmly grounded in a materialist understanding of space-and-state relations, yet his thinking is tentative and experimental, and therefore porous enough to use in relation to fiction. Lefebvre wants to '*open up* through thought and action towards *possibilities* by showing the horizon and the road' (1996: 63, original italics). Especially his description of the concept of 'the worldwide' (*mondialisation*) as a mental capacity that precedes (economic) globalisation, invites a connection between the fictional and the geopolitical, because fiction can be seen as a mental preparation for a world to come (Brenner and Elden, 2009: 20–25). Furthermore, he explicitly addresses the city as space for aesthetic and imaginative representations (see further Liinamaa, 2020). For Lefebvre, the worldwide is the horizon for any state's production of space. That is, the nation state always needs to consider both the national level and the global level, although its primary influence is over national territory. In the following discussion, I attempt to show how the state of Finland, as imagined in Luther's fiction, struggles to control the production of space through repressive and controlling measures while the global climate refugees – strangers, so to speak, to the original state – colonise its territory.

Lilja's search for home

'Fri! Ensam! Äntligen! Ingen ser mig, ingen vet var jag är och ingen, ingen, ingen kommer åt mig!' [Free! Alone! Finally! Nobody sees me, nobody knows where I am and nobody, nobody, nobody can get at me!] (Luther, 2011: 35, translations are my own), the soon-to-be sixteen-year-old Lilja Koskinen exclaims when she finally manages to escape the all-seeing surveillance system that controls every movement in her hometown Jyväskylä, the capital city of Finland in 2050. Lilja mounts her bicycle and runs away during a thunderstorm when a power cut makes it possible to pass the gates. The fact that the city is guarded and surrounded by electric barbed wire is presented by its leaders and bureaucrats as protection against the dangers of a chaotic outside world that has radically changed as a consequence of environmental disaster and the subsequent geopolitical reorganisation of the world. The EU has created an Official Immigration Zone between Helsinki and Jyväskylä, where European farmers whose land has been destroyed are allowed to cultivate the fertile fields, to the annoyance of the Finns. The coast of Ostrobothnia and large parts of the south are in ruins and flooded, including major cities like Åbo/Turku and Tammerfors/Tampere. The areas even worse off, like the 'farlig riskzon' [dangerous risk zone], have been officially emptied of people, but even the most devastated areas have been inhabited by climate refugees from 'hundra andra dränkta helveteskuster' [a hundred other submerged hellish coasts] (Luther, 2011, 12). The Helsinki area itself is no man's land.

Lilja leaves the security of her home to look for her biological mother, the marine biologist Sassa. When the Great Waves swept across the world in 2035, the survivors were evacuated from Helsinki, but Sassa stayed on to continue studying the consequences of the catastrophe for marine life. Jyväskylä and Helsinki are experienced through Lilja's eyes, and they embody opposite reactions to environmental catastrophe. The major difference between the two cities concerns state control. All leisure activities in Jyväskylä are planned by the state and include skiing (on plastic tracks, as there is no snow due to climate change), football, and music performed on the traditional instruments – the kantele and the accordion – while television offers mainly programmes consisting of music competitions and quizzes. With humour, irony, and acuity, the author takes the key aspects of today's conservative nationalist ideologies of a homogeneous Finland and its imagined traditional values and transposes them to Jyväskylä. The flood has, in a way, made it possible to create a new Finland where the elite moved and could forget about those who were left behind. In a review of the novel, Maria Lassén-Seger comments on the satirical description of Jyväskylä's focus on safety and security to mark the novel's grounding in the Finnish nationalist discourse (Lassén-Seger, 2011). Julia Tidigs, in turn, questions the plausibility of an ethnically pure Jyväskylä, as no city like this exists in Finland today. Thus, the novel's Jyväskylä is not a replication of Helsinki at the time of writing in 2011 that could be moved north and into the future (Tidigs, 2021: 69). My understanding of the novel's future Jyväskylä is more sinister than either Tidigs's or Lassén-Seger's readings are. The narrative opens for an interpretation whereby the old Finland cleansed itself of unwanted people when it relocated its capital. As it turns out, the people left behind seem to be those who were already outcasts before the catastrophe. The narrow spectrum of cultural and sports activities, the control of information and the selected history that is sanctioned by the state, and the hindrance to travel, along with the strict surveillance of inhabitants, all suggest that it is a question of more than just moving a city out of harm's way.

The cost of order is uniformity, which is stifling for an inquisitive young person; it is 'inte mer än ett halvliv' [no more than a half-life], Lilja states, while a full life 'kräver frihet och risker' [requires freedom and risks] (Luther, 2011: 18), something that she hopes to find in Helsinki. After several dangerous situations, the young woman hitches a ride on a cargo ship owned by two Indian smugglers of algae oil, which is used to produce fuel in the immigration zone near Jyväskylä. As they approach Helsinki, the Orthodox Cathedral looms over the harbour like a spooky castle, and in the ruins of the Cathedral, Lilja glimpses a troll-like figure roaming around. The description of the lively and multilingual city is as much based on stereotypes as is the image of the bland and boring Jyväskylä. Washing is blowing in the wind on lines hanging between houses, people are exotic and wear colourful clothes, and things smell and taste differently. To Lilja, everything is loud, and the people seem initially to be rather rude. The formerly well-organised life in Helsinki has been replaced by makeshift solutions to acute environmental changes; landings and suspension bridges connect houses, boats transport people and animals, and children seem to run around unsupervised. The new inhabitants have even renamed the city 'Halsingih.' Under the ship on which she travels, Lilja can see flooded neighbourhoods and rooftops that break the water's surface. The people of this city seem to be prone to risk, Lilja reasons, and they probably would not even be able to spell the words of 'trygghetsplan' [safety plan] (Luther, 2011: 76). As almost all ground floors are flooded, even in the surviving buildings the homes have become damp and mouldy, and gangs of petty criminal scavengers hunt for valuables in the debris. Although there are organised businesses and institutions such as restaurants, markets, schools, religious buildings, and banks, there is no central state control; everything is based on people's own initiatives, their diligence, joint responsibility, and agreements between different groups. In Lilja's eyes, Halsingih is of course much more exciting and dynamic than the homogenous Jyväskylä.

But there are prejudices here too between groups of people. Those who did not escape when the wave swept over the city, or those who chose to remain, as did Lilja's mother, constitute the so-called indigenous population of the city. Skin colour is pointed out repeatedly and racism prevails (Luther, 2011: 142). Some Finns are even so pale that they appear to be green. The reader also encounters another Finnish stereotype – the alcoholic. The newly arrived inhabitants shun the stinking, alcohol-infused lunatics who spend their days ranting on park benches. The Indian, Chinese, and Bangladeshis live in separate neighbourhoods. Halsingih is thus not a utopian place from where differences or inequality have disappeared; on the contrary, inequalities have re-emerged, only in different forms.

Lilja finds refuge with the large Indian Nair family of the two smugglers who brought her into the city. The family consists of a mother who is Oxford-educated and now runs an English girls' school, of intellectuals, bankers, religious leaders, and entrepreneurs. Lilja (or Leila, as they call her) makes herself a home with the Nairs, participates in everyday chores with her new-found sister Manju, and soon starts wearing a toned-down Indian dress and oiling her hair. She eventually finds her lost mother in the ruins of the Russian Cathedral, but she is nothing like Lilja expected. She too has become an indigenous outcast and turns out to be the very same troll that Lilja glimpsed at her first sighting of the city. In the social hierarchy, the mother is slightly above the others in the indigenous population, and she runs a charity for other destitute Finns. Lilja is disappointed and rather thinks her mother is an embarrassment compared to what the Nair family has achieved despite their long and tumultuous journey from India. However, Sassa has maintained her interest in marine biology and studies the sea regularly, even if there is no more infrastructure for research. The only person she can share her findings with is the scientifically minded boy Matsendra, who helps Lilja to find her mother.

Throughout the life-histories of all these uprooted people, both the refugees from other continents and the Finns who have either been moved a few hundred kilometres north or have been left behind in a radically altered city, one may wonder what the word 'home' means. Home and homesickness are recurrent themes in the novel, but it is not easy to connect home to a geographical place, and Lilja would probably answer that home is about a sense of belonging. After several months of absence from Jyväskylä, she misses her family, even though she has found her biological mother. She exclaims: 'Hem? Det här är inte mitt hem! Jag börjar tro att jag aldrig kommer hem igen' [Home? This is not my home! I'm starting to think I'll never come home again], to which Matsendra, who is a foundling rescued by the Nair family during the refugee flows, replies: 'Då är vi två om saken. Jag kommer aldrig heller hem igen! Och jag vet inte ens var mitt hem var' [That makes two of us. I'll never go home again either! And I don't even know where my home was] (Luther, 2011: 172).

Cities and capitals

Lefebvre's statement that 'the state is becoming worldwide [*se mondialise*] and at the same time opposes the worldwide' is reflected in the relationship between Jyväskylä and Halsingih (2009b: 275). The state of Finland wants to exclude from its territory all things foreign, to the extent that it retracts from its capital and locks people in as well as out. The state is identified with the capital and the city, which creates an opposition between the state and the countryside, a condition that is highlighted by the novel's emphasis on zones for different activities, such as the agricultural zone outside the city's borders or the dangerous migrant zones. The separation of peoples and global division of labour is further foregrounded by the fact that the agricultural workers are EU migrants, not Finns. Returning to Lefebvre, Jyväskylä can be described as being in direct opposition to the worldwideness as the city withdraws into nationalist isolation and creates a culturally and mentally repressive space. The backwardness of Finland's strategy is obvious to the reader.

To use Lafebvre's terminology, Halsingih carries, above and under the water's surface, the nation state's 'hierarchically ordered institutions, of laws and conventions upheld by "values" that are communicated through the national language' and the social architecture forms a 'political monumentality . . . full of symbols' (2009: 224). Each institution possesses an appropriate space for supporting political domination – but how quickly places and symbols become meaningless! While Lefebvre emphasises the monumentality and symbols of the city as essential in upholding state power, it becomes obvious that the absence of an organised state apparatus drains the symbols of power of their erstwhile meaning. What today holds a central place in the architecture of the city of Helsinki is the area in front of the Parliament (Kolbe, 2006). The architecture of the Parliament building was already debated in the 1920s, the debate revolving around the need to balance the display of newly won independence and power with humility and national unity (Hakala-Zilliacus, 2002; Niemi, 2017; Suutari, 2011). The area has more recently been the topic of another heated discussion, this time focusing on the nearby equestrian statue of Marshal Gustaf Mannerheim, who is a national war hero and former president. When Kiasma, the museum of contemporary art, was planned, debate arose about how the museum's postmodern glass architecture would dwarf the statue of Mannerheim and thereby symbolically diminish the importance of a part of Finland's history. But for Lilja, as well as for those who have recently settled in Halsingih, all these statues and buildings are completely irrelevant. The Parliament, which is now referred to simply as a stone castle, is situated in a less attractive part of the city (adjacent to a huge open cesspool) and has become the headquarters of a Chinese petroleum company called Chinopetrol. This is where Lilja goes to communicate with her family, because the Chinese have trade contacts with Jyväskylä. Statues that have been erected in tribute to the working people or bear other national cultural meanings are perceived by Lilja as 'nakna män av mörk metall med klubbor i händerna' [naked men in dark metal with clubs in their hands], and as 'ryttare till häst' [man on horseback]. Similarly, landmark buildings are strange looking 'växthus' [greenhouses]. One statue she does recognise is the Finnish mythological figure of Väinämöinen because he features in the Finnish epic *Kalevala*, a tradition highly cherished in Jyväskylä (Luther, 2011: 113–114). Without meaningful bureaucratic and political institutions, or traditions that convey historical prominence, buildings and statues completely lose their significance. As Jyväskylä no longer wants to acknowledge Helsinki on any level, the political and cultural history of the city is not taught at school, and Lilja does not know the symbolically important elements for either Finland as a state or the city in which she was born. For the migrants, the original symbolism or function of the buildings have no meaning, either.

A fictional text can create many layers of memory and recognition, and literature can create scenes that activate the reader's previous acquaintance with a specific place, without excluding readers who do not know the place. Readers familiar with the city of Helsinki will be aware of the city's vulnerable location; an environmental disaster could really mean that large parts of Helsinki would regain a shoreline that is more similar to that of the eighteenth century, and much of the current city would find itself under water.[4] For anyone else, it will still be obvious that it is a modern western harbour city that has been completely transformed through climate change, and, with descriptions of recognisable institutions that are repeated in every capital city in the world, the author creates a literary experience of the consequences of a disaster. It is worth adding that Helsinki is used as a similar, but more dystopic, setting in *Parantaja* (2014) [*The Healer*] by the Finnish author Antti Tuomainen. The novel depicts Helsinki in the near future after major climate change, when constant rain and storms haunt the city. Yet the strategies of the two authors are diametrically different. Luther draws attention to forgetting, renaming, transformation, and the unexpected, while Tuominen minutely records places, roads, buildings, addresses, and house numbers as they exist today. Tuomainen's narrative strategy can be read as a call for holding on to the city's recognisable landmarks in a period of rapid deterioration. Here, as in Luther's book, buildings are falling apart, mould is creeping up from the basements, and roofs are leaking, while affluent locals are moving north, and new people arrive from all over the world.

Literature that deals with environmental degradation often takes place in a time close to our own. In a novel's plot we can follow the consequences of our actions into the future, and this narrative strategy is crucial if the author seeks to influence the readers' behaviour, for example by linking the negative environmental changes to the daily actions of the readers.[5] The purpose of letting disasters unravel in familiar settings may be to influence readers to change their behaviour in the present. Readers are more likely to act if events happen in recognisable places which have been destroyed, radically altered, or have changed status and function. Films repeatedly use globally known cities and landmarks as proxies to create recognition. The Sidney Opera House at the end of the film *Mad Max Beyond Thunderdome* (George Miller and George Ogilvie, 1985) or the remains of the Statue of Liberty in *The Day after Tomorrow* (Roland Emmerich, 2004) symbolise the fall not only of the monuments themselves or cities they are placed in, but of the entire western civilisation. Ghosh even suggests that 'television, film and visual arts have found it much easier to address climate change than literary fiction' (2016: 83). Climate fiction as a genre can similarly rely on the wide cultural references to enhance its impact. Familiarity is thus not dependent on first-hand or local recognition as in *The City of the Homeless*, but can just as well be based on global symbols of civilisation.

To return to Luther's near-future climate fiction, we could perhaps use Michel Foucault's concept of 'heterotopia' to untangle some of the thick layers of meaning that are at stake. Heterotopia is not a well-developed concept in Foucault's thought, where it refers to actual places as opposed to a utopia as a non-existent space; heterotopia is the *other* and *actual* place rather than the *non*-place of utopia (Tonkiss, 2005: 132–133, my italics). The heterotopia of crisis in modernity is, for Foucault, a place of and for deviance such as the brothel or even the cinema. His emphasis on heterotopia as an escape from normality or the ordinary use of space does not translate straight into the conditions of Halsingih as the new inhabitants seek another place out of necessity rather than to escape normality. Yet the empty space that is created by the evacuation of the capital leads to it become vacant land to be reoccupied, whereby it becomes another place with a new normality. The capital has fallen, but it is good. In Tuomainen's mentioned novel, *The Healer*, which is also set in Helsinki during a climate crisis, the protagonist looks at the city's diverse population of destitute refugees and locals, stating that 'Helsinki oli vihdoinkin kansainvälinen. Mutta aivan tästä emme varmaan aikanaan haaveilleet' [Helsinki had finally become an international city. But this surely wasn't how we had imagined it] (2013: 229, 252).

Reading for change?

Climate change fiction is often viewed as realistic if readers can identify with the work's setting, which speaks to the value of regionally identifiable locations, and perhaps even regionally topical climate challenges (Schneider-Meyerson, 2018: 488). The interplay between what is familiar in one moment, unknown in the next, between recognition and alienation, can have an effect on readers and make them consider what consequences an environmental disaster can have in concrete places that they know of or may even have visited. Can climate fiction and cinema, stories about imagined futures, function as impulses for changing attitudes? Both reader surveys and cognitive psychology support this to some extent. Matthew Schneider-Mayerson's empirical study shows that readers of climate fiction were often already convinced that climate change was a real and concrete problem. However, many found climate fiction important for their ability to imagine futures, to visualise the ramifications of climate change or to give concrete examples of potential consequences; 'authors are able to illuminate what is otherwise invisible' (2018: 482–484). Narratives can thereby narrow the gap between spatial and temporal distance to an imaginable future, Schneider-Mayerson concludes. Thomas Morton et al. argue that 'higher uncertainty combined with a negative frame (highlighting possible losses) decreased individual intentions to behave environmentally. However, when higher uncertainty was combined with a positive frame (highlighting the possibility of losses not materialising) this produced stronger intentions to act' (quoted in Neimanis et al., 2015: 77). And it is in times of uncertainty that we seem to have a need for help in imagining probable futures. Jerry Määttä's analysis of apocalyptic literature relates the genre's peaks to dramatic historical events, and he concludes that

> [apocalyptic stories] seem to function best as pressure valves during periods of build-up to expected conflicts or crises, for letting off emotional steam, and perhaps as a way of mentally preparing for an even bleaker reality which many fear will soon be imminent.
>
> (2015: 429)

Alternatively, literature can be a forum for trying out and launching creative ideas and for imagining future-oriented solutions. Having said all this about the potential influence of literature on the mindset of readers, I am in no way suggesting that the main purpose of literature is to didactically direct human action in a pregiven way or that the effects of literature could be unequivocally determined.

Global and local

Ecological literature, to use one of the widest possible terms for fiction dealing with human/non-human relationships, is sometimes seen as localised and even nationalised, because of the connection to a place, a natural habitat, or experiences relating to a specific topography such as a mountain, a forest, or a lake. It is easy to juxtapose the local and the global as mutually exclusive or as opposites, which is discussed in Ursula Heise's influential book *Sense of Place and Sense of Planet: The Environmental Imagination of the Global* (2008). Heise's argument is that ecocritical writing needs to take stock of the global to a larger extent than it did in the early twenty-first century. Ecological catastrophes need global solutions; however, I would argue that while political solutions need to be global, it is perhaps not literature's task to follow suit. While fully sympathetic to the need to grasp the challenges of living in the age of the Anthropocene, I find literature at its best when it brings global troubles to a local and human level. Claire Colebrook has identified 'the task of literature and the humanities' as one 'to address our sentimental, affective, habitual and non-cognitive comportment toward the world' (2017: 115). Studying local solutions to global phenomena, as does *The City of the Homeless*, means that the global is not over-emphasised at the expense of the local. For example, a prerequisite for the idea of global networks is communication, Heise argues, as she stresses the importance of a 'dual vision' combining narratives of 'the Earth as a whole' with narratives of 'different earths that are shaped by varying cultural context' (2008: 210). At times, however, the aversion to the local tends towards a kind of methodological globalism (paraphrasing the much-debated methodological nationalism) according to which only that with global economic or cultural reach is regarded as relevant for study (Leppänen, 2020; Leppänen, 2022: 183–184). I would rather argue that it is exactly the attachment to a place, combined with the acknowledgement of a shifting cultural meaning of places, that is the strength of literature. In *The City of the Homeless*, for example, the global catastrophe is experienced on a local and individual level, thus capturing what Lawrence Buell terms a 'ecoglobalist affect' or 'the inextricable linkage between . . . the specific site and a context of planetary reach' (Buell, 2007: 232). Experimenting with alternative post-catastrophic futures can thus be crucial for an imaginative approach to ecological responsibility in an uncertain future.

Cities represent many forms of vulnerability. A city is a concentrate of civilisation and at the same time the most fragile of human forms of living due to its artificiality, its reliance on technology, and its resource-dependency on production of food and goods outside the city to be consumed in a concentrated area. It is not uncommon that climate fiction includes an aspect of primitivism by suggesting a return to the old ways of living and surviving, something that is the opposite of the artificiality of life in the city (Milner, 2020: 22–23). But Annika Luther's novel is wonderfully liberated from such nostalgia. Maybe the author simply wants the reader to embrace the changeability of the place and the unpredictability of the future? Because the future in Halsingih is not at all bad!

Conclusions

I will end my discussion of the future of Helsinki in Luther's novel by briefly returning to Lefebvre and the question of how fictive futures are significant in rethinking post-catastrophic cities. For Lefebvre, space is a result of social production, and it consists of several interlaced processes (2009a: 224–225). These processes are neither chronological nor separable, but they can be identified for their different roles in relation to the reproduction of or resistance to state power. The first mode relates to the location and differentiation of production and reproduction that organises everyday life. The second relates to representations of space through symbols of power, government, architecture, maps, and plans. The third level conjures up representational spaces of imagination, embodiment, and desire. Applied to the story of *The City of the Homeless*, Lefebvre's theory of the social production of space can be given very distinct, almost didactic, forms. The organisation of production and reproduction (level one) is the sole purpose of life in Jyväskylä, as work, home, and leisure all have their assigned and standardised places. Very little freedom is granted to the citizens to move or act outside the government-sanctioned spaces and activities; everything is monitored and choreographed by a bureaucratic state. The activities of citizens are limited to a monotonous reproduction of pre-determined social space with minimal opportunity for change or influence. Lefebvre's second level of maps and symbols is the creation of social space that bridges the reader's Helsinki, or any other modern metropolitan city, with the future Halsingih. The fictional story engages the reader by calling into question the meaning of the symbolic representations of power in the buildings and monuments of the capital city, while pointing out the very meaninglessness of the physical manifestations of power for the city's new inhabitants. This shatters any idea of a permanent, natural, or eternal power or symbolism of the current state. The third level, the one least developed by Lefebvre but perhaps most interesting for literary studies, speaks directly to the meaning of fiction in making sense of the city and its role in our culture. Both the acts of remembering pasts and imagining futures are required to give direction to the development of cities. Facing catastrophe, whether it come fast or slow or be small or large scale, we need to voice desires and wishes that enable us to act to avert catastrophe and to react to the inevitable change. The critical utopian power of *The City of the Homeless*, to return to Tom Moylan's term (1986: 10), is found in this open representation of the inevitable *mondialité* of the future. As the nation state has lost its defining power over institutions, languages, monuments, and people, a new world can be built on the ruins of the old. Near future fiction of the kind represented by *The City of the Homeless* exposes the challenges of ecocatastrophe in a condensed space of present-day Finland, and it offers an opportunity to rethink what makes nations and cities meaningful as a home. It is evident that values such as far-reaching security and control fuel people's complacency, which happens at the cost of another value – freedom.

*

At the end of *The City of the Homeless*, the capital city of the state of Finland is threatened yet again, this time by a nuclear meltdown. According to Määttä (2015: 415), nuclear war is the most frequently occurring reason for literary apocalypse, and as discussed by Helena Duffy (Chapter 6) and Sébastian Thiltges (Chapter 7) in this volume, a nuclear war or a nuclear disaster can create space for reconfiguring the world along new lines of alliances and values. Plants and animals, as well as places, are ascribed a new meaning in a post-catastrophic world because of the changing position of humans in relation to society and the natural world. In Jyväskylä, a nuclear power plant suddenly begins to leak, and the previously so strongly controlled society dissolves into chaos. The protection that the well-organised state offered at the cost of repression proves unsustainable in the face of yet another human-made catastrophe. The government and the administration flee in panic north towards Lapland, letting go of the Koskinen family, who instead head south to look for their daughter. It turns out that the administration of Jyväskylä never intended to allow Lilja back home because she could carry diseases, and, one guesses, could bring unwanted knowledge of the world outside. Exhausted, the family drives sixteen hours on substandard roads, across fields of mud and with great danger to themselves, to reach their old hometown. When the state withdrew, the production of a different space was made possible and it is here, in the transformed, dynamic, vibrant, cosmopolitan city of Halsingih, that they can create a meaningful future.

Notes

1 In her Swedish-language novel, Luther uses the Swedish name of the capital, Helsingfors. I have opted to use the Finnish name Helsinki throughout, as it connects to the city's English name. For a discussion of the dialectical and multilingual form of Luther's text, see Tidigs (2021).

2 Definitions of climate fiction differ. For example, Gregers Andersen (2020: 5) argues that cli-fi should not be used for just any climate change, such as that following a natural disaster, but should rather be reserved for anthropogenic (human-induced) change. This is of course not always easy to determine, as a combination of causes may be at work (global warming and a volcanic eruption), or the root cause may not be known to the reader or the protagonist.
3 The World Bank.
4 Land elevation caused by the latest ice age is still affecting the Nordic countries. A rise in water levels would reverse the effects of land elevation.
5 One example of author activism is the Norwegian authors' climate action (Klimaaksjon).

Bibliography

Ahlbäck, Pia (2019), 'From Technopastoral to Climate Idyll? The Opportunities of Disaster in Annika Luther's Novel *De hemlösas stad*', *Conference Abstract*, [Online], [Accessed 29 August 2022], available from: https://enscandotnet.files.wordpress.com/2019/11/abstracts-enscan.pdf

Ancien, Delphine (2011), 'Global City Theory and the New Urban Politics of Twenty Years On: The Case for a Geotheoretical Materialist Approach to the (New) Urban Politics of Global Cities', *Urban Studies*, 48:12, 2473–2493, https://doi.org/10.1177/0042098011411

Andersen, Gregers (2020), *Climate Fiction and Cultural Analysis: A New Perspective on Life in the Anthropocene* (London: Routledge).

Brenner, Neil and Stuart Elden (2009), 'Introduction State, Space, World: Lefebvre and the Survival of Capitalism', in Stuart Elden and Neil Brenner (eds.), *State, Space, World* (Minneapolis: University of Minnesota Press), 1–48.

The Britannica Dictionary [Online], [Accessed 20 May 2022], available from: www.britannica.com/dictionary/city

Buell, Lawrence (2007), 'Ecoglobalist Affects: The Emergence of U.S. Environmental Imagination on a World Scale', in Wai Chee Dimock and Lawrence Buell (eds.), *Shades of the Planet: American Literature as World Literature* (Princeton, NJ: Princeton University Press).

Choen, Steven and Guo Dong (2021), *The Sustainable City*. Second Edition (New York: Columbia University Press).

Colebrook, Claire (2017), 'The Twilight of the Anthropocene: Sustaining Literature', in Adeline Johns-Putra, John Parham and Louise Squire (eds.), *Literature and Sustainability: Concept, Text and Culture* (Manchester: Manchester University Press), 115–136.

Ghosh, Amitav (2016), *The Great Derangement: Climate Change and the Unthinkable* (Chicago and London: University of Chicago Press).

Goodbody, Axel and Adeline Johns-Putra (2019), 'The Rise of the Climate Change Novel', in Adeline Johns-Putra (ed.), *Climate and Literature* (Cambridge: Cambridge University Press), 229–245.

Hakala-Zilliacus, Liisa-Marja (2002), *Suomen eduskuntatalo: Kokonaistaideteos, itsnäisyysmonumentti ja kansallisen sovinnon representaatio* (Helsinki: SKS).

Heise, Ursula K. (2008), *Sense of Place and Sense of Planet: The Environmental Imagination of the Global* (Oxford: Oxford University Press).

Johns-Putra, Adeline, John Parham and Louise Squire (2017), 'Editors' Introduction', in Adeline Johns-Putra, John Parham and Louise Squire (eds.), *Literature*

and Sustainability: Concept, Text and Culture (Manchester: Manchester University Press), 1–8.
Klimaaksjon, [Online], [Accessed June 2022], available from: https://forfatternesklimaaksjon.no
Kolbe, Laura (2006), 'An Eastern or a Western Capital City? The Spirit of Helsinki', *International Review of Sociology – Revue Internationale de Sociologie*, 16:2, 329–346, https://doi.org/10.1080/03906700600708964
Laakso, Maria, Toni Lahtinen and Hanna Samola (2019), 'Young Survivors and Agents of Change: Power, Environment, and Girlhood in Contemporary Finnish Young Adult Dystopias', *Utopian Studies*, 30:2, 193–213, https://doi.org/10.5325/utopianstudies.30.2.0193
Lassén-Seger, Maria, 'Luthers vattenvärldar fungerar', [Online], [Accessed 15 June 2022], available from: http://gamla.hbl.fi/kultur/recension/2011-08-25/luthers-vattenvarld-fungerar
Lefebvre, Henri (1996), *Writings on Cities*, trans. Eleonore Kofman and Elizabeth Lebas (Cambridge: Blackwell).
Lefebvre, Henri (2009a), 'Space and the State', in Stuart Elden and Neil Brenner (eds.), *State, Space, World* (Minneapolis: University of Minnesota Press), 223–253.
Lefebvre, Henri (2009b), 'The Worldwide Experience', in Stuart Elden and Neil Brenner (eds.), *State, Space, World* (Minneapolis: University of Minnesota Press), 274–289.
Leppänen, Katarina (2020), 'Memory of Water: Boundaries of Political Geography and World Literature', *European Review*, 28:3, 425–434, https://doi.org/10.1017/S1062798719000541
Leppänen, Katarina (2022), 'Wordly Themes and Vernacular Literature: Aino Kallas on Gender, Ethicity, and Class', in Christina Kullberg and David Watson (eds.), *Vernaculars in the Age of World Literature* (London: Bloomsbury), 181–201.
Liinamaa, Saara (2020), 'The Aesthetics of Spatial Justice', in Michael E. Leary-Owhin and John P. McCarthy (eds.), *The Routledge Handbook of Henri Lefebvre, The City and Urban Society* (Abingdon: Routledge online).
Luther, Annika (2011), *De hemlösas stad* (Helsingfors: Söderströms).
Määttä, Jerry (2015), 'Keeping Count of the End of the World: A Statistical Analysis of the Historiography, Canonisation, and Historical Fluctuations of Anglophone Apocalyptic and Post-Apocalyptic Disaster Narratives', *Culture Unbound*, 7, 411–432, https://doi.org/10.3384/cu.2000.1525.1572411
Marskalk Mannerheims ryttarstaty, [Online] [Accessed 15 June 2022], available from: www.hamhelsinki.fi/sv/sculpture/marsalkka-mannerheimin-ratsastajapatsas-aimo-tukiainen
Milner, Andrew (2020), 'Utopia in Recent Climate Fiction: MaddAddam, MAEVA! and New York 2140', *Foundation*, 49:137, 19–30.
Moylan, Tom (1986), *Demand the Impossible: Science Fiction and the Utopian Imagination* (New York: Peter Lang).
Neimanis, Astrida, Cecilia Åsberg and Johan Hedrén (2015), 'Four Problems, Four Directions for Environmental Humanities: Toward Critical Posthumanities for the Anthropocene', *Ethics and the Environment*, 20:1, 67–97, https://doi.org/10.2979/ethicsenviro.20.1.67
Niemi, Marjatta (2017), 'Breaking from and Building on the Past: Helsinki and Dublin After Independence', *Irish Historical Studies*, 41:160, 238–255, https://doi.org/10.1017/ihs.2017.34

Ortman, Scott, José Lobo and Michael Smith (2020), 'Cities: Complexity, Theory, and History', *PLoS ONE*, 8 December, 1–24, https://doi.org/10.1371/journal.pone.0243621

Schneider-Meyerson, Matthew (2018), 'The Influence of Climate Fiction: An Empirical Survey of Readers', *Environmental Humanities*, 10:2, 473–500, https://doi.org/10.1215/22011919-7156848

Suutari, Inkeri (2011), 'Valta näkyväksi: Eduskuntatalohanke 1905–1926' (Master's Thesis, University of Helsinki).

Tidigs, Julia (2021), 'Från Djuvaskula till Halsingih på många språk: En flerspråkighetsläsning av Annika Luthers *De hemlösas stad*', in Jutta Ahlbeck, Judith Meurer-Bongart, Julia Tidigs and Mia Österlund (eds.), *Vill jag vistas här bör jag byta blick: Texter om litteratur, miljö och historia tillägnade Pia Maria Ahlbäck* (Åbo: Granskaren).

Tonkiss, Fran (2005), *Space, City and Social Theory* (London: Polity Press).

Tuomainen, Antti (2010), *Parantaja* (Helsinki: Helsinki kirjat).

Tuomainen, Antti (2013), *The Healer*, trans. L. Rogers (London: Harville Secker).

The World Bank, [Online], [Accessed 22 May 2022], available from: www.worldbank.org/en/topic/urbandevelopment/overview#1

5 Extreme climate and the anthropocentric conception of agency in cinematic ocean planets

Faeze Rezaii

This chapter discusses the representation of climate-related extreme events in two science fiction films, reading them as multifaceted explorations of agency in the Anthropocene. A great number of scholarly writings have addressed cinematic water imaginaries on Earth, but much remains unsaid about filmic imaginaries connected to other planets and their real-world implications. Scientific findings have constantly influenced these imaginaries. For instance, in a recent article on planetary geology, William Rapin et al. discuss climate change on Mars and examine its salt lakes that dried up '3.5 billion years ago' (2019: 889). These kinds of discoveries affect fiction and trigger the imagination of filmmakers in creating imaginary geographical spaces where climate change beyond Earth, albeit not induced by humans, is a determining factor in the narrative.

Star Wars: *Episode II – Attack of the Clones* (Lucas, 2002) and *Interstellar* (Nolan, 2014), two science fiction films that mainly take place outside Earth, are cases in point. Planet Kamino in *Attack of the Clones* and planet Miller in *Interstellar* are exemplary of the two kinds of aquatic planets that recur in cinematic climate science fiction. I have used three different terms for the taxonomy of aquatic planets: water planets, ocean planets, and wet planets. The terms 'ocean planet' and 'water planet' are used almost interchangeably in *Star Wars* books and fandom. However, in this chapter, I differentiate water planets as having both terrestrial and aquatic spaces on their surface, just as does Earth, from ocean planets which have only oceans on their surface and no dry shores. My essay focuses on two ocean planets, Kamino and Miller, but as I will argue, the latter is an instance of a wet planet that resists the binary categorisation of water and ocean planets too.

In line with the logic of projection and extrapolation that governs much speculative fiction, climate imaginaries set in the future have strong ties to how we currently perceive our place as agents in the natural world. This chapter focuses on extreme climate events and the conception of agency in the representation of these two cinematic ocean planets. The

DOI: 10.4324/9781032726953-6

representation of Kamino in *Attack of the Clones* unsettles human agency with the planet's extreme climate and shows an anthropocentric understanding of agency with the imprisoned humans of the clone army. In the *Star Wars* world, the governance of the Galactic Republic along with the Jedi knights (central protagonists holding the light side of the force) secures peace in the universe. *Attack of the Clones* begins as the separatist movement orchestrated by the Siths (the antagonists holding the dark side of the force) attempts to overthrow the Republic. A senator named Padmé Amidala (Natalie Portman) wishes to vote for creating an army to resist the separatist movement, but a bounty hunter named Jango Fett is assigned to assassinate her. Jedi Master Obi-Wan Kenobi (Ewan McGregor) thwarts the attempt and the Jedi council demands that he find the bounty hunter. After investigating, Obi-Wan realises that the hunter lives on an ocean planet named Kamino, which is inhabited by a sentient species named Kaminoans; it is the birthplace of the notorious army of human clones disguised in white armour. Kaminoans have secretly created the clone army by the order of a human Jedi but without the permission and knowledge of the Republic. The bounty hunter, Jango Fett, is the clone template of the army.

In the storyline of *Interstellar*, Earth is becoming uninhabitable in the year 2067 due to heavy dust storms, and NASA is working in secret since space exploration is deemed to exhaust resources. NASA sends a crew of astronauts to find a habitable planet. Cooper (Matthew McConaughey) and a crew of astronauts go on a mission to find a new home for humans. As they voyage through another galaxy, they come across planet Miller (named after the astronaut Dr Miller), which neighbours a supermassive black hole. Having received signals from their colleague Dr Miller, who docked on this planet before, they consider it promising. Not knowing that the planet is devoid of shores, the characters disembark the spacecraft while the planet's ocean is in the middle of a swell. On both Kamino and Miller, the anthropocentric conception of agency dominates the ocean planets.

This chapter suggests that analysing the representation of ocean planets in science fiction can help us untangle the fundamental questions about the relationship between human and non-human agencies now that the human species functions as a climate agent. While both films represent worlds beyond our planet, they still recount stories in which humans and the ideologies of the Anthropocene are central; they extend environmental issues on Earth to alien worlds. On Earth, the equation of climate change has two sides: natural occurrences along with anthropogenic activities. The environmental writer Renee Cho suggests that 'there have always been extreme weather events caused by numerous natural factors, but climate change is increasing the number and strength of these events' (2021).

In the two selected films, the representations of the oceanic worlds obscure the human ecological impact on a planet. The films' portrayals of ocean worlds undermine human agency, staging humans as weak agents. In this sense, these representations are in line with an anthropocentric conception of agency. This understanding of agency suggests that the environmental crisis on Earth is now beyond human control (Clark 2015: 6). According to Timothy Clark, 'the irony of the Anthropocene' is that even though this epoch is within the 'planet's natural history,' a time when humans become a definitive 'geological and climatological force,' it exposes itself to us as a process through and within 'the domain of natural becoming,' a procedure out of our hands, most noticeable for us in extreme weather phenomena or the destruction and downfall of ecosystems (2015: 6). Kamino and Miller reproduce the same conception of anthropocentric agency. On both planets, nature in the form of water as the main actant deprives humans of agency. Rather than bringing the human and the non-human together in an interdependent relationship, this type of agency separates 'Nature, a substance "over there"' from the human 'subjects over here' (Morton, 2016: 56, 29). These imaginaries of ocean planets revolve around human involvement with extreme weather events and natural crises like floods and tsunamis. The films' depictions of Kamino and Miller thereby largely avoid the human impact on the environment while retaining the extreme weather events and climate change. I suggest that discarding human agency from the equation while retaining its effects performs the ideological work of maintaining the status quo on Earth. Such representation of agency in these cinematic productions operates to naturalise ecocatastrophes and avoid human responsibility. This is very different from future imaginaries of places impacted by climate catastrophe in *De hemlösas stad* (2011) [*The City of the Homeless*], analysed in Chapter 4, where Katarina Leppänen discusses the transformative capacities of Annika Luther's novel in encouraging environmental responsibility in the readership.

In examining the Anthropocene and agency in the extreme climate imaginaries connected to the planets, I draw upon two optical phenomena that are best understood in and with water: reflection and diffraction. We can understand reflection, for instance, in the replication of our bodies on the surface of water. Reflection acts like a copy imitating an original. As I will argue, both films work with an aesthetics of reflection, echoing endless images of humans stuck in the Anthropocene as well as diffraction. Diffraction can be observed in water when 'waves . . . overlap and extend into one another' (Geerts and Tuin, 2016). This happens when waves face an obstacle and their patterns change from linear lines to circular ones. Conceived in aesthetic and figurative terms, diffraction can create new and different forms of agency for both the bodies of water and humans. I argue that even though *Attack of the Clones* and *Interstellar* represent extreme climate as a natural phenomenon independent of human involvement, they still render an anthropocentric conception of agency that is reflective. This type of agency serves to normalise ecocatastrophes and gloss over the accountability of humans. At the same time, however, these reflective images are ruptured by non-Anthropocentric or diffracted representations of the agency of a human in *Attack of the Clones* and the ocean in *Interstellar*.

Inundated lands: the ocean planet Kamino and the clone army

The habitability and appropriability of the planets determine the setting in both films, especially in *Attack of the Clones*, as its ocean planet Kamino is the location of the settlements for the human clone army. As Gerry Canavan points out, despite the ecological limits imposed on humans,

> late capitalism remains a mode of production that insists (culturally) and depends (structurally) on limitless expansion and permanent growth without end: into the former colonial periphery, into the peasant countryside, through oil derricks into the deepest crevices of the earth, and then, in futurological imaginings, to orbital space stations, lunar cities, Martian settlements, asteroid belt mining colonies, sleeper ships to Alpha Centauri, and on and on.
>
> (2014: 5)

This colonial capitalist outlook has consistently generated an expansive body of material in a sci-fi context. Borrowing from the title of Robert A. Heinlein's most famous novel, *Stranger in a Strange Land*, Jessica Langer suggests that one of the genre's enduring 'myths' is the figure of the faraway planet which is 'ripe for colonization' or the 'Strange Land' (2011: 3). The colonisation of space is an important context for both films, despite the differences in the story worlds of *Star Wars* and *Interstellar*. The genre of the former is science fiction mixed with fantasy, replete with various alien creatures, many of which have human qualities; the hard science fiction of the latter attempts to present a plausible universe that at times reminds the audience of space documentaries, where there is no evidence of other life forms beyond Earth.[1] Yet in both films, the oceanic agency complicates the prospect of colonisation: Kamino and Miller both unsettle human agency with extreme climate in the narratives of the films. Furthermore, both of these ocean planets are positioned as ambiguous and mysterious. While Kamino has been erased from the archive maps, planet Miller is situated near a mysterious black hole named Gargantua. The spatial positioning of the planets adds to the vague and impenetrable design of these fluid places.

In the case of Kamino, both its history of climate change and the clone army distinguish it from the rest of the planets featured in the film. I argue that Kamino represents a futuristic image of the parts of Earth tragically affected by global warming and sea-level rise. A case in point is the Marshall Islands in the Pacific Ocean, where sea-level rise caused by climate change intensifies the effects of 'extra-high tides' and 'flooding'. Therefore, it is expected that this chain of islands 'will be submerged' to some extent by 2035 (Bordner and Ferguson, 2020). Visualisations of a future Earth transformed by global warming are everywhere in contemporary media. To mention just one representative example, the maps in a 2013 *National Geographic* article entitled 'What the World Would Look Like if All the Ice Melted' project a world where many islands and coastlines have partly or entirely vanished (2013). Kamino thus exhibits thematic similarities to the way Earth is imagined as a changing planet. However, climate change on Kamino is envisioned as a natural process not caused by human activity. As the sourcebook for the roleplaying game *Star Wars: Rise of the Separatists* points out, Kamino was initially a 'terrestrial world' that transitioned from a long-lasting 'ice age' when all the mountains and fields were submerged by floods caused by the melting ice. This process made the humanoid species, Kaminoans (as they are called in this book and also by the other species in the film – for example, Dexter Jettster, an inhabitant of planet Coruscant), adapt their life to the new oceanic world. 'They continually moved to higher ground until there was little land left' and created enclosed domed towers on 'stilts' planted on the ocean floor in the city of Tipoca (Navara, 2019: 111).

Kamino's history of global warming and climate change is thereby evocative of climate change on Earth in the Anthropocene. The scenes that mostly take place indoors in the settlements built above Kamino's oceans mitigate the planet's extreme climate. The frequency of the sequences shot inside the settlements makes the profound effect of global warming less visible in the film. Kamino is represented as a place containing two contrasting modes of climate: extreme weather outdoors and clement weather inside its cities. This is evident, for instance, in the confrontation and battle between Obi-Wan Kenobi and the sought-out bounty hunter, Jango Fett. The sequence takes place in the open air and clashes with the artificially calm atmosphere within the settlement both visually and auditorily. The blindingly bright and silent interior, where the clone army of humans is engineered, contrasts with and thereby conceals the physical environment of the planet and its harsh climate with its combination of incessant rain and thunderstorms, acoustically foregrounded through sound alongside the visual presence of a dark-grey sky and clouds. The blue-hued shots (Figure 5.1) that display hundreds of clone foetuses submerged in the placid blue waters of artificial crystal wombs create a similar effect: interspersed with shots of the waters

Figure 5.1 The cloning technology of Kamino in the city of Tipoca City in *Star Wars: Episode II – Attack of the Clones* (George Lucas, US, 2002). The blue hue of the shot emphasises the water in which the clone embryos thrive.

of the violent dark floods on the surface of the planet, they create a sense of balance and order.

The clone army embodies an anthropocentric system. It is created by the order of a human Jedi character so that it fights for the Galactic Republic. The army is the result of the genetic engineering of humans in a prison-like setting. Mass production of only one species has disturbed the ecological balance of the planet. In the following episodes of the film series, the viewer will witness how this disrupts the entire universe of Star Wars when the Siths use the clone army against the Republic in never-ending wars that bring about all kinds of calamities like the slaughtering of all the Jedis. Since Kaminoans have genetically modified the soldiers to follow the orders they have been given, little to no agency is imagined for them. In the scene where Lama Su, the prime minister of Kamino, takes Obi-Wan Kenobi on a tour of the army, he introduces the clone troopers as such: 'You'll find they are totally obedient, taking any order without question. We modified their genetic structure to make them less independent than the original host.' Humans have designed a system of power from which they have removed the agency of their kind.

The representation of the clone army very much corresponds with Timothy Clark's conception of the Anthropocene. Clark describes the Anthropocene as an epoch in which 'humanity having become a geological force has implications beyond the fact of human violence against the Earth and other species' (Clark, 2015: 16). He maintains that little agency remains for 'individual rights' and for 'human decisions' due to the human-induced climate change that was generated collectively. It is a time when the environmental collapse caused by human geological agency appears to be too vast to control. The Anthropocene creates 'the demand made upon a species consciously to consider its impact as a totality upon the whole planet, the advent of a kind of new reflexivity as a species' (Clark, 2015: 16). For him, all the personal and collective human actions should be conceived at the 'level of self-reflection.' In other words, humans experience 'shared self-recognition' as the human species (Clark, 2015: 16–17). Thus, in the Anthropocene, humans become conscious of their impact as a species; they return to and examine the actions that have caused their impact on the environment.

Drawing upon Clark's conception of the Anthropocene, I regard the reflective quality of human clones in *Attack of the Clones* as a thematic and figurative representation of the Anthropocene epoch and the human species. I argue that the imagery of reflection in the form of identical human beings is a filmic representation of humans imprisoned in the age of the Anthropocene, on an imaginary, faraway planet, unable to act autonomously and freely. Reflection here means a ceaseless return to the actions of the humans and to their effects in a given space. The scenes that present the army in the film show a multitude of humans of different ages. Two actors are involved in portraying the whole army; a teenage actor plays the part of young clones, and an adult actor portrays them as they are matured. Characters in each age group are mirror images of each other. The *mise-en-scène* thereby highlights the identical and reflective features of the clones. The sequence first shows them in their teenage years wearing uniforms, then cuts to a shot of a disgruntled grown-up clone. Finally, the whole army is marching while wearing full white costumes and helmets that cover their faces, implying their dehumanisation. The scene creates a collective human image, a vision of a homogenous society that is weakened and is therefore unable to insert new possibilities and change the status quo. Regulated by the anthropocentric system, the human clone soldiers have agency, but only a diminished and limited one.

The clone army with its diminished agency is a mirror image of humanity conveying the sense that things have gotten out of control due to human sovereignty. The film nonetheless affords the possibility of disturbing and breaching this anthropocentric conception of agency by introducing a diffractive element. The conversation between Obi-Wan and Lama Su marks this difference:

Obi-Wan: And who was the original host?
Lama Su: A bounty hunter called Jango Fett. . . . Fett demanded only one thing. An unaltered Clone for himself. . . . Pure genetic replication. No tampering with the structure to make it more docile. And no growth acceleration.

The bounty hunter, the original human in the army, is a person of his own will, and even though the army has been constructed with the use of his genes, he is similar to them only physically since he is a free man. In this sense, the whole army is a reflected image of an original character, even though they all look like a 'mirror image' of him (Barad, 2007: 89), except his son, Boba Fett, whom I consider a different and diffracted depiction within the army. According to Donna Haraway, 'Diffraction patterns record the history of interaction, interference, . . . difference. Diffraction is about heterogeneous history, not about originals' (2018: 273). Haraway writes that 'Science fiction is generically concerned with the interpretation of boundaries between problematic selves and unexpected others' most visible 'in a context structured by transnational technoscience.' She proposes that the subjects that emerge as a result are 'subjects called "inappropriate/d others"' that live in SF worlds (2020: 466). In Haraway's theory, this other might be something 'other than the image of the same, something inappropriate, unfitting, and so, maybe, inappropriated' (2020: 466). Jango Fett's (the template) son, Boba Fett, is represented as this other, when a rotated shot shows him taking the helm of the spacecraft to shoot Obi-Wan in his father's defence. Boba Fett breaks the boundary of homogeneity in the clone army by acting instinctively to protect his father without any hesitation because he might have an emotional attachment towards him. He breaks the shared and collective image of the Anthropocene that only creates sameness in the army. Among all the clones that have biochips inside their heads to make them less aggressive, he is the only one who has no implanted chip and therefore is not a cyborg. His appearance is marked by his rather long and dishevelled hair, which contrasts with the helmets uniformly worn by the other clones in the same age, highlighting his difference from the rest. Among the clones that lack spontaneity, Boba Fett represents a different pattern. His mode of agency is an instance of diffraction that acts as a breach in the reflected army. On Kamino with its extreme weather climate, a diffracted instance of human agency breaks the oppressive image of the Anthropocene.

The wet planet Miller: tsunami waves and a diffracted conception of place

The ocean planet Miller of *Interstellar* is itself a diffracted planet. The film depicts the fluid agency as a misleading and manipulative force that decentres and undermines the human species and shows human perception to be limited and insufficient. Extreme climate on the planet takes the form of enormous waves that resemble tsunamis. In 'What Causes a Tsunami?,' the website of the Pacific Tsunami Museum states that tsunamis are caused mainly by natural factors like 'Plate Tectonics' and 'Volcanic Eruptions' (The Pacific Tsunami Museum). But, as the website of the International Tsunami Information Centre explains, tsunamis can also be generated by non-natural forces like 'large nuclear explosions' (Tsunami Information Centre). However, floods are different since they can be triggered more significantly by anthropogenic climate change and global warming in addition to natural factors (Hirabayashi et al., 2013: 816). On Miller, tsunami-like waves are the workings of the tidal currents, and the extreme climate exists independently of any human agency.

The oceanic agency disrupts the human agency of the astronauts on Miller. During the mission assigned to the astronauts to find a habitable planet, the crew land on Miller, from where they received signals that suggest the possibility of life. Seeing the aquatic world, Dr Laura Miller deemed the planet promising and sent a signal to NASA to indicate it, disappearing mysteriously before the arrival of the crew of the spacecraft *Endurance*. Judging by the scenes that show the wreckage of Dr Miller's spacecraft, she must have been crushed by the waves. Upon entering the atmosphere of Miller, the crew see huge masses of water that could sustain life. The astronauts calculate the shallowest waters and land on them. They walk through the waters looking for the wreckage of Miller's spacecraft in order to retrieve her data recorder and to investigate the events that took place before their arrival. A bird's-eye view shot (Figure 5.2) shows the spacecraft after it has successfully landed in shallow waters while some of the crew have just got out of it and started to walk. Then, the camera focuses on the walking astronauts in extreme long shots followed by Dr Brand's

Figure 5.2 A bird's-eye view shot presenting the aircraft on Miller in *Interstellar* (Christopher Nolan, US, 2014).

(Anne Hathaway) point-of-view shots (Figure 5.3) which reveal what seem to be tall pale-blue mountains on the horizon, making the planet appear even more Earth-like.

However, we soon realise along with the astronauts that these are not mountains, but waves. As Brand is searching for Dr Miller's spacecraft to access the data Miller might have collected on the planet, she notices the wreckage of the spacecraft along with a data recorder. She wades through the thigh-deep waters to seize the recorder and bring it to the crew but, as

Figure 5.3 Brand's (Anne Hathaway) point-of-view shot in *Interstellar* (Christopher Nolan, US, 2014) revealing their foreground of supposedly pale-blue tall mountains.

she struggles to move the recorder, she slips and accidently gets stuck under it. The camera shows Brand being pinned under the wreckage looking at the rising waves, but those waves are moving away from the astronauts who are about to be heavily hit by a gigantic wave approaching them from the rear. Cooper's lines capture the moment of understanding: 'Those aren't mountains. They're waves. That one's moving away from us.' Then he goes to the back of the Ranger to check the rear view and sees an extremely high wave approaching: 'Brand, Doyle, back to the Ranger, now! The second wave is coming, we are in the middle of a swell.' The land covered with shallow water through which the astronauts could wade transforms into an ocean within a few seconds as the camera tilts upward, apparently in Brand's point of view, to reveal the rising wave. Human perception is limited on this planet, even though the horizon is shown to be clear. 'Water is . . . an agent in our perception. And so are our eyes' (DeLoughrey and Flores, 2020: 162). Water imposes its agency by obscuring and dimming the characters' vision. The astronauts are directly looking at the gigantic waves the whole time, yet they cannot recognise them as such. We normally associate limited human perception with underwater spaces, but here both the initial bird's-eye view and the astronauts' points of view as they are walking on the planet's surface offer erroneous perceptions of the environment.

Agency on Miller is imagined as belonging only to the ocean; it is one-sided and unshared, embodied by the oceanic waves. In the discussion of material and nonhuman agency, Carl Knappett and Lambros Malafouris argue against the idea that agency should be understood as mere 'intentionality' and formed by human 'consciousness' (2008: ix). Furthermore, Veronica Strang addresses the relationship between water and agency and suggests that the idea of a 'flux' for agency resonates 'with Bruno Latour's view that agency emerges neither from people nor things but from their combination' (2014: 138–139). Such an understanding of agency is in sharp contrast to the conception of agency on Miller, where the ocean possesses all the power and humans are frail and disempowered creatures at its mercy. Latour affirms that an agent earns subject status due to being 'subjected' to 'another agent.' For him, 'to be a subject is not to act autonomously in front of an objective background, but to *share agency with other subjects that have also lost their autonomy*'(emphasis in original); subjects are all co-dependent actants (Latour, 2014: 5). Neither the subject nor the object possesses absolute 'mastery' or 'de-animation' (Latour, 2014: 11). The characters from whose perspective the film tells its story hope to colonise the planet and gain mastery over it, but instead, in less than a few minutes, their subject role is seized by the ocean which is going to devour them. In both the initial scenes that show the crew landing and the following scenes showing the giant waves, subjectivity seems to be equated with autonomy and mastery of one subject over the other.

The ocean exerts its power over humans through its deceptive nature and manipulation of human perception, which lead to a potentially fatal submersion. As the astronauts are waiting for the engines to drain so that they can leave the planet, their robot, Case, retrieves Dr Miller's data and states: 'The data Doyle received was just the initial status echoing endlessly.' It is now known that no subsequent signal ever existed and that the initial one was in fact based on misperception and a reverberating echo. We realise that humans will be stuck here endlessly, their condition/entrapment being amplified by the dilation of time on the planet; a single hour on Miller equals seven years back on Earth. The initial message of hope they received that implied they might have found a habitable water planet is nothing but a deceptive report. The planet has taken the life of Dr Miller before the arrival of the *Endurance* crew and it sweeps away Doyle, another member of the crew, upon their departure. The final scene shot on the planet shows Doyle lying face down in the water. A new wave is coming and Doyle's body will float, diffracting endlessly.

Although humans have no agency on Miller, the representation of the planet in the film paradoxically stages an anthropocentric worldview where agency can belong to only one actant; here, the human actant is simply replaced by the ocean. In this sense, this is an inverse representation of anthropocentric agency. According to Whitney Stark, agency should be understood not as a built-in characteristic of humans to be performed but as a 'dynamism of forces in which all designated "things" are constantly exchanging and diffracting, influencing and working inseparably' (2016). The anthropocentric standpoint manifests itself in the creation of a sterile ocean world devoid of life forms and their potential agencies. The planet is the flip side of the anthropocentric Earth, where humans have become the central geological force. The depiction of this planet comments on the anthropocentric understanding of agency where the only subject is human. Here it is the reverse; the only subject of power is water. Ultimately, it is important to keep in mind that this is a human-made vision of an oceanic world made mostly with the help of computer-generated imagery. Like Kamino, the planet Miller can be understood as the projection of the human fear and anxiety about the consequences of global warming in a solely oceanic planet where the oceans are the main agents. In contrast to Kamino, where diffraction patterns that break the anthropocentric agency are limited to the human body, Miller locates diffraction in the oceanic bodies and the fluid identity of the ocean itself. The ocean as the largest body of water is the best place to observe diffraction.

In *Interstellar*, Miller is presented as a diffracted entity since it is both an ocean planet and a water planet. Miller is constantly transforming from a planet that offers land with shallow waters to walk through into an ocean planet of enormous waves. It is the incarnation of both without any absolute distinction between the two. For Karen Barad, diffraction is non-binary and non-dualistic; it is what 'queers binaries' and points to a redefinition of the concepts of 'identity and difference' (2014: 171). Working with metaphors of light and darkness, Barad argues that diffraction does not lead to the production of one thing as opposed to the other (2014: 171). Miller does not have a fixed essence or an original identity. In this place, a water planet and an ocean planet are not two separate individual entities, and Miller is represented as having the potential to transform into both.

The categorisation of water and ocean planets as each other's counter-images is rooted in the theoretical opposition of land and water. This dichotomous logic reveals the colonial mindset in pursuit of fixed settlements on dry grounds that shapes these sci-fi narratives. The human geographer Kuntala Lahiri-Dutt proposes a hybrid reading of wet lands like floodplains, haors, and silts, that is of spaces 'where land and water are . . . inseparable,' producing a 'fluid environment.' She draws upon 'wet theory; a theory that can accommodate messiness and contextual variations' in order to go beyond the binary of water and land in geography (Lahiri-Dutt, 2014: 507). She uses examples of wet lands in order to break away from a rigid separation of land and water. In her discussion of the Bengal Delta, she notes that this 'aqueous land' is the result of 'fluvial action' and 'colonial interventions' (Lahiri-Dutt, 2014: 508). On a larger scale, Miller is also a hybrid and wet place that does not adhere to the binary of water and ocean planets, and thus resists categorisations. This hybridity on Miller corresponds to the entanglements of science fiction with both the 'fantasies of empire' and the 'resistant' narratives of 'decolonisation' (Canavan, 2012: 494). On Miller, the colonial pursuit of the astronauts clashes with the resistant nature of the planet. Ultimately, the astronauts have found a planet with water, a place that might be colonised, but their hopes are shattered by the planet's wet and hybrid identity. The hybrid matter is both the 'foreign' and the 'familiar' entity (Langer, 2011: 107) embodied in the oceans. The initial shots expose Miller as an Earth-like planet, another blue marble familiar to the characters, but then it transforms into a devouring ocean they have not encountered before. In this vein, Miller is the epitome of hybridity, a world that looks so much like our ocean world Earth but has too much water, a wet planet that resists any unwavering explication.

The planet Miller resists fixed figurations and categorisations, and oscillates between two modes; the ocean planet merges and overlaps with the water planet. Miller makes 'difference pattern'; it is always in the process of 'differential becoming' (Barad, 2007: 89). It does not make its different self (water planet) into its separate other, but it is the embodiment of the entanglement of two different beings that are not separate at all. There is no definite boundary between what the planet is in one moment and what it becomes in the next. Miller is the embodiment of in-betweenness, fluidity, and difference. The astronauts of the film cannot be blamed for mistakenly taking the imminent waves for solid mountains as the planet is constantly moving, rising, falling, and diffracting. Miller is constantly reshaping and redefining itself and its other. This oceanic world is a fragmented, diffracted, and fractured being that resists the dichotomy of land and ocean, as well as that of water and ocean planets.

Conclusion

We are living in times faced with anthropogenic climate change that generates extreme weather phenomena on Earth and affects the lives of many. Science fiction representations of planets with water expose much about our conception of agency. In this chapter, I have categorised planets with water as either water planets that have both land and aquatic spaces on the surface or ocean planets that are devoid of shores. I examined two ocean planets; Kamino in *Attack of the Clones* and Miller in *Interstellar*. The emphasis in *Attack of the Clones* and *Interstellar* on possibly appropriable places that have an extreme climate, which, however, is not the work of humans, disregards half of the equation of our current environmental crisis. Such a representation of the two planets works to obscure human agency and responsibility and presents nature as the sole agent producing extreme climate. I have argued that an anthropocentric conception of agency prevails in both films, but my analysis has pointed out ruptures of diffracted and resistant modes of agency as well. Kamino is a dark futuristic vision of our Earth troubled by climate change, and the settlements constructed on it conceal its actual harsh climate. The planet is a space of self-reflective anthropocentric imaginaries embodied in the clone army. The imaginaries of Kamino's clone army restate the theories of the Anthropocene through the representation of identical human clones imprisoned in a system that deprives them of agency. At the same time, there is room for divergence and diffraction in this army in the form of an unaltered human clone.

The representation of Miller's extreme climate more radically unsettles human agency. Miller lacks lands for settlement and constantly produces different oceanic patterns, evading the formation of a fixed identity; in cinematic terms, too, the images it produces are diffracted, multiplied, and differentiated. As an ocean planet, Miller resists boundaries and can transform into a water planet. Miller is a wet planet that resists the binary of water planet and ocean planets. Yet the representation of agency in Miller also reproduces a very human way of conceiving agency by simply replacing humans with the ocean as the single agent. As such, *Interstellar* holds up an inverse mirror to anthropocentric agency. In both films, ruptures of diffracted and differentiated agency surface, but finally fold back into an anthropocentric understanding of agency.

Note

1 In hard science fiction, writers attempt to restrict themselves to 'what can be extrapolated directly from known science' (Samuelson, 1993: 201).

Bibliography

Barad, Karen (2007), *Meeting the Universe Halfway: Quantum Physics and the Entanglement of Matter and Meaning* (London: Duke University Press).
Barad, Karen (2014), 'Diffracting Diffraction: Cutting Together-Apart', *Parallax*, 20:3, 168–187, https://doi.org/10.1080/13534645.2014.927623
Bordner, Autumn and Caroline E. Ferguson (2020), 'Marshall Islands Could Be Wiped Out by Climate Change – and Their Colonial History Limits Their Ability to Save Themselves', *The Conversation*, 11 December, [Online], [Accessed 29 October 2022], available from: https://theconversation.com/marshall-islands-could-be-wiped-out-by-climate-change-and-their-colonial-history-limits-their-ability-to-save-themselves-145994#:~:text=By%202035%2C%20the%20U.S.%20Geological,migrate%20away%20from%20their%20homelands
'Can Asteroids, Meteorites or Man-Made Explosions Cause Tsunamis?', *Tsunami Information Center*, [Online], [Accessed 17 October 2022], available from: http://itic.ioc-unesco.org/index.php?option=com_content&view=article&id=1161:can-asteroids-meteorites-or-man-made-explosions-cause-tsunamis&catid=1340&Itemid=2057
Canavan, Gerry (2012), 'Decolonizing the Future: Review of Jessica Langer's *Postcolonialism and Science Fiction* and Ericka Hoagland and Reema Sarwal's *Science Fiction, Imperialism and the Third World*', *Science Fiction Studies*, 39:3, 494–499, https://doi.org/10.5621/sciefictstud.39.3.0494
Canavan, Gerry (2014), 'Introduction: If This Goes On', in Gerry Canavan and Kim Stanley Robinson (eds.), *Green Planets: Ecology and Science Fiction* (Middletown: Wesleyan University Press), 1–21.
Cho, Renee (2021), 'Attribution Science: Linking Climate Change to Extreme Weather', *State of the Planet Columbia Climate School*, 4 October, [Online], [Accessed 18 October 202], available from: https://news.climate.columbia.edu/2021/10/04/attribution-science-linking-climate-change-to-extreme-weather/
Clark, Timothy (2015), *Ecocriticism on the Edge: The Anthropocene as a Threshold Concept* (New York: Bloomsbury).
DeLoughrey, Elizabeth and Tatiana Flores (2020), 'Submerged Bodies: The Tidalectics of Representability and the Sea in Caribbean Art', *Environmental Humanities*, 12:1, 132–166, https://doi.org/10.1215/22011919-8142242
Geerts, Evelien and Iris van der Tuin (2016), 'Diffraction and Reading Diffractively', *New Materialism*, 27 July, [Online], [Accessed 17 October 202], available from: https://newmaterialism.eu/almanac/d/diffraction.html
Haraway, Donna J. (2018), *Modest_Witness@Second_Millennium.FemaleMan©_Meets_OncoMouseTM:Feminism and Technoscience* (New York: Routledge).
Haraway, Donna J. (2020), 'The Promises of Monsters: A Regenerative Politics for Inappropriate/d Others', in Geffrey Andrew Weinstock (ed.), *The Monster Theory Reader* (Minneapolis: University of Minnesota Press), 459–521.
Hirabayashi, Yukiko, Roobavannan Mahendran, Sujan Koirala, et al. (2013), 'Global Flood Risk Under Climate Change', *Nature Climate Change*, 3 September, 816–821.
Knappett, Carl and Malafouris, Lambros (2008), 'Material and Nonhuman Agency: An Introduction', in Carl Knappett and Lambros Malafouris (eds.), *Material Agency: Towards a Non-Anthropocentric Approach* (New York: Springer), ix–xix.
Lahiri-Dutt, Kuntala (2014), 'Beyond the Water-land Binary in Geography: Water/Lands of Bengal Re-visioning Hybridity', *ACME: An International Journal for Critical Geographies*, 13:3, 505–529.

Langer, Jessica (2011), *Postcolonialism and Science Fiction* (Basingstoke: Palgrave Macmillan).
Latour, Bruno (2014), 'Agency at the Time of the Anthropocene', *New Literary History*, 45, 1–18, https://doi.org/10.1353/nlh.2014.0003
Lucas, George (2002), *Star Wars: Episode II – Attack of the Clones* (California, Los Angeles: 20th Century Fox).
Luther, Annika (2011), *De hemlösas stad* (Helsingfors: Söderströms).
Morton, Timothy (2016), *Dark Ecology: For a Logic of Future Coexistence* (New York: Columbia University Press).
Navara, Andrew (2019), *Star Wars: Rise of the Separatists* (Roseville: Fantasy Flight Games).
Nolan, Christopher (2014), *Interstellar* (California, Los Angeles: Paramount Pictures).
Rapin, William, Bethany L. Ehlmann, Gilles Dromart, et al. (2019), 'An Interval of High Salinity in Ancient Gale Crater Lake on Mars', *Nature Geoscience*, 12, 889–895, https://doi.org/10.1038/s41561-019-0458-8
Samuelson, David N. (1993), 'Modes of Extrapolation: The Formulas of Hard SF', *Science Fiction Studies*, 20:2, 191–232.
Stark, Whitney (2016), 'Intra-action', *New Materialism*, 15 August, [Online], [Accessed 18 October 202], available from: https://newmaterialism.eu/almanac/i/intra-action.html
Strang, Veronica (2014), 'Fluid Consistencies: Meaning and Materiality in Human Engagements with Water', *Archaeological Dialogues*, 21:2, 133–150, https://doi.org/10.1017/S1380203814000130
'What Causes a Tsunami?', *The Pacific Tsunami Museum*, [Online], [Accessed 17 October 2022], available from: http://tsunami.org/what-causes-a-tsunami/
'What the World Would Look Like if All the Ice Melted', *National Geographic*, [Online], [Accessed 17 October 2022], available from: https://.nationalgeographic.com/magazine/article/rising-seas-ice-melt-new-shoreline-maps

6 The radiant future or the end of history?

The (eco)politics of Antoine Volodine's novel *Radiant Terminus*

Helena Duffy

Radiant Terminus (2017) (originally published in French in 2014 as *Terminus radieux*) is set in the future, after the fall of Orbise, the last stronghold of communism and the capital of the Second Soviet Union. Capitalists have now gained control of the few areas remaining habitable after a series of nuclear disasters had struck the Second Soviet Union. In a quasi-suicidal gesture, the novel's central character, Elli Kronauer, and Kronauer's two fellow revolutionaries, Ilyushenko and Vassilissa Marachvili, leave the beleaguered capital for the irradiated territories. Soon enough they begin to suffer the effects of radiation: 'Malaises, affaiblissement, dégoût de l'existence, pour ne pas parler de vomissements et de diarrhées' [Sickness, weakening, disgust at existence, not to mention vomiting and diarrhoea] (Volodine, 2015: 11; Volodine, 2017: 8). To seek help for the dying Vassilissa, Kronauer leaves his two comrades and trails through the steppe and the taiga until he reaches an isolated *kolkhoz* called 'Radiant Terminus.' On his arrival, he is interrogated by Grandma Ugdul, the self-appointed guardian of the collective farm's nuclear core which, during the accident of the power plant, sank below the surface of the earth. Kronauer is eventually imprisoned by Solovyei, Grandma Ugdul's common-law husband and the president of the *kolkhoz*. Pathologically possessive of his three beautiful daughters, Solovyei subjects Kronauer to a humiliating show trial on drummed-up charges, and sentences him to centuries of incarceration, mental and physical pain, and aimless wandering through the taiga. The novel ends hundreds of years later, with Kronauer meandering through the irradiated and contaminated boreal forest, either alone or in the company of one of the former inhabitants of the titular *kolkhoz*.

Radiant Terminus, which in 2014 was awarded the highly prestigious Prix Médicis, is the work of an original and prolific author who writes under several pennames, including 'Antoine Volodine.'[1] The plurality of the novelist's authorial persona reflects his self-perception as a spokesperson for a group of revolutionaries who, having been imprisoned for their

DOI: 10.4324/9781032726953-7

beliefs, continue to resist political oppression by founding a literary movement called 'post-exoticism.' Rather than written in a traditional way, the post-exotic texts are whispered, chanted, or yelled from cell to cell (Volodine and Wagneur, 2003: 40). Just as the label 'post-exotic,' the generic categories – romånces, interjoists, narracts, and shaggås – that Volodine has invented to classify his own writings are designed to counter the efforts of literary critics to pigeonhole his prose using existing taxonomies.[2] Volodine's defiance of the literary tradition and establishment is further suggested by his rejection of the notion of authorship, by the deliberate confusion of narrative and authorial identities, and by the self-definition of his work as 'world literature in French' rather than as 'French literature.'[3]

As part of its contestation of conventional narrative practices, which, as Markku Lehtimäki (Chapter 1) and Georgiana Bozîntan (Chapter 2) suggest in this volume, may be a response to the perceived inadequacy of traditional narrative forms to address ecocatastrophe, Volodine's oeuvre also unsettles the teleological impulses characteristic to narrative, which reassure readers by satisfying their desire for order, causation, and closure. These impulses include a rectilinear succession of events in a story of development and transformation, which ends with some final and conclusive point (Shinn, 2018: 10). For example, although the title of *Radiant Terminus* suggests a definitive closure and the novel's diegetic present is situated on a timeline of historical phenomena both familiar and unfamiliar – the Scythian Empire, the Mongols' thirteenth-century conquest of the Kievan Rus', the Russian Revolution, and the fall of the Second Soviet Union – the novel replaces the idea of history as progress with the idea of history as a cyclical construct where revolution and counterrevolution alternate. In Lucas Hollister's words, 'the post-exotic narrative stands as a refusal to reveal the truth of history, to close history in a final meaning,' thus remaining in 'an indecisive but highly politicised state of indeterminacy and openness' (2020: 198). Similarly, Annie Epelboin describes the post-exotic world as 'une entreprise eschatologique privée de toute orientation vectorielle, d'aspiration, d'issue. Il désigne une impasse, totale, définitive, circulaire voire sérielle, où se ressasse un discours de survie, sur l'au-delà de la fin' [an eschatological enterprise deprived of any vectorial orientation, of aspiration, of a way out. It constitutes a dead end, total, definitive, circular or even serial, where a discourse of survival beyond the end is rehashed] (2013: 365). In *Radiant Terminus*, Volodine's conception of history can be gleaned from the representation of the Second Soviet Union as a revival of the USSR, or from Grandma Ugdul's fleeting idea that she might start a new communist state. The circularity of history is also illustrated by Kronauer's never-ending trudge through the boreal forest in the final part of the novel, whose events, significantly, echo the protagonist and his comrades' wanderings in Part I. Kronauer's trek is, in addition, a re-enactment of the fate of his own parents who, in the protagonist's imagination, are forever lost in the taiga, as well as of the lifestyle of the prehistoric nomads mentioned in the novel, who endlessly travelled the steppe.

Volodine's rejection of stabilising teleologies on both the structural and thematic levels of his narrative bears upon my examination of *Radiant Terminus*, whose aim is to gauge the novel's ideological standing in the face of the environmental catastrophe that forms its backdrop. To achieve this, the first part of the chapter analyses the novel's post-apocalyptic aura and its linked critical engagement with the concepts of 'the end of history,' as prophesised by the American Right, and of 'the radiant future' (*svetloe buduschchee*) coined by Soviet demagogues. Having examined Volodine's satirisation of both the projection of a worldwide triumph of liberal capitalism and market economy, and of the promise of social equality and financial prosperity made to Soviet citizens, in the second part of the chapter I discuss the novel's figuration of the post-catastrophic landscape. I focus here on the character of Kronauer, from whose perspective we view the ravages wrought on the natural world by humanity's activities, and whose Marxist-Leninist commitment, non-violent sexuality, and loyalty to his comrades may prove at odds with his implication in humanity's masterful violence towards the non-human world. I will inquire whether the ambivalence marking Kronauer betrays Volodine's scepticism about the revolution and its promise of egalitarianism, which, as implied by his characters' nostalgia for their communist youth, his oeuvre valorises positively, if not champions.[4] By considering *Radiant Terminus* in dialogue with critical readings of the first man's relationship with the vegetal and animal world, as recounted by the Bible, I will propose that, rather than any specific economic or political system, Volodine sees our culture's anthropocentrism as responsible for the ecocatastrophe that we are currently facing. I then comment on the effectiveness of *Radiant Terminus* to counter anthropocentric logic, before, in conclusion, returning to its ecopolitics. I suggest that it is at humanity, regardless of its ideological positioning, that *Radiant Terminus* points an accusatory finger, and that, consequently, it is humanity's extinction or at least its dramatic depletion that the novel identifies as a solution to the environmental crisis.

The end(s) of history and the last men

Set in a post-catastrophic world, Volodine's novels, including *Radiant Terminus*, appear to fall under the rubric of apocalyptic fiction. As theorised by Frank Kermode in his seminal study *The Sense of an Ending* (1967), apocalyptic fiction serves to stabilise and circumscribe the dynamic realm of history by enabling us to project ourselves 'past the End, so as to see the structure whole, a thing we cannot do from our spot of time in the middle'

(1967: 8). According to Kermode, much of literature follows the paradigm established by the Book of Revelation; it does so simply by ending and by thus offering readers a sense of 'consonance,' as the critic calls the impression that past and present events can be correlated to knowable future outcomes. To the apocalyptic narrative model, which promotes 'the idea of history as a coherent developmental process bounded by clear origin and termination events,' Connor Pitetti opposes 'post-apocalyptic' fiction (2017: 439). The prefix 'post' marks here a break with the apocalyptic literary tradition in the same way as postmodernism 'departs from a specific set of ideas about the nature of the present that scholarship designates as modernist,' and as posthumanism 'is a way of thinking about humanity that [Cary] Wolfe wishes to distinguish from the older discourses of humanism' (Pitetti, 2017: 447). This is to say that post-apocalyptic novels 'oppos[e] the idea that history proceeds by way of clean breaks and the advent of new worlds,' instead offering scenarios where the end of the world is never total' and highlighting 'the various forms of adaptive continuity and survival that link the pre- and post-disaster worlds' (Pitetti, 2017: 447).

In *Radiant Terminus*, which, like Volodine's other novels, is situated 'sur un lieu impossible . . . entre la fin et après la fin' [in the impossible space between the end and its aftermath] (Epelboin, 2013: 365), history is a cyclical rather than rectilinear construct, and the nuclear meltdown and the ensuing collapse of communism lead neither to the end of the world nor to humanity's regeneration. The novel therefore clearly inscribes the post-apocalyptic rather than apocalyptic narrative pattern. Indeed, in its final pages, humans, animals, and plants continue to exist, even if in radically decreased numbers. After physically adapting to living with radiation, which has seemingly rendered them immortal, the few surviving humans dwell endlessly and aimlessly in the bleak and timeless post-cataclysmic world. Either alone or in the company of a fellow human being, they wander through a landscape dotted with ruins of concentration camps, nuclear plants, *kolkhozes*, railway tracks, and roads. What remains of these structures is testament not only to humanity's dream of technological and social progress, but also to the coercive measures used to achieve it. The nomadic women and men featured by Volodine's novel spend their time reminiscing about their life in the Second Soviet Union or in its last surviving outpost, the *kolkhoz* 'Radiant Terminus.'

The post-apocalyptic circularity and open-endedness of Volodine's novel is consistent with its critical questioning of the concept of the end of history, as theorised by Francis Fukuyama who in 1989 projected the imminent worldwide triumph of liberal democracy and market economy. By announcing the end of historical time, the American political scientist did not mean, however, that there was to be no further events. Rather, he meant that, following the near-extinguishment of monarchy, the demise of fascism, and, finally, the collapse of communism, liberalism has prevailed over other forms of governance.[5] Contrary to Fukuyama's anticipation of an eternal liberal democratic idyll, Volodine's post-historical world is a place dominated by greed, rivalry, and violence. As perceived by two staunch supporters of Marxism-Leninism, Kronauer and Ilyushenko, capitalism is synonymous with fascism, sadism, nihilism, savagery, monstrosity, misogyny, and barbarism.[6] As for capitalism's proponents, Kronauer and Ilyushenko see them as a de-individualised mass of 'fanatiques de l'exploitation de l'homme par l'homme' [zealots for exploitation of man by man] (2015: 31, 2017: 25, translation modified). That the capitalists, labelled 'ennemies à tête de chien' [dog-headed enemies] (2015: 31, 2017: 25), are also strongly associated with sexual violence is illustrated by their final storm on Orbise, during which they rape and murder the female patients of a cancer ward, including Kronauer's wife.[7] The capitalists' rampant brutality is offset by the posture of the failed revolutionaries whom Volodine invariably portrays as courageous, principled, generous, companionate, loyal to their comrades, and unreservedly committed to communist ideals. Their behaviour is attributed to their upbringing and education by Soviet institutions dedicated to – to paraphrase Stalin's words – the engineering of human souls and the production of '"the cogs" who keep [the] great state machine in motion' (Cheng, 2009: 23). The upbringing and education of Kronauer and other communists populating Volodine's novel are indeed evidence of the excessive political socialisation that 'was conducted solely to achieve acceptance of the party's authority and identification of the party with the state and nation to the extent that . . . the ideological doctrine of the party comprised the norms of everyday life' (Cheng, 2009: 27). For example, Kronauer's schooling is described as 'alignée sur le destin des Communes ouvrières et paysannes' [focused on the future of Communes for workers and countrymen] and his conduct as illuminated by proletarian morality based on 'abnégation, altruisme et combat' [sacrifice, altruism, and confrontation] (2015: 32, 2017: 26). In the same vein, Kronauer's comrade, Vassilissa Marachvili, is figured as a woman 'dure à la douleur' [impervious to pain] who would put on a brave face despite the bleak prospect and who 'acceptait l'adversité sans faire d'histoires' [accepted adversity without making a fuss] (2015: 19, 2017: 15, translation modified).

Bordering on a parody of the idealistic image of Soviet men and women peddled by Stalinist propaganda, Volodine's characterisation of his protagonists undermines a reading of *Radiant Terminus* as conduit for communist nostalgia and instead marks the author's distance from the ideology of the Left. As implied by his novel's title, alongside Fukuyama's prediction of the worldwide domination of liberalism, *Radiant Terminus* subverts the competing Soviet narrative of 'radiant future.' According to this narrative, a paradise on earth was to follow a transitional period marked by food shortages, lack of civil liberties, and cramped living conditions of the communal apartments (*kommunalki*). In Alastair Renfrew's words, 'The future utopia became a device for the rhetorical and practical neutralisation of the manifest deficiencies of the present "utopia"' (2016: 289). The propagandist concept has been famously lampooned by Alexander Zinovev's novel *The Radiant Future* (1978) in which gigantic portraits and slogans conceal the dystopic reality of the Soviet Union made up of, on the one hand, waste, rubbish, and physical decay and, on the other, apathy and moral corruption. While in Zinovev's novel the promised metaphorical radiance has turned into 'the reality of the wasteland' (Gottlieb, 2001: 246), in Volodine's narrative it has become the 'unhappily literal radiance' emanated by radionuclides (Ellis, 2019: 224). Correspondingly, the communal march towards a better tomorrow has become 'une marche commune vers la mort' [a communal march towards death] (Volodine, 2015: 14, 2017: 12).

The adulteration of the Soviet concept of the radiant future finds embodiment in the titular *kolkhoz* which, together with the surrounding village of Levanidovo, is set in an area so badly affected by radiation that the authorities have declared it unsuitable for human habitation for the next four to five centuries. Reminiscent of the exclusion zone of a nineteen-mile radius established around Chernobyl after the 1986 nuclear disaster, the area has been corded off with barbed wire and erased from official maps. To the high level of radioactive caesium and iodine bear witness the waters in the nearby lake, which resemble lukewarm mercury, and the drastic depopulation of the collective farm itself. Among the handful of survivors is Solovyei whose resistance to the gamma rays is attributed to his shamanic status. There are also Solovyei's three daughters – Hannko Vogulian, Myriam Umarik, and Samyia Schmidt – who have miraculously become immune to radiation, and a group of lethargic and emasculated *kolkhozniks*. These farmhands, who hover on the verge of death, must be periodically revived by the sorcery of Solovyei or Grandma Ugdul. Indeed, Grandma Ugdul, who first arrived in Levanidovo as the head of a team of liquidators, is portrayed in Volodine's novel as a priestess who guards the nuclear core as if it were a hearth, feeding it with incantations, irradiated objects, and, ultimately, her own body.

Described as resembling more closely 'un repaire de brigands qu' . . . une institution agricole' [a den of thieves than . . . an agricultural establishment] (Volodine, 2015: 48, 2017: 38), the *kolkhoz* 'Radiant Terminus' is figured as a monstruous perversion of the communist utopia. Since all the farm animals were killed by the nuclear explosion, the farm's sties and stables are empty, and most of the houses have fallen into ruin. One of the exceptions is the building that once housed the soviet (the local council) and whose façade is supported by a row of Ionic columns, a grotesque memento of the Second Soviet Union's imperial past. The tattered red flag flying above the building indicates in turn the farm's ideological deviation from Marxism-Leninism initiated by its president himself. Solovyei's character is an amalgam of several historical, fictional, and mythical figures: a *kulak* in a Tolstoy novella, a totalitarian ruler, and the Primal Father in Sigmund Freud's account of human prehistory in *Totem and Taboo* (1913). Indeed, like the latter who, before being slain and devoured by his sons and brothers, terrorised the other males of the horde and monopolised all the women, Solovyei engages in incestuous relations with his daughters and persecutes their suitors and husbands. Solovyei's name also gestures towards the figure of the predatorial half-man, half-bird in the *bylina* (a traditional form of Russian heroic and orally-transmitted narrative) that constitutes one of the novel's intertexts.[8] Like Solovey Razboinik in the *bylina* 'Ilya Muromets and Nightingale the Robber,' the president of the *kolkhoz* reigns over the nearby forest, deafening anyone who enters his realm with his ear-splitting whistle. His shriek can also be interpreted as a parodic echo of 'the Russian Woodpecker,' the nickname given to the hugely powerful Cold War-era Soviet radar system Duga-3, constructed in the woods near Chernobyl. Designed to detect foreign ballistic missiles directed at the Soviet territory, the radar interfered with radio and television signals with the characteristic staccato 'rat-tat-tat' sound (Hewitt, 2019: 101–102). Solovyei is therefore cast by the novel as a personification of the panoptic, controlling, and abusive Soviet state. As illustrated by his (vain) efforts to revive Vassilissa Marachvili, he fancies himself a master of the life and death of other people. More specifically, Solovyei's Oedipal jealousy of his daughters makes him an echo of Stalin, who had his daughter's Jewish lover, Aleksei Kapler, sent to exile before locking him up in a hard-labour camp. Similarly, Solovyei banishes Hannko Vogulian's husband, Aldolay Schulhoff, a runaway deportee whose appearance and extraordinary musical and linguistic talents are tokens of his veiled Jewishness.[9] Later, unfairly suspecting Kronauer of having raped or intended to rape his daughters, the president of the *kolkhoz* subjects him to a humiliating trial on the drummed-up charge of the sabotage of a hydrant, before sentencing him to years of solitary confinement followed by centuries of atrocious headaches, nightmares, and hallucinations. If the legal proceedings call to mind the show trials of the Stalinist era, the use of waterboarding as the method of torture invokes the CIA's interrogations of Al-Qaeda suspects at Guantánamo Bay.[10] The conflation of two different imageries in this episode confirms Volodine's scepticism of both forms of governance that dominated the twentieth century.

The novel's critical distance from communist ideology can be further detected in its representation of the Second Soviet Union which, with its institutions (the Komsomol, the Red Army, the Pioneer movement), insignia (the red flag), and Marxist-Leninist values, uncannily resembles the USSR as we remember it. The Second Soviet Union is, like its predecessor, an intrusive and repressive state whose territory is dotted with hard-labour camps and prisons, such as the one from which Kronauer's parents escape only to perish in the taiga. Among the novel's many characters who spend time in the gulag is Solovyei, who was jailed for forty-five years for having refused to embrace, as a poet, socialist realist aesthetics. He also sabotaged 'ses autocritiques en y insérant des narracts ésotériques, des considérations sur l'apocalypse et des discours politiquement incorrects sur la sexualité et les rêves' [his self-criticisms by inserting esoteric narracts, considerations of the apocalypse, and politically incorrect discourses on sexuality and dreams] (Volodine, 2015: 50, 2017: 40). Despite her wholehearted commitment to Marxism-Leninism and her self-sacrifice as a heroic, tireless, and bemedalled liquidator of exploded nuclear reactors,[11] Grandma Ugdul, too, fell under suspicion of the authorities puzzled by her extraordinary resistance to radiation.[12]

> [E]lle devait . . . affronter les soupçons du Parti et des services, se plier à des procédures humiliantes, réécrire son autobiographie pour la millième fois, reprendre à zéro ses autocritiques, et pour couronner le tout, elle devait se rendre à des convocations à l'Académie de médecine, justifier de son état organique et idéologique devant des embryologues, devant des xénologues, devant des commissions ouvrières spéciales qui n'hésitaient pas à l'accuser de l'individualisme petit-bourgeois face au trépas, voire de sorcellerie.
>
> [[S]he was . . . forced to deal with the Party, and the suspicion of its teams, undergo humiliating procedures, rewrite her autobiography for the thousandth time, do her self-criticisms over again, and, on top of all that, she had to appear at the Medical Academy's meetings, justify her natural and ideological state in front of embryologists, in front of xenologists, in front of special work councils that didn't hesitate to accuse her of petit-bourgeois individualism in the face of death, and even of witchcraft.]
> (2015: 43, 2017: 35)

Interestingly, the persecution that Grandma Ugdul and Solovyei suffered in the Second Soviet Union has not precluded their redeployment of Stalinist coercive methods at 'Radiant Terminus.' Having tested Kronauer's ideological position on his arrival in Levanidovo, Grandma Ugdul plans to subject him to 'des mois d'enquête et de réclusion, accompagnées de multiples autobiographies écrites pendant des moments de privation de sommeil' [months of investigation and imprisonment, along with multiple autobiographies written during sleep deprivation] (2015: 120–121, 2017: 99).

Apart from enabling Volodine's satire on the USSR, the character of Grandma Ugdul is instrumental in thematising the nuclear meltdown whose disastrous consequences form the background of Kronauer's story. *Radiant Terminus* resolutely lays the blame for the environmental catastrophe at the door of the Second Soviet Union and, equally significantly, links it to the colonial expansionism of the communist state. For it was to settle, urbanise, and industrialise the newly captured and often remote and underpopulated territories that the Soviet authorities heavily relied on nuclear energy. In the Second Soviet Union, nuclear reactors powered the new cities, factories, and collective farms. It is accidents such as those that struck 'Radiant Terminus' and the nearby *sovkhoz* 'Red Star' that drastically reduced habitable areas, weakened the country's economy, and, consequently, opened it to counter-revolutionary attacks: 'D'immenses régions devaient être évacuées et laissées à l'abandon. La marche triomphale vers le communisme, déjà fortement retardée par les agressions extérieures, avait dû ralentir encore son rythme' [Massive regions had to be evacuated and left to ruin. The triumphant march toward communism, already hampered by outside attacks, found its pace slowed down even further] (2015: 40, 2017: 32). To mitigate the implications of the nuclear meltdown, the authorities sent out teams of liquidators prepared to immolate themselves 'pour l'avenir radieux de [leurs] enfants et [leurs] petits-enfants' [for the radiant future of [their] children and [their] grandchildren] (2015: 40, 2017: 33). However, ironically, the liquidators would be rendered infertile by radiation.

It is therefore evident that, despite its indebtedness to the long tradition of Russian utopianism (Heller and Niqueux, 1995), *Radiant Terminus* counters utopian literature's enthralment with technology as a tool 'used in the "fight against nature"' (Dalton-Brown, 1995: 104).[13] For instance, in Vladimir Maiakovskii's play *Klop* (1928) [The Bedbug], metallic trees 'indicat[e] the transformation of nature into the useful and subservient.' In contrast to Russian utopian fiction where the sun often labours on behalf of the people or, in the form of electricity, 'is the basis for the creation of utopia' (Dalton-Brown, 1995: 105), in Volodine's novel displaced women and men huddle around nuclear reactors regardless of the deadly radiation they emit, and of whether they still give off heat or not. For instance, after leaving the *kolkhoz*, Hannko Vogulian moves into a cottage constructed around the ruins of a small nuclear core. Similarly, the failed revolutionaries who wander through irradiated landscape in search of a *gulag* that would accommodate them camp in the reassuring vicinity of defunct nuclear plants. In this sense, *Radiant Terminus* both shows affinity with the Soviet literature and film concerned with the twin cataclysms of Chernobyl and the dissolution of the USSR, and promotes the prevailing understanding of the 1986 nuclear accident as what sealed the fate of the already ailing Soviet Union.[14] According to Marjolein van der Veen, it was the exorbitant cost of the clean-up, of the relocation and compensation of inhabitants, and of the provision of ongoing healthcare that became the proverbial last straw for the USSR (2013: 127). Such a reading of *Radiant Terminus* is further encouraged by the glaring likeness of Levanidovo and the surrounding forests resonating with Solovyei's ear-splitting shriek to the exclusion zone around Chernobyl with its Duga-3 radar system.

The day after: naming the steppe

The description of Kronauer's journey through the steppe situated at the beginning of the novel provides the reader with insight into the extent of the environmental fallout of the multiple nuclear disasters. As he walks, Volodine's protagonist listens to birds that belong to a species 'déjà pratiquement éteinte' [nearly extinct] (2015: 68, 2017: 55) and watches crows fly over the irradiated zones without stopping. Considering that these ominous scavengers have traditionally allegorised death, danger, and transformation from physical to spiritual world, their presence befits the novel's post-catastrophic tone. The crow's symbolism illuminates especially the predicament of Kronauer and his two comrades who, having crossed into the irradiated areas, expect 'une totale absence d'avenir' [a completely empty future] (2015: 10, 2017: 15). But, even though Volodine's novel espouses an anthropocentric perception of the ill-boding birds by recognising in their appearance a warning to humans lost in the steppe, it undermines the human/animal divide undergirding this perception by deciphering the appearance of the crows as a manifestation of their solidarity with their fellow (human) animals. Other rare creatures encountered in *Radiant Terminus* are individual wolves, rabbits, reindeer, and bears. During his journey, in the vicinity of the *sovkhoz* 'Red Star,' Kronauer also spies two cows grazing in the wild. Their presence is explained with the fact that, after the explosion of the *sovkhoz*'s reactor, all farm animals were set free so that they may die on the open steppe. This suggests that the nuclear meltdown has liberated animals from domestication or – to do away with the misleading euphemism – from enslavement and exploitation to which humans had subjected them for centuries.[15]

As for cultivated species of plants such as buckwheat or rye, which, like animals, have been genetically modified to suit human purposes, since the nuclear catastrophe they have reverted to their prehistoric forms. Other plants have undergone a different sort of genetic modification: 'après une brève période de mutations où l'on avait observé des apparitions baroques et spectaculaires, la stérilité avait été la règle, et la planète a retourné à un état essentiellement végétal' [after a short period of mutations when baroque and spectacular apparitions could be seen, sterility had regained, and the planet had returned to an essentially vegetal state] (2015: 533, 2017: 445). In the depopulated irradiated areas, flora has proliferated abundantly, colonising the ruins of cities, villages, and camps, and covering roads, and railway tracks. In an interview following the publication of *Radiant Terminus*, Volodine describes the storyworld of his novel as one where 'La nature a pris le dessus et . . . l'espèce humaine est en voie d'extinction' [Nature has taken over and . . . the human species has become endangered] (2014a).[16] As a consequence, humans are no longer masters of nature; rather, as in premodern times, the natural environment can once again be a source of anxiety. For instance, Kronauer is apprehensive about entering the taiga, his fear being rooted in his childhood trauma of losing his parents to the boreal forest. The very word 'taiga' brings to Kronauer's mind

> les épaisseurs ténébreuses, les sapins morts tombés de leur belle mort depuis trente ou quarante ans. . . . La taïga, ça peut pas être un refuge, une alternative à la mort ou aux camps. C'est des immensités où l'humain a rien à faire.
>
> [gloomy thickness, the dead pines fallen to their natural death thirty or forty years earlier. . . . The taiga can't be a refuge, an alternative to death or the camps. It's vastness where man has no place.]
>
> (2015: 13, 2017: 10–11)

As opposed to the endless, treacherous, eerily silent, and dark taiga, whose representation by Volodine tallies with the forest's traditional association with evil and threat, the open steppe provides Kronauer with a sense of safety, comfort, and aesthetic pleasure. Its vast expanse offers him and his two comrades a respite from the violence of the city they have fled, and its tall grasses shield them from potential aggressors. With recourse to religious imagery, Kronauer marvels at the openness and tranquillity of the steppe; he compares the overarching sky above it to 'Une *voûte* uniformément et magnifiquement grise' [A uniformly and magnificently grey *vault*], and invokes human desire to 'chanter des *hymnes* à la nature, à sa force communicative, à sa beauté' [sing *hymns* to nature, to its inexhaustible force, and to its beauty] (2015: 12, 2017: 10, my italics). Volodine's protagonist reifies this desire by reciting long lists of plants, which, relying on the rhetorical figure of anaphora, invoke the litany, a prayer that is a series petitions to a saint alternating with invocations pronounced by a religious congregation. The novel itself encourages such a reading of these enumerations by referring to them as 'interminables litanies post-exotiques' [endless post-exotic litanies] (2015: 31, 2017: 25).[17]

The quasi-religious bedazzlement of Volodine's protagonist by the steppe is enmeshed with his enduring love for Irina Echenguyen, who was a botanist charged with identifying and naming new species of steppe grasses and who, during the final attack of the counterrevolutionaries on Orbise, was raped and murdered at a cancer clinic. It is to revive the memory of his wife that Kronauer names plants using the botanical terms she coined: 'La mole-guillote, la malveinée, l'ashrang, la captive-petite-gloire, la benaise-de-saules. La demoiselle-en-fuite, la mascaratte, la belle-de-quatre-heure, la pituitaine, la douce-lieuse ou Jeanne-de-minuit' [Molle-guillotes, malveinés, ashrangs, smallglory captives, willow benaises. Damsels-in-flight, masquerats, four-o'clock-beauties, pituitaines, sweetbalers, or midnight Jeannes] (2015: 31, 2017: 25). Such a form of remembrance is hardly surprising, considering that in Volodine's oeuvre, the list constitutes a form of political resistance. As Gaspard Turin notes, a list is what Volodine's oppressed protagonists offer their interrogators instead of a confession or silence: 'elle empêche l'interrogatoire de se poursuivre, étant monologique et monopolisant la parole; . . . un blocage syntactique qui, en embrayant constamment sur lui-même, empêche le récit d'advenir' [it prevents the interrogation from continuing, being monological and monopolising speech; . . . a syntactic blockage which, constantly self-engaging, stops the narrative from developing] (2013: 87).

The association between the lists recited by Kronauer and the violent death of his wife reasserts the connection between oppressed and silenced femininity and the invisibility and muteness that our culture has thrust upon the vegetal world, established in Volodine's earlier novels. Correspondingly, it reposits efforts to overcome plant blindness as an echo of women's erstwhile struggle for emancipation. For example, *Herbes et golems* (2012), which, significantly, Volodine published under his female heteronym, Manuela Draeger, assimilates imprisoned women with plants whose names the women recite as a form of resistance against the political oppression they suffer:

> Les actrices sont réelles mais déjà mortes, les herbes sont vivantes mais imaginaires. . . . Nous redonnons voix à celles qui l'ont perdue, nous donnons sève à celles que nulle n'a connues. . . . Des femmes disent la prairie et multiplient ici inventions et néologismes.
>
> [The actresses are real but already dead, the grasses are alive but imaginary. . . . We are restoring the voice to those who have lost it and the sap to those whom no one has known . . . Women are speaking the meadow, multiplying linguistics inventions and neologisms.]
>
> (2012: 11–12)

The identification of women with grasses is invited by the feminine gender of the French noun 'l'herbe' [grass] and is reinforced by the suggestion of the vulnerability of the grasses by the terms uttered by Kronauer. It appears therefore that Volodine inscribes the long tradition of feminising plants and especially flowers (Artman, 2013: 108), illustrated by their description by botanists as seductive women, as women needing protection, and as sisters or maids carrying for a male without sexual connotation (Le-May Sheffield, 2004: 43).[18] However, in seemingly following this tradition, Volodine personifies and, more specifically, feminises plants to rather different ends; notably, he draws attention to the centuries-long exploitation of the vegetal world by humans as food, medicine, and clothing, and to our culture's simultaneous neglect of it, which, as we shall see later, *Radiant Terminus* attempts to counter by figuring plants as ontologically independent and agential beings. In this sense, *Radiant Terminus* shows affinity with the work of the Polish artist, Agata Siniarska, and with Barbara Kingsolver's novel *Prodigal Summer*, which, as Chloé Bour--Lang (Chapter 9) and Katarzyna Bojarska (Chapter 10) demonstrate in this volume, intertwine oppression of women and nature.

The enumerations of plants, mushrooms, and animals that disrupt the flow of Volodine's prose have intrigued his commentators, some of whom have read these enumerations (eco-)politically.[19] Among other interpretations, critics have deciphered the lists as a rejection of traditional taxonomical systems (Hermetet and Posthumus, 2020: 98) and as an attempt to individualise plants rather than to reduce them to 'the mass of non-human inanimate green,' to unsettle conventional generic distinctions and hierarchies, and to oppose the correlated drive to separate what is useful to humans from what is not (Stamm, 2019: 48, 51). It has also been noted that these enumerations decentre humans and invite a non-anthropocentric reading of Volodine's prose (Hermetet and Posthumus, 2020: 95), which means that they are 'the first step in an ethical stance that would consider plants in and of themselves, escaping from . . . the disease of "plant blindness"' (Stamm, 2019: 48). Coined in 1999, the term 'plant blindness' denotes a cognitive bias against the vegetal world (Wandersee and Schussler, 1999), that is its active 'repress[ion] in the semantic texture of urban, technological consciousness' (Laist, 2013: 10).[20] Since the lists of plants and animals combine fabulated and real terms or, as in *Radiant Terminus*, contain only made-up names, they can be seen as intended to unsettle the principle of referentiality (Hermetet and Posthumus, 2020: 96), to 'introduc[e] instability and indeterminacy in the world-building work of reading,' to resist 'the progressive logic of storytelling' (Cf. Turin, 2013: 87), and to draw attention to the lack of transparency of the vegetal world (Brunau-Zaragoza, 2021: 234).

As already suggested, enumerations of plants are enmeshed in Volodine's prose with opposition to authoritarian political regimes, as in *Herbes et golems* where 'the poetic practice of naming plants creates a larger community of resistance metaphorically linked to a liberty attributed to plants themselves' (Stamm, 2019: 49). Similarly, in *Radiant Terminus* Kronauer uses lists of botanical terms in the same way as he recites the names of his fallen comrades, which he does to memorialise a community of revolutionary fighters, even if, having forgotten the actual names, he makes them up. The invented character of both the botanical terms and the names of Kronauer's fellow soldiers points to the arbitrariness, subjectivity, and cultural and political contingency of the taxonomical work carried out by geographers, botanists, and zoologists who name the non-human world. This is exemplified by the visible influence of various constructions of femininity, including politicised ones, on the names of steppe grasses coined by Irina Echenguyen: 'Jeanne-des-communistes' [Jeanne-of-the-Communists] (2015: 14, 2017: 11), 'la dame-exquise' [dame-exquise], 'la folle-en-jouise' [folle-en-jouise], 'la mère-du-lépreux' [mère-du-lépreux] (2015: 27, 2017: 22). While the term 'la dame-exquise' may be an allusion to the 1932 novella 'The Lovely Lady' by D.H. Lawrence, translated into French as 'La Dame exquise,' 'la mère-du-lépreux,' which can be rendered as 'mother-of-the-lepers,' is how Mother Teresa and similar figures have been called. In his discussion of the lists proliferating in Volodine's oeuvre, Hollister goes even further by arguing that the act of naming and localising something in a system of meaning betrays a quest for control and dominance. Following Jean-Luc Nancy's assertion that 'naming equals domination' (2016, 45), the critic considers the 'semantic stabilisation' performed by Volodine's protagonists as an expression of their wish to master both the human and the non-human world. Indeed, in Volodine's novels, prisoners under interrogation enumerate names of plants and animals as confessions or as strategies of encryption and evasion (Hollister, 2020: 217). In *Radiant Terminus* the blurring of the boundary between interhuman violence and our masterful violence towards plants finds embodiment in Grandma Ugdul's interrogation of Kronauer. To test the newcomer's class consciousness, the old woman quizzes him on the botanical terms used in the Second Soviet Union. Since it was Irina Echenguyen who forged the terminology that he is being tested on, Kronauer feels that to supply the names would constitute a betrayal of her memory. However, despite his initial defiance, in the end, unheroic as he is, he provides answers to Grandma Ugdul's questions.

Tallying with the typically postmodern ethical undecidability marking Volodine's oeuvre, the taxonomical work undertaken by Kronauer is stamped with ambiguity. As well as a means of resistance, including against humanity's oppression of the vegetal world, naming and categorising is a tool of mastery and exploitation. By naming steppe grasses Kronauer extends to the vegetal world Adam's relationship with the animal kingdom in Genesis. His gesture thus, on the one hand, foregrounds and rectifies the 'zoocentrism' or 'zoochauvinism' marking our perception of our natural environment or, to quote James Wandersee and Elizabeth Schussler, 'the misguided anthropocentric ranking of plants as inferior to animals and thus as unworthy of consideration' (1999: 82). On the other hand, read in the terms of Jacques Derrida, by identifying and categorising plants, Volodine's protagonist asserts his power over them just as the first man asserted his power over the animals (1999: 265–268).[21] Echoing Derrida, Julietta Singh claims that 'mastery is tied to language' and that 'in its power to name the human . . . gains authorisation to particular forms of masterful consumption: because I have named you, I can consume you' (2018: 12). Singh illustrates her position with reference to the ecological writings of the Antiguan American novelist Jamaica Kincaid, which intersect colonialism with humanity's domination of the vegetal world. For Kincaid, 'this naming of things is so crucial to possession . . . that it is a murder, an erasing' (1999: 122). Yet, by putting in his protagonist's mouth made-up rather than real names, and by linking naming to a quest for control, Volodine distances himself not only from humanity's rational, scientific, and violent approach to the non-human world, but also, more generally, from anthropocentric thinking. This is borne out by Grandma Ugdul's cited redeployment of a tool of humanity's mastery over nature as one of interhuman violence. All this situates Volodine's protagonist in an ambiguous position: while being a champion of egalitarianism and a loyal husband and comrade, he is implicated in humanity's conquest and abuse of the non-human world. He is therefore indirectly responsible for the nuclear meltdown and the ensuing environmental catastrophe, which, as I argued in the first part of the chapter, Volodine's novel clearly posits as an outcome of our exploitative attitude to the natural world.

Volodine's paradoxical representation of humanity's relationship with the vegetal world is further illustrated by the passage describing Kronauer's painstaking trudge through the steppe. The passage seems to be aimed at increasing the visibility of plants, contesting their perception as silent and non-agential beings, and individualising them by challenging their status as fungible commodity and their traditional reduction to visual décor. The agency of plants is manifest in their gradual reclaiming of lands and structures – rail tracks, roads, and even buildings – abandoned by humans. This lends support to Monika Bakke's view of plants as 'territorial beings' capable of a degree of 'self-recognition, decision-making and communication' (2012: 17). Personified by Volodine's narrative, steppe grasses whisper, rustle, murmur, sigh, and rasp, each specific species speaking its own tongue. Their language is a form of resistance to the environmental damage that has been inflicted upon them by humans and that has resulted in, among other repercussions, the disappearance of pollinators and the consequent condemnation of plants to sterility. It is noteworthy that plants are thus once again identified with women who, like Grandma Ugdul, have become barren following their exposure to radiation. Volodine further empowers the steppe grasses by figuring Kronauer's march as a struggle against dense and resilient vegetation:

> Herbes qui font barrage contre les mollets, contre les genoux, contre les cuisses. Herbes rarement cassantes. . . . Herbes dures, élastiques, violentes. . . . Herbes que le pied n'écrase pas quoi qu'il arrive. Herbes qui répandent des senteurs fortes et désagréables . . . Herbes rassemblées en haies difficiles à franchir.
>
> [Plants that obstruct his calves, his knees, his thighs. Plants that rarely snap. . . . Hard, elastic, violent plants. . . . Plants that feet could never crush. Plants that give off strong and disagreeable scents. . . . Plants that look like thick hedges.]
>
> (2015: 27, 2017: 22)

Radiant Terminus also challenges the anthropocentric and utilitarian view of the flora championed by, among other texts, the Bible, where, as opposed to animals that predate Adam's appearance, plants fully emerge only once man is there to toil the soil.[22] As Rhona Trauvitch remarks, in Genesis, the 'growth [of vegetation] hinges on the active participation of man. . . . Man's active will draws the plants from the seeds that are waiting at the gate of the ground. Man "activates" the plants via his collaboration' (2013: 168). In contrast, in *Radiant Terminus* the steppe grasses are figured as autonomous beings that do not require human presence. Rather, their peaceful and internally harmonious existence is hampered by humanity's harmful intervention in the form of urbanisation and industrialisation, or even of Kronauer's march through the steppe, which is described as laborious and noisy. It is therefore unsurprising that plants do not welcome Kronauer's intrusion into their territory; what is more, they *wish* him dead:

> Le bruit de cette avancée. Sa violence crissante. Un homme avance à allure forcée au milieu d'une végétation qui ne lui témoigne aucune bienveillance. Un homme traverse la steppe au lieu de dormir à jamais sur la terre. Un homme casse le silence des herbes.
>
> [The noise of this progress. Its screeching violence. A man pushing at full speed through vegetation that doesn't welcome him at all. A man crossing the steppe instead of sleeping forever on the ground. A man breaks the silence of the plants.]
>
> (2015: 27, 2017: 22, translation modified)

Volodine's efforts to overcome the plant blindness marking our culture are, however, compromised by the focalisation of his description of the steppe through a human character and by his corresponding recourse to anthropomorphism. Even if well-intentioned, such a strategy implies that we can engage with the vegetal world only by endowing it with recognisable human characteristics such as agency, articulate speech, territorialism, and resistance. That we cannot escape anthropocentric thinking and that our understanding of nature is inevitably conditioned by science and culture is confirmed by Kronauer's perception of different biomes. While his dread of the taiga is rooted in his filial loss and possibly also in the traditional construction of the forest as a perilous and alien space, his love of the steppe results from the metonymic transferral of his love for his wife on to it. In addition, Kronauer's relationship with the steppe is shown to be invariably mediated by the discourses of culture, science, and religion. Before Irina Echenguyen encouraged him to see the steppe through a taxonomical lens, Volodine's protagonist, who had been raised in an urban environment, could relate to it only through culture. In films, books, and *byliny*, the steppe served as a backdrop for battles, conquests, and fantastic events such as the struggle between Ylya Muromets and Solovey Razboinik: 'La steppe avait fini par lui être aussi familière et indispensable que les rues de la capitale' [The steppes eventually became as familiar and essential to him as the capital's streets] (2015: 30, 2017: 24). Finally, as we have seen, as he is approaching death, Kronauer conceives of the steppe and of nature in general in religious terms.

Conclusions

In an interview, Volodine stated that a novelist does not necessarily have to be 'more insightful than others about political and ethical problems' (Volodine and Wagneur, 2003: 13). Yet, on another occasion, he described his oeuvre as 'entièrement liée à un engagement politique' [entirely linked to political commitment] and the experience of his protagonists as 'totalement à voir avec une pensée politique, une idéologie' [totally to do with political thought, with ideology] (Nicolino et al., 2002). Indeed, this chapter has shown that *Radiant Terminus* takes a political and ethical stand, even if, as my discussion has equally demonstrated, it shirks definitive commitment to the values of either the Left or the Right. By associating Kronauer, who is a staunch communist, with humanity's abuse of nature, Volodine invalidates attempts to read his novel as an apology of the Left. This is confirmed by the novel's exposure of the Soviet Union as a paranoid and coercive state, and by its distance from the concept of the radiant future, which was to be achieved through, among other means, colonial expansion and technological and scientific progress. As I have contended, in *Radiant Terminus* the Soviet utopia is ridiculed through the portrayal of a *kolkhoz* where the metaphorical radiance has been reified as fatally toxic radiation. Equally disparaging is the novel's portrayal of capitalism, whose greed Volodine hyperbolises as the murderous and sexually predatorial behaviour of its proponents.

If *Radiant Terminus* is equally dismissive of the two main politico-economic paradigms of the twentieth century, it adopts a clear ethical position by condemning teleological constructions of human history, be they the product of the Right or the Left. The novel's position is evident from the embodiment of the teleological end in the titular *kolkhoz*, a dystopian space governed by a psychotic, abusive, and oedipal ruler. Solovyei's kinship with the mythical primal father, as imagined by Freud, points to our civilisation's return to its original social organisation as a horde ruled by a despot. The concept of the end is replaced with the bardic state between life and death, in which Volodine's protagonists are eternally caught and which is mirrored by the circular structure of *Radiant Terminus*. The endless continuation of life is paralleled by the protracted consequences of human activity, illustrated by the persistence of radioactive fallout and other forms of pollution such as crude oil stagnating in the taiga centuries after the near disappearance of humans. It is therefore the technological and scientific advancement motivated by future happiness of humanity that *Radiant Terminus* isolates as the source of the ecological disaster that we are presently witnessing. This means that it is only in the (near) absence of humans that nature will have a chance to regenerate and that, paradoxically, it is the man-made ecocatastrophe and the predicted resulting drastic reduction in human population that can liberate nature from our yoke and solve the environmental crisis.

Notes

1 Volodine has also published as Lutz Bassmann, Elli Kronauer, and Manuela Draeger, and has anticipated further heteronyms. 'Antoine Volodine' is the author's principal penname and one that he uses in public settings. According to Sabrinelle Bedrane, the novelist's real name is Jean Desvignes. The critic adds that she is unsure of the name's exact spelling (Bedrane, 2010: para. 30). *Radiant Terminus* was also awarded the Prix de la Page 111.

2 Volodine has offered many definitions of 'post-exotic literature.' He has defined it as 'A literature of elsewhere, from elsewhere, headed elsewhere,' as 'An internationalist, cosmopolitan literature with memories rooted in the tragedies of the twentieth century: the wars, revolutions, genocides and defeats of the twentieth century,' and as 'A foreign literature written in French' (Volodine, 2014a). Volodine has published the manifesto of the literary movement he initiated (Volodine, 1998a), whose title – *Le Post-exotisme en dix leçons, leçon onze* [*The Postexotism in Ten Lessons, Lesson Eleven*] – parodies the title of Jean-François Lyotard's *Le Postmodernisme expliqué aux enfants* (1986) [*The Postmodern Explained to Children*]. Originally, Volodine's writings were published by Denoël as part of its science fiction series 'Présence du futur'. As for the generic categories Volodine has invented, they function, as Jean Hippolyte

observes, 'only intratextually, within a closed signifying economy,' deliberately hindering 'topological elucidation' (2006: 148).

3 For example, Elli Kronauer is the main character of *Radiant Terminus* and the autodiegetic narrator of some sections of the novel. He is also a post-exotic writer, yet Volodine has stressed that Kronauer the writer and Kronauer the character must not be conflated (Volodine, 2014b). Volodine's texts move freely between first- and third-person narration, and are strewn with neologisms and archaisms, which unsettle the standard use of the French language.

4 For a discussion of Volodine's nostalgic relationship with communism, see Duffy (2021) and Lamarre (2013).

5 Fukuyama considered the potential rise of destructive forms of extreme nationalism and religious fundamentalism, but found them unlikely to prevail.

6 As Mélanie Lamarre rightly notes, Volodine never gives capitalists a voice and caricatures them in the style of the roman noir (2013: 291).

7 In Volodine's other novels, capitalists are described using equally if not more offensive labels. These include 'pogromistes' [pogromists] (Volodine, 2002: 26), 'loups fauteurs de guerre' [lupine warmongers], and 'mafieux'[mafiosi] (Volodine, 1999: 71).

8 In 1999, Volodine published under the penname of Elli Kronauer *Ilia Mouromietz et le rossignol brigand* [*Ilya Mouromets and Nightingale the Robber*]. In *Radiant Terminus*, the *bylina* is performed by Aldolay Schulhoff at his wedding to Hannko Vogulian.

9 Traditional Jewish multilingualism has been the result of the historical experience of the Jewish diaspora in Eastern Europe. Jews would normally speak between five and seven languages, including Yiddish, Polish, and Russian. Among other languages spoken by Jews were Hebrew, German, and French (*The Yivo Encyclopedia of Jews in Eastern Europe*).

10 If in public consciousness waterboarding is strongly linked with the interrogation practices of the CIA, the torture dates back to ancient times. It has been practised in many different contexts, including by the Spanish Inquisition, by the Japanese and the Nazis during World War II, by the Khmer Rouge in Cambodia, by the British in Palestine, and by the French during the Algerian War. See Weiner (2007).

11 She was decorated as 'Valeureuse Combattante de l'atome' [Valiant Combatant of the Atom], 'Héroïne rouge' [Red Heroine], 'Glorieuse Liquidatrice' [Glorious Liquidator], 'Doyenne rouge intrépide' [Intrepid Red Doyenne], and 'Vétérane, de Grande Sœur rouge' [Veteran, Red Big Sister] (Volodine, 2015: 42, 2017: 34).

12 Grandma Ugdul's unwavering loyalty to communism is reminiscent of the characters of Volodine's earlier novels: Gloria in *Le Port intérieur* (2010) [*The Inner Harbour*] and Maria Samarkande in *Vue sur l'ossuaire* (1998b) [*View over the Ossuary*] whose commitment to the Party remains unchanged despite the persecutions they suffer.

13 For a discussion of the engagement of Volodine's writing with the Russian tradition of utopianism, see Epelboin (2013).

14 For the study of this trend, see Banerjee (2016). It is noteworthy that it is *Piknik na obochinie* [*Roadside Picnic*], a 1972 novel by Arkady and Boris Strugatsky whose work Volodine has translated into French, that has provided a template for visualising zones of exclusion (Banerjee, 2016: 71). Banerjee extends her comment to Andrei Tarkovsky's cinematic adaptation of the Strugatsky brothers' novel as *Stalker* (1979).

15 For a discussion of the role of animals in Volodine's oeuvre, see Légeron (2019).

16 We find a similar scenario is Volodine's earlier novel *Des anges mineurs* (1999) [*Minor Angels*] which depicts 'la fin de la civilisation humaine, l'extinction de l'espèce sur une planète dévastée par les humains' [the end of human civilisation, the extinction of the species on a planet devastated by humans] (Volodine, 1999: 97).

17 The religious rhetoric does not have to be at odds with Kronauer's communist outlook. Marcel Gauchet defines totalitarian ideologies as forms of 'réligions séculières' [secular religions] which attempt to restore religiosity in the aftermath of the social and cultural upheavals of the second half of the nineteenth century (2005).

18 Suzanne Le-May Sheffield notes that there were no intellectual women or divorcées in the flora settings of Erasmus Darwin (Darwin's grandfather).

19 Other novels featuring lists of animals and plants are *Le Non des singes* (1994), *Dondog* (2002), *Lisbonne, dernière marge* (1990), and *Nos animaux préférés* (2006). For an analysis of these lists, see Turin (2013).

20 Wandersee and Schussler define plant blindness as 'a) the inability to see or notice the plants in one's environment; b) the inability to recognize the importance of plants in the biosphere and in human affairs; c) the inability to appreciate the aesthetic and unique biological features of the life forms that belong to the Plant Kingdom; d) the misguided anthropocentric ranking of plants as inferior to animals and thus as unworthy of consideration' (Wandersee and Schussler, 1999: 82).

21 Rhona Trauvitch's reading of Genesis and its commentaries foregrounds important differences between man's relationship with animals and with plants. While there is a finality in Adam's naming of animals, his relationship with nature is ongoing and physical, rather than cerebral. Trauvitch would also take issue with Volodine's indication of the arbitrariness of the naming process since, in line with the kabbalistic understanding of Hebrew, biblical commentators have noted that, to name the animals, Adam had to be able to comprehend fully their nature before attributing to them names which expressed it (2013: 169–172).

22 'And God said, "Behold, I have given you every seed-bearing herb, which is upon the surface of the entire earth, and every tree that has seed-bearing fruit; it will be yours for food. And to all the beasts of the earth and to all the fowl of the heavens, and to everything that moves upon the earth, in which there is a living spirit, every green herb to eat", and it was so' (Gen. 1:29–1:30).

Bibliography

Artman, Stacey (2013), 'Refusing Form: A Reading of Art, Americanism, and Feminism through Plant Imagery in Susan Glaspell's *The Verge*', in Randy Laist (ed.), *Plants and Literature: Essays in Critical Plant Studies* (Amsterdam: Brill), 103–122.

Bakke, Monika (2012), 'Art for Plant's Sake? Questioning Human Imperialism in the Age of Biotech', *Parallax*, 18:4, 9–25, https://doi.org/10.1080/13534645.2012.713196

Banerjee, Anindita (2016), 'From Fallout Fantasy to Bunker Bildunsgroman: Nuclear Imagination after Utopia', *Foundation: The International Review of Science Fiction*, 45:2, 70–84, https://doi.org/10.4000/resf.11670

Bedrane, Sabrinelle (2010), 'L'Engagement poétique d'Antoine Volodine: le communisme de la parole', in Marc Dambre and Richard J. Golsan (eds.), *L'Exception*

et la France contemporaine: Histoire, imaginaire et littérature (Paris: Presses Sorbonne Nouvelle), 181–192.
Brunau-Zaragoza, Sophie (2021), 'Ecospirituality: Vegetal Ghosts in Antoine Volodine's *Le Nom des singes* (1994)', *Contemporary French and Francophone Studies*, 25:2, 231–240, https://doi.org/10.1080/17409292.2021.1878693
Cheng, Yinghong (2009), *Creating the 'New Man': From Enlightenment Ideals to Socialist Realities* (Honolulu: University of Hawaii Press).
Dalton-Brown, Sally (1995), 'Signposting the Way to the City of Night: Recent Russian Dystopian Fiction', *Modern Language Review*, 90:1, 103–119.
Derrida, Jacques (1999), 'L'Animal que donc je suis (À suivre)', in Marie-Louise Mallet (ed.), *L'Animal autobiographique* (Paris: Galilée), 251–301.
Draeger, Manuela (2012), *Herbes et golems* (Paris: Olivier).
Duffy, Helena (2021), 'Back in the USSR: The Prose of Andreï Makine and Antoine Volodine', in Anna-Louise Milne and Russell Williams (eds.), *Contemporary Fiction in French* (Cambridge: Cambridge University Press), 132–151.
Dundes, Alan (1997), 'Binary Opposition in Myth: The Propp/Lévi-Strauss Debate in Retrospect', *Western Folklore*, 56:1, 39–50.
Ellis, Suzannah (2019), 'Messianic Fiction in Antoine Volodine's Nuclear Catastrophe Novel *Minor Angels*', *Paragraph*, 42:2, 223–237, https://doi.org/10.3366/para.2019.0300
Epelboin, Annie (2013), 'L'Utopie de la fin et la fin de l'utopie', in Frédérik Detue and Lionel Ruffel (eds.), *Volodine, etc.: Post-exotisme, poétique, politique* (Paris: Garnier), 365–382.
Fukuyama, Francis (1989), 'The End of History?', *The National Interest*, 16, 3–18.
Gauchet, Marcel (2005), *La Condition historique* (Paris: Gallimard, folio essais).
Gottlieb, Erica (2001), *Dystopian Fiction East and West: Universe of Terror and Trial* (Montreal: McGill-Queen's University Press).
Heller, Leonid and Michel Niqueux (1995), *L'Histoire de l'utopie en Russie* (Paris: PUF).
Hermetet, Anne-Rachel and Posthumus, Stéphanie (2020), 'Retour du végétal dans *Herbes et golems* (2012) de Manuela Draeger (Antoine Volodine) et *Ruines-de-Rome* (2002) de Pierre Senges', *Esprit Créateur*, 60:4, 94–106, https://doi.org/10.1353/esp.2020.0046
Hewitt, Cameron (2019), 'A Visit to Chernobyl: Travel in the Postapocalypse', in Alexandra Fuller (ed.), *The Best American Travel Writing 2019* (Boston: Mariner Books), 95–108.
Hippolyte, Jean-Louis (2006), 'Antoine Volodine: Inside Wars', in *Fuzzy Fiction* (Lincoln: University of Nebraska Press), 144–187.
Hollister, Lucas (2020), *Beyond Return: Genre and Cultural Politics in Contemporary French Fiction* (Cambridge: Cambridge University Press).
Kermode, Frank (1967), *The Sense of an Ending: Studies in the Theory of Fiction* (Oxford: Oxford University Press).
Kinçaid, Jamaica (1999), *My Garden (Book)* (New York: Farrar, Straus and Giroux).
Kronauer, Elli (1999), *Ilia Mouromietz et le rossignol brigand* (Paris: École des Loisirs).
Laist, Randy (2013), 'Introduction', in Randy Laist (ed.), *Plants and Literature: Essays in Critical Plant Studies* (Amsterdam: Brill), 9–17.
Lamarre, Mélanie (2013), 'Antoine Volodine et la fiction idéologique', in Frédérik Detue and Lionel Ruffel (eds.), *Volodine etc.: Post-exotisme, poétique, politique* (Paris: Garnier), 287–299

Légeron, Karine (2019), 'Les Animaux post-exotiques: "ce qui reste quand il ne reste rien"', *Postures*, 29, 1–16, [Online], [Accessed 12 May 2022], available from: http://revuepostures.com/sites/postures.aegir.nt2.uqam.ca/files/legeron_30_0.pdf

Nancy, Jean-Luc (2016), *Que faire?* (Paris: Galilée).

Nicolino, Sylvain, Laurent Roux and Sébastien Omont (2002), 'L'Humour du désastre. Entretien avec Antoine Volodine', *La Femelle du Requin*, 19, [Online], [Accessed 12 May 2022], available from: http://lafemelledurequin.free.fr/intervenants/volodine/entretien_volodine/droite_antoine_volodine_entretien.htm

Pitetti, Connor (2017), 'Use of the End of the World: Apocalypse and Post-Apocalypse as Narrative Modes', *Science Fiction Studies*, 44:3, 437–454, https://doi.org/10.5621/sciefictstud.44.3.0437

Renfrew, Alastair (2016), 'Lenin and Utopia', *Poetics Today*, 37:2, 269–294, https://doi.org/10.1215/03335372-3481943

Sheffield, Le-May Suzanne (2004), *Women and Science: Social Impact and Interaction* (Santa Barbara, CA: ABC CLIO).

Shinn, Abigail (2018), *Conversion Narratives in Early Modern England: Tales of Turning* (Basingstoke: Palgrave Macmillan).

Singh, Julietta (2018), *Unlinking Mastery: Dehumanism and Decolonial Entanglements* (Durham: Duke University Press).

Stamm, Gina (2019), 'Inventing a Vegetal Post-Exotic in the Work of Antoine Volodine', *Ecozone*, 1–2, 42–57, https://doi.org/10.37536/ECOZONA.2019.10.2.3134

Trauvitch, Rhona (2013), 'The Bible's Paradise and *Oryx and Crake*'s Paradice: A Comparison of the Relationships between Humans and Nature', in Randy Laist (ed.), *Plants and Literature: Essays in Critical Plant Studies* (Amsterdam: Brill), 165–180.

Turin, Gaspard (2013), 'La Stratégie du silence dans "La Stratégie du silence dans l'oeuvre de Robert Malipiero" d'Antoine Volodine', in Frédérik Detue and Lionel Ruffel (eds.), *Volodine etc.: Post-exotisme, poétique, politique* (Paris: Garnier), 85–100.

van der Veen, Marjolein (2013), 'After Fukushima: Revisiting Chernobyl and the Collapse of the Soviet Union', *Rethinking Marxism: A Journal of Economics, Culture and Society*, 25:1, 121–129, https://doi.org/10.1080/08935696.2012.711103

Volodine, Antoine (1998a), *Le Post-exotisme en dix leçons, leçon onze* (Paris: Gallimard).

Volodine, Antoine (1998b), *Vue sur l'ossuaire* (Paris: Minuit).

Volodine, Antoine (1999), *Des anges mineurs* (Paris: Seuil).

Volodine, Antoine (2002), *Dondog* (Paris: Seuil).

Volodine, Antoine (2010), *Le Port intérieur* (Paris: Gallimard).

Volodine, Antoine (2014a), 'Interview', [Online], [Accessed 12 March 2022], available from: www.youtube.com/watch?v=wystdZmsFac

Volodine, Antoine (2014b), 'The Fringe of Reality', trans. Jeffrey Zuckerman, *The White Review*, September, [Online], [Accessed 12 March 2022], available from: www.thewhitereview.org/fiction/the-fringe-of-reality/

Volodine, Antoine (2015 [2014]), *Terminus radieux* (Paris: Seuil. Collection Points).

Volodine, Antoine (2017), *Radiant Terminus*, trans. Jeffrey Zuckerman (New York: Open Letter).

Volodine, Antoine and Jean-Didier Wagneur (2003), 'Let's Take That from the Beginning Again . . .', trans. Roxane Lapidus, *Substance*, 32:2, 12–43.

Wandersee, James, H. and E. Schussler Elizabeth (1999), 'Preventing Plant Blindness', *The American Biology Teacher*, 61:2, 82, 84, 86, https://doi.org/10.2307/4450624

Weiner, Eric (2007), 'Waterboarding: A Tortured History', *NPR*, 3 November, [Online], [Accessed 1 February 2022], available from: www.npr.org/2007/11/03/15886834/waterboarding-a-tortured-history

The Yivo Encyclopedia of Jews in Eastern Europe, [Online], [Accessed 1 February 2022], available from: https://yivoencyclopedia.org/article.aspx/language/multilingualism

7 The nuclear disaster as metaphor for the impending ecocatastrophe in anticipatory fiction from Luxembourg

Sébastian Thiltges

In *Down to Earth: Politics in the New Climatic Regime*, Bruno Latour states that:

> The question of landing somewhere did not occur earlier to the peoples who had decided to 'modernize' the planet. It arose – ever so painfully – only for those who for four centuries had been subjected to the impact of the 'great discoveries,' of empires, modernization, development, and finally globalization. They knew perfectly well what it meant to find oneself deprived of land. And they even knew quite well what it meant to be chased out of one's land. They had no choice but to become experts on the question of how to survive conquest, extermination, land grabs.
> (Latour, 2018: 7)

The loss of territory described by Latour is a fate that, historically, the colonised cultures knew all too well. However, with climate change and its impending local, regional, and global catastrophes, the fight for your land will not spare the 'pays modernisateurs' [modernising countries] (Latour, 2018: 7). In addition to the protection of the biosphere, Latour sees some potential for mobilisation here, as people would rather defend their threatened 'territory' than endangered but abstract 'nature' that they feel disconnected from (2018: 8). With enough historical evidence to prove its point, risk theory (Beck, 1986; Horn, 2020) has shown that accidents and catastrophes are an inevitable price to pay for technological progress independent of whether located in developing or industrialised countries, even if social, racial, and environmental injustices always play a major role (Nixon, 2011). In countries and regions that do not see themselves directly threatened by natural disasters, a major industrial, technological, or nuclear incident may already convey a consciousness of an analogous territorial and biological threat which Latour sees in the 'New Climatic Regime,' the term he uses 'to summarize the present situation, in which the physical framework that the Moderns had taken for granted, the ground

DOI: 10.4324/9781032726953-8

on which their history had always been played out, has become unstable' (Latour, 2017: 3). This, according to Latour, makes nature political, because every decision and action needs to take in account its environmental consequences. It also has cultural implications, which means that the stories we tell about nature have to be profoundly reshaped (2017: 3).

One of the concerned countries is Luxembourg, a small landlocked state between France, Germany, and Belgium. It is one of the richest countries in the world and has the second-highest ecological footprint per habitant, only behind Qatar.[1] This chapter analyses two contemporary anticipatory fictional texts from Luxembourg, the short story 'Eng Stëmm an der Stëllt' [A Voice in the Silence] (2018) by Yorick Schmit, and the novel *Luxembourg Zone rouge* (2019) [*Red Zone Luxembourg*] by Pierre Decock.[2] Both narratives depict the once familiar environment as unrecognisable due to a major nuclear accident, and challenge readers to imagine its multifaceted consequences for the territory and population of Luxembourg, mainly through the eyes of an individual protagonist. New ecocritical modes of reading, sensitive to the non-human and to much larger spatial and temporal frames, analyse how a literary portrayal of catastrophic scenarios addresses 'la nécessité d'une nouvelle esthétique, au sens d'un renouvellement de nos modes de perception, de notre sensibilité, pour pouvoir répondre à ce qui est en train de nous arriver' [the need for a new aesthetic in the sense of renewing our modes of perception, our sensitivity, in order to be able to respond to what is happening to us] (Hache, 2014: 13).[3] I argue that to address this need, the localisation of global risk scenarios in the authors' and readers' own geocultural context may be essential to highlighting the cultural significance of ecocatastrophes. As opposed to Switzerland, for example, where the Alpine dangers play an important role in the national imaginary (Utz, 2013) or the Netherlands' constant exposure to sea floods, in Luxembourg, the natural environment seldomly carries representations of a risk.[4] Despite the increasing manifestations of global climate change and the upheavals associated with it, literary representations of ecocatastrophe are here seldom depicted as natural disasters, but rather as anthropogenic hazards, such as industrial, technological, or nuclear incidents.[5] Due to this local cultural specificity as well as the New Climatic Regime on a planetary scale, the depictions of nuclear catastrophes in these examples can thus also be read as metaphors for a much more global catastrophe that englobes phenomena like global warming, ocean acidification, sixth mass extinction, climate exoduses and conflicts, etc. One catastrophe hints at another: Nuclear disasters are a threat to human society as well as to non-human creatures and the natural environment, but, in contemporary literary texts, these plots are also a way to imagine risk scenarios and coping mechanisms that become urgent while humankind faces the impending ecocatastrophe.

Indeed, in Luxembourg's literary heritage, which dates back to the nineteenth century, 'nature' mostly appears as 'landscape' and is closely linked to the importance of the soil in the emerging national literature: In the context of the nineteenth-century nation-building project, the idea of a Luxembourgish national identity was partly established through the aestheticisation of the natural environment into an authentic Luxembourgish landscape, which encouraged people's identification to it (Sunnen, 2008). Because this nature is essentially cultural, there are few literary representations of sublime nature, freed from human presence, nor is there the idea that the natural environment could be detrimental to its inhabitants.[6] To illustrate my argument, I will first examine Schmit's and Decock's texts in the context of the literary representations of nuclear power stations in Luxembourgish literature, before briefly discussing the relationship between ecology and the nuclear, both from a general point of view, as well as by providing arguments for a metaphorical reading of the two literary texts.

Nuclear power plants in the literature of Luxembourg

Besides being a relatively young national literature, Luxembourgish literature has two distinct characteristics: It is multilingual and it is small (Gilbertz, 2019). 'Small' should not be understood in Deleuze and Guattari's 'revolutionary' sense coined in *Kafka: Pour une littérature mineure* [*Kafka: Toward a Minor Literature*] (1975), where the expression 'minor literatures' is used to stress its opposition to a 'major or established literature' (Deleuze and Guattari, 1985: 591). I use the adjective 'small' in the more neutral sense of other contemporaneous researchers (Gilbertz, 2019; Glesener, 2012), to indicate a literary field with quantitatively limited production and reception, emerging professional structures, and limited public visibility, including in school and university curricula. This sociological context influences the (self-)perception of authors, often leading to a 'discourse of smallness' (Glesener, 2015: 161) and even of 'self-devaluation' (Dirkx, 1995: 78). Yet, small literatures attract interest in a comparatist, world-literary framework, particularly because of their often-prominent intercultural dimension and peripheral situation, which highlights asymmetrical literary relations (Glesener and Kohns, 2022: 20–24).

The presence of nuclear power plants in Luxembourgish literature is closely interwoven with both national and international history: On the one hand, fictional power plants or those located near Luxembourg become an important motive in the national literature; on the other hand, the great nuclear disasters, like those of Chernobyl and Fukushima, influence literary scenarios all around the globe. In and near Luxembourg, two nuclear power plant projects have been and are still being discussed on the social, political, and cultural stages. The power plant in Cattenom, near

Thionville in France, ten kilometres from the Luxembourgish border, has been in operation since 1987 and has regularly been at the centre of cross-border debates. In the 1970s, there was also a project of a Luxembourgish nuclear power plant which was to be built in Remerschen, on the Moselle River (as is Cattenom), but which was abandoned in 1978 due to public opposition and political indecision (Pulli, 2020).

Authors from the Greater Region which, apart from Luxembourg, includes the German province of Saarland and the French department of Lorraine, participated in the protests against these two nuclear stations, generally in private capacity, but also as writers. In the 1970s a first wave of texts emerged accompanying the nuclear projects: an engaged literature in the form of *poésie engagée* (René Welter, Jos Jacquemoth, and Nico Graf), *Kabarett* (Jemp Schuster and Jhemp Hoscheit), and protest songs was written and performed, for instance, by Guy Rewenig and Fernande Kimmel (*Dem Fiisschen säin Fest*, 1979) or the Lorraine Franconian bands *Tutti Futti* and *Geeschtemat?*. Over time, the motif has become a literary *topos* metonymically linked to other themes, such as political power, inequity and injustice, ecocidal economic behaviour, etc. Literature then highlights the impact of the power plants on the environment and local populations rather than trying to confront nuclear projects directly. Finally, another form of resistance – which will be at the centre of this chapter – is expressed in texts that depict a nuclear power plant as an important motif, but not necessarily as the main topic, thus in contrast to the first category. These texts are about nuclear energy and nuclear accidents, but at the same time a metaphor for something else, namely the global impending ecocatastrophe and individuals' or societies' responses to it. Those two figurative meanings, the first a metonymic one, the second metaphoric, show that the trope of the nuclear power plant has become an important part of Luxembourg's literary imaginary.

The first text under analysis in this chapter is the Luxembourgish-language short story 'Eng Stëmm an der Stëllt,' published in the collection *Der Geruch der Erde nach dem Regen* [*The Smell of the Earth after the Rain*]. The author is Yorick Schmit, born in Luxembourg in 1986, who writes fiction, children's literature, and critical essays in German and Luxembourgish. The short story opens with an epigraph stating that it is inspired by the life of Naoto Matsumura, a farmer who stayed in the contaminated zone around Fukushima to look after abandoned animals (Pagnotta, 2013). The narrator depicts the protagonist as an 'alen Hiert' [old shepherd] (Schmit, 2018: 41) who hides in the contaminated and evacuated south of Luxembourg after a major explosion at the Cattenom nuclear power plant. The text never mentions the accident itself; readers only learn about the evacuation and official recommendations given on the radio from the protagonist's reminiscences. Nor is Cattenom nominally mentioned, even if precise descriptions of the protagonist's geographical location on a hilltop in the Luxembourgish town of Dudelange, from which the towers of Cattenom can be seen, leave no doubt about the referent (2018: 12–13). Having lost his wife soon after the accident, the old man refuses to abandon his home region, but leaves his house, unable to bare the silence and the memories of his former life: 'Dat Schlëmmst war d'Stëllt. Déi verdaamte Stëllt. Se huet alles ënnert sech erdréckt, huet engem d'Loft fir ze ootmen ewechgeholl. Eng Welt, wéi a Watt agepaakt. De Mënsch ass net dofir gemaach, ouni Kaméidi, ouni Geräischer ze liewen' [The silence was the worst. This damn silence. It crushed everything under it. It took all the air to breathe. A world wrapped in cotton wool. Men are not made to live without noise and sound] (2018: 11). He then looks for houses and farms where he can find animals to take care of: a dog he names Argos (which in the *Odyssey* was the name of Ulysses's faithful dog), a cat, a calf, and even a whole flock of sheep. The voices and noises of animals are generally represented as the 'voices in the silence' hinted at by the title. At the end of the short story, fatally ill, the old man is taken in by a military patrol and brought to a medical camp, from which he escapes as soon as he can to get back to the exclusion zone and to the animals he was looking after:

> Fir d'éischt war do just e bronge Fleck op der Wiss, net wäit vu sengem Schapp ewech. Hien ass stoe bliwwen, huet eropgekuckt. Wéi en d'Gebills fir d'éischte Kéier héieren huet, ass eppes an him gebrach. De bronge Fleck huet gewibbelt, ass ëmmer méi grouss ginn. Kaum een hätt deen ofgemouerte Mann a sengem verknaschten Nuetshiem mat de groen Hoer an dem laange Baart erëmerkannt, mee den Argos wousst direkt, wien do den Hiwwel eropkoum. Säi Gebills huet iwwert d'Koppe bis erof an déi enk Gaasse vum Duerf geschaalt.
>
> [At first, there appeared just a brown spot on the lawn not far from the shed. He stopped and looked up. When he heard the barking for the first time, something broke inside him. The brown spot got bigger and bigger. Almost no one would have recognised the emaciated man in his soiled nightgown, with his grey hair and long beard. But Argos knew immediately who was coming up the hill. Its barking sounded along the hilltops and into the narrow streets of the village.]
>
> (2018: 42).

The second text to be considered here is a short novel, *Luxembourg Zone rouge*, by French-language author Pierre Decock. Born in Brussels in 1959, Decock is a Luxembourgish national and publishes mostly in Luxembourg. He is also known for his historical works, as an illustrator and as a writer of crime fiction. The homodiegetic narrator of Decock's novel is a young man named Mike Olinger (a typical Luxembourgish name). He is almost twenty years old, which we know because he repeatedly mentions the 'grande catastrophe' [major disaster] (Decock, 2019: 9) which took place fifteen years earlier when he was still at a nursery. Through Mike's narration, we know that the French nuclear power station 'Mortange' exploded, leaving most of Luxembourg and some of Germany contaminated and uninhabitable. 'Mortange' is an invented name, composed of the word 'mort,' which means 'death' or 'dead,' and the suffix '-ange,' which is common to a lot of towns in this French and Luxembourgish region, like Hayange and Hettange in France, or Dudelange in Luxembourg. The morpheme is also a homonym with 'angel,' hinting at the angel of death, but it could also be drawing intertextually on Gilles Ortlieb's *Tombeau des Anges* [*Tomb of Angels*], where the author portrays the steel towns of Lorraine as fallen or dead 'anges,' meaning both the towns as well as the metaphorical angels. In *Luxembourg Zone rouge*, this regional characteristic as well as the numerous geographic indications leave little doubt that the French nuclear station of Cattenom serves as a model for Mortange. After the disaster, Mike and his family, except his father who has died from radioactive exposure, live along with many Luxembourgers in a camp in the Marne department. Mike describes his new home: 'Le pays lulu en terre étrangère. Une réserve d'Indiens où nous sommes parqués, généreusement hébergés par un État français pris de remords' [Lulu [slang for 'Luxembourgish'] land on foreign soil. An Indian reserve where we are packed in, generously lodged by a French state seized by remorse] (Decock, 2019: 9). Because of a promise made to his dying father, Mike goes on a long journey on foot back to Luxembourg and to the contaminated zone, back to his childhood home, to gather some secret documents which will finally reveal the French government's responsibility for the nuclear accident. Mike's journey structures the book, as most of the chapter titles name the towns he crosses and indicate the number of former inhabitants, obviously contrasting with the now deserted land.

Natural landscape and nuclear energy

Published in Luxembourg, and one of them being written in Luxembourgish, *Luxembourg Zone rouge* and 'Eng Stëmm an der Stëllt' clearly situate themselves in the context of the small literary field of Luxembourg. Furthermore, in the diegesis of the texts, which both multiply geographic references and detailed descriptions of the landscape, the country is represented as a natural, rural, and urban setting. Both narratives thus participate in the tradition of landscape description that has had an important role in the emerging national literature of Luxembourg. Landscapes, be they natural or not, also play an important role in other works of both authors: Schmit shows a regular interest in nature writing, and Decock is primarily known as a writer of regional thrillers, where landscape descriptions, combined to toponymical references, serve to localise the plot. An interesting twist, as opposed to traditional forms of nature or landscape description, is, of course, that these anticipatory fictional texts are set in a familiar environment that, however, has been rendered unrecognisable and inhospitable by the nuclear catastrophe. The environment, which traditionally has been a natural resource, and the landscape, which has been a cultural resource, are now critically endangered, even though, due to the imperceptibility of radioactivity, the illusion of a peaceful and abundant nature persists (more on that later).

This brings me to the constantly changing and complex relationship between ecology and nuclear energy.[7] They are most obviously ontologically linked because radioactivity poses a major threat to human and environmental health. On a geological scale, this makes the first nuclear explosions and fallouts into potential markers of the beginning of the Anthropocene (Subramanian, 2019). On another level, the evolution of nuclear energy and ecological thought is historical, because (at least in Europe), the emergence of political ecology is embedded in anti-nuclear movements.[8] Consequently, in his book *Les Ecofictions: Mythologies de la fin du monde* [*Ecofictions: Mythologies of the End of the World*], Christian Chelebourg sees a strong cultural continuity between the nuclear and the ecological end-of-world narratives: 'Le spectre d'une conflagration nucléaire des deux blocs, qui hantait des consciences frappées par le martyre d'Hiroshima et de Nagasaki, a progressivement cédé la place à l'angoisse écologique' [The spectre of a nuclear conflagration of the two blocs, which haunted the minds stricken by the martyrdom of Hiroshima and Nagasaki, has gradually given way to ecological anxiety] (Chelebourg, 2012: 7). Indeed, in literary and cultural studies, a conceptual bond occurs regarding nuclear and eco- or climate criticism, with some stating that ecocriticism has replaced nuclear criticism, while others see a closer coevolution.[9] In line with this chapter's claim regarding a strong link between the literary depictions of both nuclear and environmental catastrophes, I agree with Daniel Cordle that a cultural analysis of all things nuclear still seems very relevant. This is particularly true in the light of the most recent historical developments, with the Russian invasion of Ukraine blurring the frontiers between war and ecology and giving birth to the concept of 'écologie de guerre' [war ecology], which describes energy reduction efforts not out of environmental concern, but as 'une arme pacifique de résilience et d'autonomie' [a pacific weapon of resilience and autonomy] to harm a hostile state (Charbonnier, 2020, quoting Salomon).

Local nuclear disasters as metaphors for global ecocatastrophe

Two characteristics of ecocatastrophes are important for considering, in Decock's and Schmit's literary texts, the depiction of these fictional nuclear disasters as metaphors for the much more global impending ecocatastrophe and environmental crisis. First, an ecocatastrophe has multiple impacts – physical and environmental – as well as with socio-political and cultural effects on a very large and even global scale. In this sense, an ecocatastrophe means a major threat to the natural environment and its inhabitants. It also refers to climate change or the destruction of the biosphere. Taking Timothy Morton's well-discussed concept of 'hyperobjects,' I contend that nuclear power is also a hyperobject because of its imperceptible fallout, nonlocal quality (it transcends both geographical and physical boundaries), and a temporality that is beyond the human scale, with some particles being radioactive for millions of years (Küpper, 2020: 145). Somewhat paradoxically, a nuclear power plant would then be the localisation of the hyperobject, both as a physical structure and as a place on a map, belonging to a national territory with its specific nuclear politics and regulation.

Second, as 'eco' comes from 'oikos,' which means 'household,' and subsequently forms 'ecology,' the science which studies the interdependence between an organism and its environment, but also the (partly political) movement of thought and action to defend the environment against human destruction, 'ecocatastrophe' refers to a major event that says something about the mostly flawed relationship between humans and the environment or between humans and non-humans. In that sense, quoting Eduardo Kohn and Baptiste Morizot, anthropologist Sophie Houdart states that: 'La radioactivité a compliqué, ou est venue grossir ou densifier 'la toile de pensées vivantes' dont parle Eduardo Kohn' [Radioactivity has 'complicated, magnified or densified' the ways we live along other beings and things] (Houdart, 2019: 83).

The descriptions of the nuclear disaster in the literary texts considered in this chapter contain these two features of ecocatastrophe. Radioactivity is characterised by its disregard for national territorial borders and its transgression of bodily boundaries, as it destroys almost imperceptibly all life. The old man in Schmit's text laments:

> Et war sou einfach, et ze vergiessen. Et gesäit een se net, et rich teen se net, et héiert een se net. Et war sou einfach, sou ze maachen, wéi wann näicht wier. Mee se sinn do, a fréier oder spéder bréngen se dech ëm.
>
> [It was so easy to forget. You can't see the radiation; you can't smell it and you can't hear it. It was so easy to pretend nothing happened. But it is there, and sooner or later it will kill you.]
>
> (Schmit, 2018: 29)

And Mike, in Decock's novel, says at the beginning of his journey:

> Je suis content d'avoir trouvé ces solides bottines, car je me méfie des sols; même si mon compteur Geiger n'affiche qu'un petit 1.0, un micro-Sievert par heure, je ne peux m'empêcher de penser sans cesse à cette sourde menace. Incolore et inodore. C'est dommage, parce que l'air est si pur, d'une pureté que je n'ai jamais connue. Le vent porte des odeurs qui ne me sont pas familières: des parfums d'herbe, de fleurs, de terre. Quel étranger pays ! C'est un monde désolé, mais la nature y est rayonnante.
>
> [I'm glad I have found these sturdy boots, as I'm wary of the ground beneath my feet. Even if my Geiger counter only displays a small 1.0, one micro-Sievert per hour, I can't help thinking about this deaf threat all the time. Colourless and odourless. It's a shame because the air is so pure, of a purity I've never known. The wind carries unfamiliar smells: scents of grass, flowers, and soil. What a strange country! It's a desolate world, but nature is flourishing.]
>
> (Decock, 2019: 19)

Furthermore, chiming with the second definition of ecocatastrophe, which is the problematic relationship between nature and human, 'Eng Stëmm an der Stëllt' is basically about resilience that goes hand in hand with a reestablishment of the connection between humans and the non-human. Rather than seeing his text as a disaster story, Schmit claims it to be a Robinsonade (Schmit, 2020): a story of a man confronted with solitude and with the need to reinvent his relationship with his environment. As for *Luxembourg Zone rouge*, the novel is about the relationship between a nation and its territory, and about the tragic loss of land in the sense of Latourian thought evoked earlier.

One reason for considering the nuclear disaster as a metaphor for eco-catastrophe is obviously contextual: In the current era of the Anthropocene, we cannot read stories about nature, the environment, or the non-human the same way as we did a century ago. Ecocriticism claims that the present context, defined by mass extinction and climate change, deeply affects literary representations of nature. Firstly, it becomes more and more impossible to write about nature without depicting it as something endangered by human activity, with the consequence that the destruction of the environment is a threat to all of humankind. To say it again with Bruno Latour, in the New Climatic Regime what was once 'nature,' external to human activity, is now deeply political, affecting every decision and urging for a new narrative. This new context also affects the way we write and read about risk, catastrophes, and accidents, which become increasingly related to a much more global instability. Secondly, in this ecological turn, stories set in a local environment must be put into a larger global context, as they show the multiple relations that exist between various temporalities and localities (ZoneZadir, 2021: 10). Quoting Borges's *Aleph* (1949), Serenella Iovino states that as scholars of environmental humanities, we should not consider places as they appear on maps, but as something dynamic, as 'points in the space that contain all points' (Iovino, 2016: 2). This is particularly interesting concerning nuclear catastrophes, because they affect environments well beyond their local context, and because they expose risk scenarios and fears around the globe.

Due to this embeddedness of the local and the global, the main question raised by 'Eng Stëmm an der Stëllt' and *Luxembourg Zone rouge*, written within the small cultural context of Luxembourg, concerns their ability to 'bring the ecocatastrophe home.' How can the two authors' choice to relocate the nuclear catastrophe into their own ecocultural context be explained other than by referring to the main identification argument, which supposes that readerly interest could be stirred and maintained only through this proximity? Relocating the Chernobyl or the Fukushima disaster to Cattenom near Luxembourg turns the memorial function of literature into a preventive scenario and highlights the interplay between fiction and reality, between imagination and information. Localising stories can also have political implications, as has been shown by the 2006 film *Die Wolke* [*The Cloud*], an adaptation of Gudrun Pausewang's best-selling but controversial 1987 novel of the same title, which realistically describes the brutal losses of a young protagonist due to the explosion of a nuclear power plant near her hometown: On one hand, the controversy was political, dividing the proponents and detractors of the nuclear. On the other, it was literary, with some hailing the courage of the author and others castigating the traumatic (Liere, 2014) and moralising aspects of the novel (Boie, 2011). The German nuclear industry tried to prevent the adaptation, but managed only to have the name of the exploding power plant changed, turning the real nuclear power plant in Grafenrheinfeld in the novel into the fictive 'Markt Ebersberg' in the film (von Glasenapp, 2013: 76–77).

One of the main focal points of ecocritical reading is to show, according to Ursula Heise, 'how new risk perceptions are shaped by already existing cultural tropes and narrative templates' (Heise, 2008: 13). In that sense, representing the catastrophe is always cultural. As for Luxembourg, the Cattenom power plant plays such an important role in the collective imagination of the catastrophe that the risk scenarios due to environmental and biological collapse can also be expressed through existing yet constantly evolving tropes and narratives. What is more, the small territory and culture of Luxembourg, threatened by the neighbouring power plant, add an important idea to the representation of nuclear catastrophe because the latter could – at least by way of imagination – mean the eradication of a whole country. Thus, the smallness of Luxembourg becomes a metaphor for globality.

Yet, independently of the specific Luxembourgish context, an apocalypse in itself is always a cultural revelation. In 'What Is the Plural of Apocalypse? An Essay on the Ends of Time' (2019), Matthew Carey states that there is no end, but only ends in the plural. The imagined apocalypse is never the universal end of time, but how a culture imagines itself coming to an end. According to Carey, there lies 'the poetic force of the apocalypse':

> This idea of the loss of the world is, obviously enough, inherent to all ideas of the apocalypse, but it is also critical to stress the sense of loss of homeland and home (*Heimatverlust*) that very often accompanies and shapes the apocalyptic imaginary. . . . One cannot speak of denaturalisation without touching upon defamiliarisation. The society which we comfortably inhabit as a *Heim* or as a *Heimat*, never really calling it into question, suddenly becomes uncanny, unfamiliar or strange. . . . It is always at the very moment that we raise the possibility of a world's annihilation that we become aware of its contours, or even its existence.
> (Carey, 2019: 23–28)

In both 'Eng Stëmm an der Stëllt' and *Luxembourg Zone rouge*, the loss of home and homeland is indeed a recurrent motive. In Decock's novel, the homeland has a very strong, almost religious meaning to the protagonist: 'Normal qu'à mon âge je tienne à revoir mon pays natal. Ou du moins ce qu'il en reste. Un pèlerinage en quelque sorte' [It is normal that, at my age, I want to see my native country again. Or at least what's left of it. A pilgrimage in a sense] (Decock, 2019: 10). In Schmit's text, the protagonist leaves his house, but is unable to abandon his homeland and settles in a hut in nearby woods (2019: 10–11). When he finds Argos, he promises: 'Mir sichen dir lo en neit Doheem' [Now, we will find you a new home] (2019: 15).

As aforementioned, the many precise geographic references situate the plot in the Luxembourgish context. They are justified in a narratological manner through the two protagonists' knowledge of the region and with recourse to maps and even travel guides they use. *Luxembourg Zone rouge* particularly highlights the discrepancy between the idealised natural and cultural representation of Luxembourg and the now abandoned land:

> Tout aussi indispensable, ce vieux guide touristique que j'ai pu dégoter chez les voisins: 'Découvrir le Grand-Duché. Tous les hôtels et restaurants.' Pas certain que ce soit encore d'actualité. De toute façon, pour compléter ce bouquin, il me fallait une carte, je me la suis déchargée en plusieurs morceaux sur internet, morceaux que j'ai ensuite

méticuleusement assemblés à la façon d'un puzzle. Le Luxembourg du début des années 2000. Un document historique, une pièce de musée, une vue idyllique et colorée où les petites villes et modestes villages alternent avec les champs et forêts. . . . que reste-t-il de ce charmant pays?

[This old travel guide which I was able to dig up from the neighbours is equally indispensable: 'Discover the Grand Duchy. All the hotels and restaurants,' it says. I doubt that this is still relevant. Anyway, to complete this book, I needed a map, and so I downloaded it in several pieces from the Internet, and then meticulously assembled them like a puzzle. Luxembourg at the beginning of the 2000s. A historical document, a piece out of a museum, an idyllic and colourful view where small towns and modest villages alternate with fields and forests. . . . [w]hat remains of this charming country?]

(Decock, 2019: 12)

Here, Decock's novel clearly builds on the nature/culture dualism rather than deconstructing it. At the end of the text, Luxembourg is no longer a cultural territory but rather a natural land or, to put in in the terms Heise uses to describe the environments of Chernobyl and Pripyat: a 'remarkable nature sanctuary . . . a "radioactive wilderness" of indisputable beauty' (Heise, 2008: 201). Starting as a warning against a nuclear apocalypse, the novel ends on an ecological note, picking up the idealised, utopian depiction of a world without humans, where nature could thrive:

Quant au Luxembourg, ce pays est perdu pour toujours. Enfin, pas pour tout le monde, car en un certain sens les écolos ont réalisé leurs rêves au-delà de toute espérance . . . oublié le Luxembourg paradis fiscal: le Grand-Duché est devenu une immense réserve naturelle au cœur de l'Europe, un Eden dans lequel prolifèrent le gibier et les grands prédateurs. Loups, lynx, mouflons, cerfs, chevreuils, sangliers, rongeurs de toutes sortes; toutes ces espèces s'ébattent joyeusement dans un parc de trois mille kilomètres carrés ou lentement se réinstaure un nouvel équilibre dont l'homme est absent. Ou presque. Bien sûr, il y a ces radiations, bien trop dangereuses pour que l'on songe avant longtemps à revenir s'installer, mais il y a plus là, ni voitures, ni industries, ni pesticides. Et la nature est florissante . . . une renaissance. Peut-être cet univers préfigure-t-il le monde de demain, quand les hommes auront disparu, laissant la planète à son destin?

[As for Luxembourg, the country is lost forever. At least not for everyone, because in a way the ecologists have realised their dreams beyond their hopes . . . Luxembourg as a tax haven is gone: the Grand Duchy has become a huge natural reserve in the heart of Europe, an Eden in which game and large predators proliferate. Wolves, lynx, moufflons, deer, wild boar, and rodents of all kinds; all these species frolic happily in a park three thousand square kilometres wide, where a new balance is slowly being re-established without humans. Almost. Of course, there is radiation, much too dangerous for resettlement any time soon, but there are no more cars, no industries, and no pesticides. And nature is flourishing . . . A true rebirth. Perhaps this world prefigures a future world of tomorrow, in which humans will have disappeared, leaving the planet to its destiny.]

(Decock, 2019: 97–98)

In 'Eng Stëmm an der Stëllt,' despite the protagonist's attachment to his home, a landscape is described as something that needs to be rediscovered according to new needs: 'An den nächste Woche sinn ech mat menge Schof all Dag iwwert d'Minetter Koppe gezunn, ëmmer op der Sich no engem neie gréngen Teppech, dee meng Béischte géif sat maachen' [Over the next few weeks, I wandered through the hills of the Minette with my sheep, always looking for a new green field to feed my flock] (Schmit, 2018: 23). The homeland's landscape often appears paradoxically unfamiliar, which is symbolised by the description of soundscapes or silentscapes (Schmit, 2018: 8–9, 21–22), and even of one smellscape, signifying death and rotten bodies (Schmit, 2018: 25), rather than of landscapes which are traditionally grasped with the eye. This sensorial shift expresses the rediscovery of the post-apocalyptic environment, but also contrasts with the imperceptible radioactive pollution that will finally kill the old man.

Both plots thus explore the consequences and the ways to 'live on a damaged planet' (Tsing et al., 2017). Decock's novel imagines the disappearance of a whole country in the centre of Europe. Again, what Latour has described as a new situation due to the New Climatic Regime had already been experienced in the past, not only because of military conflicts between countries, but also in the case of major technological accidents with which modern societies pay the price for their modernisation (Horn, 2020: 422). Even if the described situation is not less dramatic, Schmit offers a more optimistic suggestion of how, to borrow Donna Haraway's formulation, to 'stay with the trouble' (Haraway, 2016). Haraway writes:

> The task is to make kin in lines of inventive connection as a practice of learning to live and die well with each other in a thick present. . . . Staying with the trouble does not require such a relationship to times called the future. In fact, staying with the trouble requires learning to be truly

> present . . . as mortal critters entwined in myriad unfinished configurations of places, times, matters, meanings.
>
> (Haraway, 2016: 25–26)

Clearly, the two narratives are neither a manual for survival of a nuclear catastrophe nor a literary pendant to official flyers and broadcasts on what to do in such scenarios. In fact, the old shepherd in Schmit's text describes the unbearable briefings and instructions:

> Am Hannergrond huet een déi midd Stëmm vum Radiospeaker héieren, dee wéi e Roboter, Stonn ëm Stonn, déi neisten Entwécklungen erofgeleiert huet: Rettungsmesuren, Stralewäerter, Appeller un d'Populatioun. De Perimeter, deen evakuéiert huet misse ginn, ass Dag fir Dag an d'Luucht gaangen. Fir d'éischt waren et just zéng, dunn zwanzeg, schlussendlech fofzeg Kilometer. Iergendwann hunn ech déi monoton, ëm Professionalitéit beméite Stëmm, dee gespillten Optimismus, einfach net méi ausgehalen. Ech wollt näicht méi héieren. De Radio ass wider d'Mauer geflunn.
>
> [The tired voice of the radio speaker could be heard in the background. Like a robot, it rattled off the latest developments hour by hour: rescue measures, radiation levels, appeals to the population. The perimeter that had to be evacuated was widening day by day: initially ten kilometers, then twenty, and finally fifty. At some point I just could not stand the monotonous voice trying to be professional and the feigned optimism anymore. The radio hit the wall.]
>
> (Schmit, 2018: 9–10)

On the contrary, both protagonists display a quasi-suicidal behaviour, as they do not want to survive, but to live in the perilous situations they are in:[10] Mike Olinger's promise to his father and his urge to visit his homeland prevail over the risks posed by radioactive exposure and groups of dangerous looters in the contaminated zone, where he will even live a short romance with a girl named Léa: 'Ce sera sans doute la meilleure nuit de tout mon séjour en zone nucléaire. Ma meilleure nuit tout court, en fait' [It will undoubtedly be the best night of my entire stay in the nuclear zone. My best night ever, actually] (Decock, 2019: 84). Schmit's text perfectly illustrates Haraway's 'thick present': Resisting the unreality of the factual official communication, his protagonist neither leaves his home region nor abandons the ill-fated animals. He thus becomes an allegorical figure of an 'entwined' humankind forced by a major catastrophe to rediscover ways to live along the non-human – meaning both the environment and animals – and with the imperceptible danger, which is radioactivity in the first place, but which also hints at the still largely unnoticed global ecocatastrophe. Finally, it is important to mention that Schmit draws upon a real biography and a real event as a main inspiration for his story. The connection between Fukushima and Luxembourg suggests that there is a 'becoming with,' in the Harawayian sense of describing tangled existences with the non-human (Haraway, 2008), but also with other cultures and fates (Houdart, 2019: 79) from which we cannot feel detached.[11]

Conclusion

Even if set in a local setting and adopting its specific tropes and narratives, like the smallness of Luxembourg and the absence of an imaginary of natural disasters, Schmit's and Decock's stories contain multiple openings by recognising the international ramifications of the production, consumption, and risks of nuclear energy. Furthermore, in the context of global 'climate uncertainty' (Caracciolo, 2022), these catastrophic scenarios also convey the authors' and readers' imaginations and fears regarding a troubled future where actual modes of living can no longer be sustained. Locally specific yet ecoglobally informed, these texts draw upon literary traditions, like nature writing, homeland literature, or the Robinsonade, as well as upon the deep-rooted fear of a nuclear disaster, to express aesthetically their concern regarding humanity's interaction with nature. The nuclear disaster is an ecocatastrophe, because it is man-made yet results in long-lasting effects on the environment, the recognition of which has made it a spur for ecological thought and political action. Fears and hopes triggered by new risk scenarios are built using existing narratives and images. As nuclear criticism has shown, that was once the case regarding nuclear warfare and energy. Now the nuclear disaster becomes the reference for imagining the consequences of global ecological disturbance, particularly in regions like Luxembourg, which seem or imagine themselves less vulnerable to it. By inventing stories in which individuals must make crucial choices in unfamiliar environments, Schmit's and Decock's literary texts confront their readers intellectually and emotionally with the impending ecocatastrophe that they might still feel unconcerned by.

Notes

1 The ecological footprint measures how much in natural resources a population or, as in this case, an individual needs for consumption and production. 2022 Global Footprint Network.
2 Anticipatory fiction anticipates an event and its consequences in a contemporary setting and in a realist mode.
3 Unless otherwise indicated, all translations from Luxembourgish, German, and French are my own.

4 There are some exceptions, like the depictions of the floods of the Moselle River in Nikolaus Hein's novel *Gertrud. Ein Moselroman*, written between 1916 and 1930 (see Thiltges, 2020: 225).
5 One recent notable exception would be Maxime Weber's (2021) novel *Das Gangrän*, where an all-destroying gangrene also affects Luxembourg.
6 Exceptions, from a different context, are the numerous deaths in the mines, suffered mostly by immigrant workers, as described by Jean Portante's (1993) novel *Mrs Haroy ou la Mémoire de la baleine*.
7 Even if they are historically opposed, the borders between ecology and nuclear energy seem porous, if we look, for instance, at the taxonomical project of the European Commission to consider nuclear energy as a sustainable source of power, because it contributes to climate neutrality (European Commission, 2022). For further details, see Thiltges (2022).
8 For the Luxembourgish context, see Pauly (1989).
9 'Ecocriticism – literary and cultural studies' engagement with environmentalism – has appeared since nuclear criticism's disappearance and is institutionally established in crucial forms that eluded nuclear criticism' (Williams, 2009: 253). 'Nuclear literature might usefully therefore be considered a special subcategory of climate fiction' (Cordle, 2019: 282).
10 In the French equivalent, 'vivre dans un monde abîmé,' Marielle Macé uses the word play between 'survivre' [survive] and 'vivre' [live] (Macé, 2019: 4).
11 I develop this idea further in Thiltges (2022).

Bibliography

Beck, Ulrich (1986), *Risikogesellschaft. Auf dem Weg in eine andere Moderne* (Frankfurt-am-Main: Suhrkamp).

Boie, Johannes (2011), 'Moralin. Gudrun Pausewangs *Die Wolke* schwebt in die Bestsellerlisten', *Süddeutsche Zeitung*, 15 March.

Caracciolo, Marco (2022), *Contemporary Fiction and Climate Uncertainty: Narrating Unstable Futures* (London: Bloomsbury).

Carey, Matthew (2019), 'What Is the Plural of Apocalypse? An Essay on the Ends of Time', *Terrain*, 71, 6–25.

Charbonnier, Pierre (2020), 'La naissance de l'écologie de guerre', *Le Grand Continent*, [Online], [Accessed 17 August 2022], available from: https://legrandcontinent.eu/fr/2022/03/18/la-naissance-de-lecologie-de-guerre/

Chelebourg, Christian (2012), *Les Ecofictions: Mythologies de la fin du monde* (Bruxelles: Les Impressions Nouvelles).

Cordle, Daniel (2019), 'Climate Criticism and Nuclear Criticism', in Adeline Johns-Putra (ed.), *Climate and Literature* (Cambridge: Cambridge University Press), 281–297.

Decock, Pierre (2019), *Luxembourg Zone rouge* (Esch/Sauer: Op der Lay).

Deleuze, Gilles and Félix Guattari (1975), *Kafka. Pour une littérature mineure* (Paris: Minuit).

Deleuze, Gilles and Félix Guattari (1985), 'Kafka: Toward a Minor Literature: The Components of Expression', trans. Marie Maclean, *New Literary History*, 16: 3, 591–608.

Dirkx, Paul (1995), 'Describing Literature in Belgium: Some Reflections on Evolution, Domination and Dependence', *Dutch Crossing. A Journal of Low Country Studies*, 19:2, 73–92.

European Commission (2022), 'EU Taxonomy: Commission Begins Expert Consultations on Complementary Delegated Act Covering Certain Nuclear and Gas

Activities', [Online], [Accessed 27 June 2022], available from: https://ec.europa.eu/commission/presscorner/detail/en/ip_22_2
Gilbertz, Fabienne (2019), *Wortproduzenten: literarische und ökonomische Professionalisierung im Luxemburger Literatursystem der 1960er und 1970er Jahre* (Heidelberg: Universitätsverlag Winter).
Glesener, Jeanne E. (2012), 'On Small Literatures and Their Location in World Literature: A Case Study on Luxembourgish Literature', *Interlitteraria*, 17, 75–92.
Glesener, Jeanne E. (2015), 'The Periodical as a Strategy of Recognition for Small Literatures', *Interlitteraria*, 1, 159–174.
Glesener, Jeanne E. and Oliver Kohns (eds.) (2022), *Weltliteratur und kleine Literaturen* (Würzburg: Königshausen & Neumann).
'Global Footprint Network' (2022), [Online], [Accessed 15 June 2022], available from: https://data.footprintnetwork.org/#/
Hache, Émilie (2014), 'Introduction. Retour sur terre', in Émilie Hache (ed.), *De l'univers clos au monde infini* (Bellevaux: Éditions Dehors), 11–25.
Haraway, Donna J. (2008), *When Species Meet* (Minneapolis: University of Minnesota Press).
Haraway, Donna J. (2016), *Staying with the Trouble: Making Kin in the Chthulucene* (Durham: Duke University Press).
Hein, Nikolaus (1999), *Gertrud. Ein Moselroman* (Luxembourg: Sankt Paulus Druckerei).
Heise, Ursula K. (2008), *Sense of Place and Sense of Planet: The Environmental Imagination of the Global* (New York: Oxford University Press).
Horn, Eva (2020), *Zukunft als Katastrophe* (Frankfurt-am-Main: Fischer Verlag).
Houdart, Sophie (2019), 'Fukushima, l'expérience en partages', *Critique*, 860–861:1–2, 70–86.
Iovino, Serenella (2016), *Ecocriticism and Italy: Ecology, Resistance and Liberation* (London: Bloomsbury).
Küpper, Achim (2020), 'Der Reaktor als grenzüberschreitende Risikozone in der Großregion: Ökologische, interkulturelle und didaktische Herausforderungen der Kernenergie', in Sébastian Thiltges and Christiane Solte-Gresser (eds.), *Kulturökologie und ökologische Kulturen in der Großregion* (International: Peter Lang), 143–159.
Latour, Bruno (2017), *Facing Gaia: Eight Lectures on the New Climatic Regime*, trans. Catherine Porter (Cambridge: Polity).
Latour, Bruno (2018), *Down to Earth. Politics in the New Climatic Regime*, trans. Catherine Porter (Cambridge: Polity).
Liere, Judith (2014), 'Pädagogische Horrorshow', *Spiegel Online*, 27 February, [Online], [Accessed 17 August 2022], available from: www.spiegel.de/geschichte/die-letzten-kinder-von-schewenborn-horror-schocker-kinderliteratur-a-959122.html
Macé, Marielle (2019), 'Présentation', *Vivre dans un monde abîmé. Critique*, 860–861:1, 3–4.
Morton, Timothy (2013), *Hyperobjects: Philosophy and Ecology after the End of the World* (Minneapolis: University of Minnesota Press).
Nixon, Rob (2011), *Slow Violence and the Environmentalism of the Poor* (Cambridge, MA: Harvard University Press).
Ortlieb, Gilles (2011), *Tombeau des Anges* (Paris: Gallimard).
Pagnotta, Antonio (2013), *Le Dernier Homme de Fukushima* (Paris: Don Quichotte).
Pauly, Michel (1989), 'Die grün-alternative Bewegung', in Martin Gerges (ed.), *Mémorial 1989. La Société luxembourgeoise de 1839 à 1989. Les Publications mosellanes*, 28 (Esch-sur-Alzette: Editpress/Schortgen), 90–92.

Pausewang, Gudrun (2006), *Die Wolke. Jetzt werden wir nicht mehr sagen können, wir hätten von nichts gewusst* (Ravensburg: Ravensburger Taschenbuch).

Portante, Jean (1993), *Mrs Haroy ou la Mémoire de la baleine* (Echternach: Phi).

Pulli, Sasha (2020), *Das gescheiterte Jahrhundertprojekt. Die Geschichte der Atomzentrale in Remerschen von 1973–1979* (Luxembourg: Fondation Lydie Schmit).

Schmit, Yorick (2018), 'Eng Stëmm an der Stëllt', in *Der Geruch der Erde nach dem Regen* (Luxembourg: Saint-Paul).

Schmit, Yorick (2020), 'Bachelor-Seminar "Natur an Ëmwelt an der Lëtzebuerger Literatur"', *led by Sébastian Thiltges*, University of Luxembourg, 24 April.

Schnitzler, Georg (2006), *Die Wolke* (Munich: Clasart Film & Fernsehproduktion).

Subramanian, Meera (2019), 'Anthropocene Now: Influential Panel Votes to Recognize Earth's New Epoch', *Nature*, [Online], [Accessed 27 June 2022], available from: www.nature.com/articles/d41586-019-01641-5

Sunnen, Myriam (2008), '*Den Zolverknapp as kê Parnas*. Le paysage dans la littérature luxembourgeoise', in Claude D. Conter and Germaine Goetzinger (eds.), *Identitäts(de)konstruktionen* (Mersch/Differdange: Centre national de littérature/Éditions Phi), 33–50.

Thiltges, Sébastian (2020), 'Frontières liquides: l'eau dans les romans de la terre', in Sébastian Thiltges and Christiane Solte-Gresser (eds.), *Kulturökologie und ökologische Kulturen in der Großregion* (International: Peter Lang), 213–233.

Thiltges, Sébastian (2022), 'La représentation interculturelle du désastre nucléaire entre appropriation et écart: deux exemples luxembourgeois', *RELIEF. Revue électronique de Littérature française*, 16:1, [Online], [Accessed 17 August 2022], available from: https://revue-relief.org/issue/view/587

Tsing, Anna, Heather Swanson, Elaine Gan and Bubandt Nils (eds.) (2017), *Arts of Living on a Damaged Planet: Ghosts and Monsters of the Anthropocene* (Minneapolis: University of Minnesota Press).

Utz, Peter (2013), *Kultivierung der Katastrophe. Literarische Untergangsszenarien aus der Schweiz* (Munich: Wilhelm Fink).

von Glasenapp, Gabriele (2013), 'Akopalypse Now! Formen und Funktionen von Utopien und Dystopien in der Kinder- und Jugendliteratur', in Hans-Heino Ewers, Gabriele von Glasenapp and Claudia Maria Pecher (eds.), *Lesen für die Umwelt. Umwelt und Umweltschutz in der Kinder- und Jugendliteratur* (Baltmannsweiler: Schneider Verlag Hohengehren), 67–86.

Weber, Maxime (2021), *Das Gangrän* (Luxembourg: Kremart Edition).

Williams, Paul (2009), 'Nuclear Criticism', in Mark Bould, Andrew Butler, Adam Roberts and Sherryl Vint (eds.), *The Routledge Companion to Science Fiction* (London and New York: Routledge), 246–255.

ZoneZadir (Collectif) (2021), 'Pour une écopoétique transculturelle: introduction', *Littérature*, 201:1, 10–23.

8 Speculating on ecological futures

Narratives of hope and multispecies justice in contemporary ecofiction

Elizabeth Tavella

Contemporary discourses on climate change, mass extinction, and systemic injustices tend to promote narratives of collapse and end-of-the-world scenarios, a symptom of a widespread failure of the imagination. We are in fact currently experiencing an epistemic blindness to ways of being that exist beyond fatalistic futures, especially in western imaginaries, which consequently generates paralysing anxiety and resignation, and even outright denial.[1] While it is important to acknowledge and legitimise the reality of environmental grief and eco-despair, we need to be alert to the ways in which neoliberal fatalism interferes with political mobilisation as much as corporate-funded climate denialism.[2] The dominance of apocalyptic thinking in current political debates concerning the climate crisis demonstrates a fixation on a politics of low expectations as well as on a weak rhetoric of welfare reform that pervades virtually all climate discussions. Both frameworks, I argue, should be considered active contributors to what Rob Nixon has termed 'slow violence,' in that they limit the possibilities of radically changing the social structures and ideologies responsible for this crisis in the first place.[3]

A similar attitude of resignation manifests itself also in the stories we tell. In fact, post-apocalyptic stories of (solely) human survivors, colonial fantasies, and white saviours are prevalent in dominant cultural narratives, including in many ecotopias, which are often built on heteronormative, racist, and elitist premises, exacerbated by an anxiety-driven tone (de Haan, 2019: 55–72). Besides reproducing oppressive social dynamics, depictions of the future that are anthropo-focused generate a scenario that, echoing Vandana Singh's words, 'is a sign of a deep malaise, apart from being utterly boring and unrealistic' (2014). Among the factors that contribute to our limited imaginative horizons is a vision of future urban environments as biodiversity wastelands depicted as the main stage for violent conflict across species for survival. While it is true that the city remains a shared space of conflict due to pervasive speciesist ideologies and infrastructures,

DOI: 10.4324/9781032726953-9

this does not mean that the literary imagination must reproduce such conflictual dynamics and ignore the richness of urban ecologies, which are far from being void of more-than-human presence.

Promisingly, recent publications dedicated to speculative writing reframe futurity from the lens of the oppressed, and at times even embed in ecological thought. Nevertheless, more-than-human animals tend to remain at the outskirts of solidarity discourses, especially those deemed consumable and who remain stuck within a chain of production supported by a hierarchical ideology of violent domination.[4] In particular, the fracture between individuals considered 'wild' and those who have been 'domesticated' remains strikingly unchallenged, thus revealing visions of ecological futures built on speciesist premises. As a consequence, in contexts not explicitly attentive to undoing normative ideas about the subordinate positioning of non-human animals in society, ambivalent forms of care and control arise, particularly within animal consumption discourses, which limit the possibilities for radically dismantling interlinked systems of oppression.

Given these premises, now more than ever, a critical shift in our imagination is key in disrupting historic and ongoing practices of colonialism, capitalism, industrialisation, and anthropocentrism responsible for enabling exploitative and extractive relations with the more-than-human world.[5] As José Medina affirms, 'the imagination can be both empowering and disempowering. It can create and deepen vulnerabilities, but it can also make people stronger and able to resist' (2013: 253). In response, then, to the widespread obituaries of nature, such as the one for Iceberg A68a and for the Bramble Cay melomys, the first mammal to go extinct directly because of human-caused climate breakdown, radical hope within the literary page can function as a reconfiguration of our conceptual toolbox aimed at actively investing in the future as a response to the problems of the present. Of course, hope must not be interpreted as weaponised optimism or toxic positivity, especially given the political economies that manifest themselves through uneven distributions and experiences of hope, but rather, as a rejection of the notion that we are unable to change the current flow of events, or, as Andreas Malm has so perceptively put it, 'a hope that acknowledges the immensity of the defeats and the desolation. Yet it refuses to capitulate' (2021: 255).[6]

Thus, by recuperating the creative, liberating potency that lies in speculative imagination, this chapter turns to speculative writing, promoting solution-oriented approaches to socio-ecological adversity with the aim of daring to temporarily materialise forms of existence not yet realised, but nevertheless ripe enough in the present to become an actuality.[7] In particular, I take the literary genres of solarpunk and ecopoetry as my point of departure to celebrate the potential for (re-)emergence from ruins: the first, in the anthology of short stories *Multispecies Cities: Solarpunk Urban*

Futures (Rupprecht et al., 2021), the second in Tamiko Beyer's collection of poems *Last Days* (2021). More specifically, drawing on abolitionist frameworks, urban ecology, and multispecies studies, I investigate how these works may contribute to transforming the conditions rooted in white supremacy, heteropatriarchy, corporate power, and speciesism fuelling the present global crises as well as how they may – or may not – designate more-than-human actors as allies and worldbuilders in facing uncertain futures. Likewise, I look at how these works participate in destabilising the presumption that cities are exclusively populated by humans, while helping to re-envision interspecies solidarity against structural violence and oppression within the cityscape.

As a result of a comparative analysis of speculative literary narratives by writers who centre the experiences of the Global Majority, more-than-human perspectives, and queer imaginaries, I engage with the following questions: how are marginalised individuals – of all species – (re)imagined within these texts as thriving in multispecies communities? What shape do such alliances take, and in whose interests might they work? What does ecological and social healing look like when, and if, practised from the perspective of interspecies liberation? How can we avoid perpetuating anthropocentric ecologies in charting paths towards more just futures? By addressing both the speculative potentials and epistemic limitations inherent to the selected texts, the aim of this chapter is twofold: to engage with narratives framed as hopeful in a strategising effort to catalyse our imaginaries for present action in the political struggle for collective liberation; and to reclaim the central role of radical imagination in dismantling interlocking systems of oppression affecting disenfranchised human communities as well as the more-than-human world.

Multispecies kinship and worldbuilding in solarpunk storytelling

Over the last decade, solarpunk sprouted online as a subgenre of cli-fi. As a response to grimdark narratives, this evolving lifestyle and storytelling trend quickly took off through crowd-sourced publications, that is, outside the economic logics of centralised publishing houses.[8] Its emphasis on community-building through cooperation is reinforced by the aim for a unity of theory and practice reliant on a forward-looking aesthetic, and grounded in high technology coexisting with or even complementing nature.[9] Foundationally, however, this rejection of dystopia does not mean embracing utopia. Rather, by combining realism with hope, the goal of solarpunk is to arise from a political mood of resistance against capitalist logics leading to the destruction of the environment. While the aspiration of this literary movement is to function as counter-hegemonic media by intertwining issues of race, gender, sexuality, class, and colonialism with an ecological ethic, the genre's lack of racial diversity has been highlighted by writers within the community, as well as its still-too-often anthropocentric approach, reinforced by sanitised futuristic cityscapes that rely on civilisational tropes.[10]

A step in an auspicious direction is, however, exemplified by the anthology *Multispecies Cities: SolarPunk Urban Futures*, published by World Weaver Press in partnership with the Research Institute for Humanity and Nature in Kyoto, Japan. In the introduction, the editors accurately recognise that 'what sets these stories apart is the foregrounding and focus on examining and including these [interspecies] relationships as a core element of world-building' (Rupprecht et al., 2021: 17). By platforming authors from and stories set in the Asia-Pacific region, and by showcasing abundant queer representation, this collection grapples with issues concerning systemic oppression towards marginalised populations through tales of overcoming and redemption. In addition, sustainable practices such as composting and tiered gardens are narratively intertwined with AI systems, holographs, and biomimicry, thus establishing the imaginative conditions for urban resilience. Among the main ecological concerns are questions of de-extinction and coexistence with 'wildlife,' which proliferate in the collection by giving space to characters who are often excluded from discussions of interspecies care on the literary page, including seagulls in search of nest sites and eels inhabiting underground water systems.

A thrilling example of multispecies inclusivity can be found in the depiction of interspecies coalition in Amin Chehelnabi's short story 'Wandjina,' inspired by the 2020 Australian bushfires. In response to the government's lack of action to help the native species of Australia, a group of human rescuers sets out in a blasted landscape to help the non-human animals at risk of dying. When the crew breaches the resource clause using resources and equipment for helping animals instead of Australians, an act that harbours jail time and even torture, a police vehicle approaches them to halt their mission. Using their bodies as literal shields 'against [their] oppressors' (Chehelnabi, 2021: 256), the crew of humans, bandicoots, wallabies, and dingoes fiercely fights for survival against the police officers. This literary scene of multispecies resistance echoes the many real-life stories of nonhuman animals who either escape, come to the aid of other animals, or fight back for their liberation.[11] Thus, instead of reading this act simply as an anthropomorphic gesture, this moment of interspecies solidarity needs to be understood as a manifestation of non-human agency and personhood as well as an expression of the desire to access the phenomenological world of our more-than-human kin.

Similarly, in 'Old Man's Sea' by Meyari McFarland, an act of resistance beyond the human species is imagined in a flooded world devasted by wars: an orca once exploited for military operations and now living with PTSD comes in defence of a human sea-dweller who is being targeted by a military boat. In the future outlined in this story, while land-dwellers continue to fight brutally over the little land that is left, the human-animal boundary is deliberately subverted beneath the ocean's surface as a sign of evolutionary adaptability to climate change and militarism. In fact, the relationship that is established between a new version of human who 'wasn't a fish and wasn't an orca and wasn't, really, a land-dweller, either' and an orca with cybernetic limbs and a human voice because 'the military wanted their sea-going conscripts to sound like any other soldier' underscores not only histories of trauma shared by marginalised identities across species, but also the strength of liminality as a space of possibility where ecological resilience and new social configurations may arise (McFarland, 2021: 39–40). Thus, by stretching the imagination without losing credibility and touch with reality, a sense of hope is instilled in the multispecies refusal to surrender to anthropogenic acts of violence.[12]

Besides devoting considerable attention in the collection to rethinking our relationships with 'wildlife,' the ethical ambivalence that characterises our current relationships with species deemed as food also comes to the surface. In fact, by illustrating the enduring difficulties inherent in grappling with the long history of human domination over the more-than-human world, which includes the internalised speciesism legitimising exploitation per se, the narratives raise crucial questions regarding bodily autonomy and values surrounding definitions of multispecies justice. So, for instance, intensive farming and overfishing are indeed recognised as problematic practices and the consumption of artificial meat is occasionally normalised within the narrative space for environmental reasons (Spires, 2021: 71).[13] Yet, at the same time, non-human animals as food, either as flesh or secretions, appear in a scattered way throughout the collection: maguro (tuna) sashimi, ham, smoked salmon, and mac and cheese, among others. All of these iterations imply – both in real life and at an epistemic level – the commodification of animal bodies, a value system that relies on the existence of raising and processing facilities, which in these futuristic visions at times continue to be concealed geopolitical spaces whose ties to interspecies violence, capitalist ideologies, and environmental racism are overlooked.[14] An example that puts the spotlight on these cultural contradictions within a single story, occurs in 'It Is the Year 2115' by Joice Chng, in which a future without caged animals is conceived as a reality, to the point where reminiscing a past when chickens were kept in cages leaves the protagonist horrified (2021: 110). However, soon after, the same character enjoys a stir-fry with 'strips of chicken' advertised as Kampung Chicken, a particular breed known for being raised using free-range production techniques:

> I turn my attention to the stir-fry, some of the kai lan and Chinese kale from row 11. I toss in some strips of chicken. It is touted to be Kampung Chicken, because the fowl run free in the farms. The *you tiao* uncle talked about caged chicken . . . Imagine that. Either way, it is going to be my delicious dinner.
>
> (Chng, 2021: 112–113)

While the elimination of cage systems represents an important milestone to aspire to, this dinner scene signals a future investment in welfarist approaches preoccupied with chasing a mode of production still based on ownership and killability of other animals, thus exposing the contradictory impasse between animal consumption and multispecies kinship. In other words, only what is considered excessive exploitation activates an emotional response, leaving unchallenged other forms of violence that continue to be regarded as 'natural' and therefore as socially acceptable.[15]

The short story that more directly exposes our troubling consumption practices is 'A Life with Cibi' by Natsumi Tanaka, which takes place in a future where humans live alongside genetically modified beings known as Cibi (from *cibus* – food in Latin) designed to be consumed by humans. Blurring the line between plants and animals, this 'new crop of edible creatures' (Tanaka, 2021: 160) wanders around the city and freely offer their flesh to be eaten. Suffering is not an issue, they do not feel any pain when sliced; on the contrary, they actually die sooner if not consumed. In the introduction to the collection, the editors interpret the story as a parable on reciprocal consumption, intended to support the claim of humans as eater and potentially eaten, and as a demonstration of 'how our need to consume can conflict with any universalist notions of bodily autonomy and individual freedom.' At the same time, they argue that multispecies justice 'does not simply mean straightforward notions of care, guardianship, and living and letting live, but rather acting both responsively and responsibly in both life and death' (Rupprecht et al., 2021: 7). Yet, if read as a cautionary tale, just like with the previous literary example, this story forces us to consider who benefits from this type of interspecies dynamic and from whose perspective we approach the question of multispecies justice.[16]

In fact, besides illustrating our relations with other animals as guided primarily by our consumption interests, the editors' statement defends an idea of justice rooted in notions of anthropocentric ecologies that safeguard the comfort of the human species while supporting oxymoronic practices of sustainable exploitation. If the aim is to envision a future community that crosses the species barrier and honours the ethical and political subjecthood of more-than-human animals, it is crucial, in the words of Fayaz Chagani, to embrace 'an openness to the question of the animal as a question of (social) justice' that finally comes to terms with 'uninterrogated anthropocentrism and the representational and physical violence which it legitimises' (2016: 634). Additionally, the humans in the story are never at risk of being eaten, hence never in the position of becoming potential prey. As part of their privileges, they also take advantage of the Cibi's ability to shift shape, since they 'prefer not to eat [Cibi] in its human form' (Tanaka, 2021: 160). However, once the human protagonist starts living with one of them, the leap into companionship makes him uncomfortable to slice his 'dear Cibus' (Tanaka, 2021: 162). This reaction outlines a common psychological mechanism emerging from speciesist categorisations and hierarchies that is often triggered by consumption practices. It must also be noted that none of the stories in the collection depict a future with humans as both eater *and* eaten, thus leaving the prospect of a primitivist human-animal relationship based on mutual edibility unexplored, which further demonstrates a disconnection between the aspired philosophical premises outlined in the introduction and the actual imaginative worldbuilding of the collection.[17] By preserving, then, normative relations of subjugation upheld by claims of 'necessity' versus the right to bodily liberty, the reading proposed by the editors underestimates the ethical conundrum that the humans in the story constantly face, while restricting the potentiality of new epistemological and ontological paradigms that could lead to expanding our vision of what multispecies kinship may signify.

The contribution that perhaps most strikingly displays an all-encompassing anti-oppressive stance is Priya Sarukkai Chabria's 'Listen: A Memoir.' The story alerts us to an incredibly populated world by bringing to life the pulse of the more-than-human, which functions as a reminder that 'everything speaks' (Chabria, 2021: 21).[18] This near-future story draws in part from the writer's response to the pandemic and the global forced isolation from the natural world, and in part from her childhood in Madras (now Chennai) where she lived with her grandparents and believed she could understand some of the languages of our more-than-human kin. The garden of her childhood mapped into her consciousness increasingly found its echoes in the city, 'in the rip of sky between towers, damp petals on the pavement; light that changes tar roads into glistening rivers, in the sleep of stray dogs' (Sengupta, 2021). Driven by a desire for interspecies connection, enabled by a sharpened capacity for listening, the author transfers a multisensory embodied experience informed by vibrant multispecies soundscapes to the page that 'reverberate with affective, gestural significance' (Abram, 1997: 55).[19] In the attempt to translate species-specific knowledge, emotions, and concepts, a rich language of poetic wonder and speculation permeates the story: dragonflies spoke to her 'in their language'; two squirrels befriended her, 'barking into [her] face in their loud voices'; she overhears conversations between spiders, 'though [she] can't decipher the exact words'; and she speaks in 'snail tongue' to her lover (Chabria, 2021: 14–18). While the risk of misinterpreting more-than-human languages remains a real possibility, the employment of 'strategic anthropomorphism' (Iovino, 2015: 16) destabilises anthropocentrism, thus urging readers to extend their empathy and creative imagination to our fellow earthlings. Moreover, acknowledging the value of more-than-human languages entails honouring their political voices as well as reimagining accepted views of language – namely logo/human-centric frameworks.[20] To further stress the semiotic interconnectedness between all lifeforms, the writer mentions that the protagonist and her grandmother speak in 'preserved Tamil' (Chabria, 2021: 13), conscious of the links between the flourishing of linguistic and biological diversity.[21] The writer's recording of habitually backgrounded voices deeply resonates with Alexis Pauline Gumbs's formulation of listening for trans-species communion, which she defined as 'a transformative and revolutionary resource that requires quieting down and tuning in' (2020: 15). Yet, the author's attunement with the more-than-human is not limited to 'wildlife' creatures, but rather optimistically extends to the species used for human consumption. In fact, while reminiscing the past, the young narrator notes that 'people used to own, grow and eat animals. As a child I wonder why people wanted to eat their friends. Perhaps because people didn't hear them sing, gossip and lament; didn't know the stories of their lives, big, small, and even smaller, which are much like our own, but oh-so frail' (Chabria, 2021: 16).[22] It is then the longing for more inclusive dialogues and non-exploitive relations that sets the foundations for the emergence of a multispecies society where all its members are actually given true respect and dignity. Through the speculative lens of this story, it is also possible to leave behind the imaginary of survival due to someone's extinction or killing: everyone survives, everyone lives. And this gives flickering hope.

Embodied cross-species solidarity in Tamiko Beyer's ecopoetry

On her personal website, Tamiko Beyer defines *Last Days* as 'a poetic practice of radical imagination for our current political and environmental crises.'[23] Inspired by adrienne maree brown's principles of 'emergent strategy' (2017) as well as by Robin Wall Kimmerer's 'grammar of animacy' (2013), her poems give voice to inequalities and troubled histories among human populations alongside the urgent needs of the more-than-human world, while centring immanent resilience and ecocentric inclusiveness grounded in hopeful speculation. Despite the apocalyptic tone of its title hinting to an end, a temporal end, the poems actually vibrate on the notes of hope by celebrating the beauty of interconnectedness between all organisms. From the very first pages, a resilient spirit takes over by recognising the necessity for 'a system of recovery, of remedy' (Beyer, 2021: 11) in order to thrive in the ruins, both literal and metaphorical, caused by current eco-social catastrophes. At the same time, the poems offer space for futurity and for possibility, such as in the poem titled '[ ]' in which neoliberal attempts to stifle political imagination are countered with an embodied awareness of collective and personal trauma, leading to an empowered response necessary to concretise the desired future: 'I carry my weight on my heels. I am always leaning forward' (Beyer, 2021: 66–70).

Throughout the collection, the process of mapping the way towards a more just future is always marked by a 'we,' a collectivity bounded by ecological proximity and precarity. In fact, collective solidarity is hereby extended to all those affected by systems of oppression: from Black and Brown people, to migrants, and the more-than-human world. In doing so, Beyer compels her readers to see their complicity in racism, environmental destruction, and a general permissiveness for violence in all its forms. Additionally, traditional notions of the poet standing apart from nature to observe it or to use it as a trope are abandoned. Invoking what she defines a 'queer::eco::poetics' (2010), the poet expands the very idea of 'nature' by writing from within, conscious of being inextricably related to all surrounding elements. It is within this context that new possibilities for attachment, kinship, and care are imagined. Each poem is meant to uniquely destabilise our perspectives, starting with the very first one titled 'What It Means To Be Human,' which sets the tone for deep ontological questioning and necessary deconstruction. The imagery presented in this opening poem establishes a thematic thread rooted in body politics and what I define a poetics of light, breath, and flame: a match is lighted, a fist is raised, animal muscles gallop, and fingers become wings (Beyer, 2021: 1). The body becomes the vehicle for this porousness of form, for the dissipation of all artificial dichotomies and categories, violently constructed to separate, to delegitimise one's essence. The trans-species fluidity envisioned through human-to-bird bodily mutations and metonymic galloping are a clear demonstration that, in David Harvey's words, 'the body is not a closed and sealed entity, but a relational "thing" that is created, bounded, sustained, and ultimately dissolved in a spatiotemporal flux of multiple processes' (2000: 98). Coming full circle, in the last poem, the author identifies her 'ambiguous skin' as 'mammal' (Beyer, 2021: 76), thus rejecting the historical weaponisation of animality towards marginalised populations, while creating the conditions for cross-species kinship based on a shared lived experience of embodiment.

The poems also boldly support an abolitionist approach, from prisons to police and geographical borders, and intersecting oppressive systems, such as colonialism and capitalism, which affect also the lives of other animals through commodification practices, objectification, biopolitical control, systemic violence, and erasure.[24] Social justice is thus framed ecologically and extended to the more-than-human world: trees 'planted when white/men drafted a constitution' participate in the collective struggle for liberation by 'showing their roots,' as a sign of resistance to a violent history of white supremacy and colonial genocide to which their existence is coercively tied (Beyer, 2021: 7).[25] To stress further the intersectional ecologies speaking to dynamics of privilege, the poet understands that 'the odds stack up by species, neighborhood, race, and wealth' (Beyer, 2021: 5), thus including, along with race and class, species discrimination among the social markers that generate in-group favouritism. In response to these hierarchical social structures, the liminality of the body once again functions as a site of protest amplified by ecological mergings: 'If we flood/the streets with salt water, we can/flood the sky with wings' (Beyer, 2021: 72). The syntactic parallelism between the flooding of both skies and streets creates a metaphorical, yet experiential, continuity between the urgency for collective (re)action, traceable in the 'we' taking to the streets, and the resulting opportunity for social and ecological change, for dreaming winged flights of freedom. This semantic continuity is further reinforced by the entangled nature of human bodies with bodies of water, thus stressing once again the interconnectedness between vibrant elements in the fight for liberation. In the same poem, addressed to 'a child of the near future,' an uplifting speculative tone conveys the vision of a shared meal: 'One day, I hope to set a table, invite you/to draw up a chair. Greens steaming garlic./Slices of bread, still warm' (Beyer, 2021: 72). While the invitation is specifically extended only to a generalised human child, the absence of animal flesh on the table leaves open the possibility of including more-than-human animals as active interlocutors driving ecological recovery rather than as consumed objects, shaping a future where 'we climb/out of our greedy mouth.'

A similar illustration of potential cross-species solidarity based on non-oppressive reciprocal relationships is explored in the poem 'Wintering' (Beyer, 2021: 12). Inspired by a real-life encounter with a 'four-legged creature' whose species remains unknown to the poet, the 'I' establishes a connection with them through a shared moment of blindness caused by the overwhelming presence of whiteness in the surrounding landscape. By establishing a lexical synchronisation between their lived experiences, 'I'm blinded by so much white. . . . skinny creature/blinded by so much white,' the individuality of the mysterious animal person is not only recognised, but it is also elevated to characterhood within the literary sphere. The snow, the sound of ice, the frozen river, all contribute to giving form to a wintering process that engages metaphorically with whiteness, a power structure that thrives on homogeneity, which is just one of the many ways in which it contributes to the annihilation of (bio)diversity. As a reaction to their shared vulnerability, at the end of the poem, their bodies synchronise: 'Our spines show./We cross the frozen river.' In the act of collectively crossing a precariously dangerous surface affected by the wintering process, there lies not only the possibility for interspecies coalition but also – and most importantly – the potential for envisioning multispecies flourishing beyond the frozen river.

By the end of the collection, a renewed sense of connection with the natural world is established, and it is at this point that the poet's voice surfaces more vigorously, such as in the poem 'I Vow To Be The Small Flame' in which she invites the readers to commit to a promise, to repeat after her: 'I vow/a ravenous undoing./I vow to love the fire always' (Beyer, 2021: 63). So, while these are certainly narratives of loss and struggle informed by racial violence, abusive immigration practices, and ecological disasters, they are also crafted to animate hope, or at least to support an awareness of a future that is partly open and promising. In order then to fully embrace a 'ravenous undoing' (Beyer, 2021: 63) of the engrained social structures and cultural assumptions impeding collective liberation, it is necessary to rethink the meaning of multispecies kinship through an anti-speciesist lens that challenges us to include the more-than-human world in pressing political, social, economic, and cultural issues. Despite an overall representational absence of 'domesticated' species, which may indicate an epistemic lack of reconciliation between their struggles for liberation and intersectional conceptualisations of justice, the poems clearly create a space for a process of unlearning to occur. In the uplifting spirit of 'constructing strategies, drafting poems' (Beyer, 2021: 29), Beyer views poetry

> as a means to help us re-think and re-imagine our relationships – our relationships to language, to meaning, to each other, and to what surrounds us . . . Poetry (that is, a text that takes nothing for granted and requires the reader to actively participate in meaning making) can help to raise questions, and to open up new possibilities for interactions, to the benefit, I hope, of the earth's survival.
>
> (Chin-Tanner, 2013)

It is precisely by celebrating these possibilities for growth and by embracing discomfort leading to collective healing and liberation that we will make it to the other side of the frozen river.

Conclusions

Collective visions have power – *stories* have power. And because of this, they hold the transformational potential necessary for fuelling hope for the future; in other words, 'humans need stories and art to engage in imaginative play to develop alternative perspectives on reality' (Räpple, 2019: 123). Thus, literature, and speculative writing in particular, with its capacity to undermine the dichotomy of practical use and aesthetic value (Felski, 2008: 8), can become a vehicle with which to activate and cultivate a political and ethical imagination that can project on reality. Viewed through this lens, the narratives analysed in this chapter may function as ecological models, that is, as openings for alternative strategies that can guide us towards meeting our aspirations going forward. They can also give us access to fragments of lives that we can model from while viscerally feeling their impact. The enduring value of speculation lies precisely in its insightful critiques of the limitations of present structures and modes of being, which inevitably requires shattering the perceived immutability of the 'Now,' an epistemic practice that, as a result, demonstrates that other worlds are indeed possible. By exploring how the awareness of the current ecological crisis has influenced the selected texts, an alternative etymology of the Greek *katastrophē* emerges: an 'overturning' or 'sudden turn' in individual or cultural consciousness. This is the only kind of catastrophe we really need: the formation of a new cultural paradigm in solidarity with our more-than-human kin; a relationship with the future that involves attunement to environmental change.

As Alexis Shotwell candidly admits,

> it is hard for us to examine our connection with *unbearable pasts* with which we might reckon better, our implication in *impossibly complex presents* through which we might craft different modes of response, and our aspirations for *different futures* toward which we might shape different worlds-yet-to-come.
>
> (2016: 8, original italics)

Yet, embracing futurity through a multispecies justice lens, rooted in an infrastructure of hope that critically and radically challenges anthropocentric values, holds the power to shift the cultural practices and social structures responsible for the ecological rupture affecting our relationships with the more-than-human world. Nonetheless, if the ramifications of anthroparchy are not fully problematised, there is a risk of collapsing into anthropocentric ecologies that still put human interests first, drastically limiting the scope of our ethical responses and epistemic frameworks.[26] In order, then, to overcome the obstacles preventing us from extending solidarity across species and liberation movements, it is necessary to welcome an expansive and omni-comprehensive idea of multispecies kinship that accepts the uncomfortable challenge of ceasing to benefit uncritically from anthropocentric power structures.[27] It is also necessary to reject explicitly and actively the use and exploitation of other animals, which entails 'choosing to negotiate a world stacked against the nonhuman without giving up on the possibility of systemic cultural and institutional transformation' (Quinn, 2021: 267).[28]

Besides revitalising our imaginary, the narratives presented in these works encourage us to notice the cultural and social constructs that currently prevent futures of multispecies flourishing not dependent on ideals of selective justice. At the same time, what the selected works all have in common is that they promote a sense of interconnectedness that pushes the readers to rethink the ontological split between Human and Nature that underpins apocalyptic discourse. The adoption of such a stance lays the foundations for more radical re-imaginings of systems as well as visions of multispecies kinship based on notions of justice that transcend logics of commodification, instrumentalisation, exploitation, and extraction, rather than rely on unquestioned systems of violence and domination. 'Before the realisation comes the dream, the conception,' as Vandana Singh reminds us (2008). Ultimately, end-of-the world stories will unlikely be as useful as restorative narratives, which is exactly the subversive epistemological project promoted in these literary works. To conclude with the visionary, yet pragmatic, words of Priya Sarukkai Chabria,

> Listening to others – other species, other ways of imagination, other cultures, is just the beginning in a long process of loving, sharing, and emerging from the shackles with which we imprison ourselves, while reaching towards an abundance, a boundarylessness. A major shift in our thinking is required, whether or not we hear birdsong and insect songs, fish bubbles and eggs hatching, the sound of flowers opening. However, it will be much more amusing if we do listen, if we do accept the old Jaina thought – to a large degree – that all life wants to live, all living beings fear death, all life dreams.[29]

Notes

1 See, for instance, Vogelaar et al. (2018); McFarland (2021). Amitav Ghosh has notably associated the climate crisis to a cultural and imaginative failure in *The Great Derangement* (2016). On the ideology and rhetoric of denial in environmental discourse, see Almiron and Xifra (2020).

2 On the links between mental health and climate crisis, see Wray (2022); Frumkin (2022).

3 According to Rob Nixon's definition, slow violence is 'a violence that occurs gradually and out of sight, a violence of delayed destruction that is dispersed across time and space, an attritional violence that is typically not viewed as violence at all' (Nixon, 2011: 2). Among the examples, Nixon includes climate change, toxic drift, deforestation, and other 'slowly unfolding environmental catastrophes' (2011: 2). On weak sustainability, see Ergas (2021: 21–72).

4 Examples include Ring (2020); Sjunneson-Henry et al. (2018).
5 On the connections between colonialism and climate change, see White (2017: 153–162).
6 Controversies of hope are discussed in greater detail in Shewry (2015: 11–16). On the political economies of hope, see Cook and Cuervo (2019: 1102–1117); Out of the Woods Collective, 2020.
7 'Potency is the energy that transforms the possibilities into actualities' (Berardi, 2017: 1). The imaginative force of utopian econarratives is stressed also by Alberro (2021: 273–288).
8 Among the most notable solarpunk collections are Lodi-Ribeiro (2018), Ulibarri (2018), Grzyb and Sparks (2017), and Wagner and Wieland (2017).
9 For an overview of the aesthetics of solarpunk, see Johnson (2020). A text considered foundational for the movement is Flynn (2014).
10 On the lack of racial diversity, see Cameron (2019). Robyn Eckersley discusses at length the anthropocentric nature of social ecology (1992), an ideology universally upheld by solarpunk literature. See also Kopnina et al. (2018: 109–127).
11 Animal resistance has been thoroughly theorised by Hribal (2011) and, more recently, by Colling (2020).
12 Besides tackling the impacts of militarism on multispecies ecologies, this story also denounces institutional racism in the police 'No resource for sea-dwellers because they were floating trash, that was what the cops always claimed. Mama said that people with dark skin or poor people or just those passing through used to be treated the same. Cops were cops, dedicated to serving the rich and powerful instead of protecting everyone' (McFarland, 2021: 37).
13 For instance, in 'The Exuberant Vitality of Hatchling Habitats' by D.A. Xiaolin Spires, intensive human fishing is called out for depleting the food sources of seagulls. However, no explicit reference is made to the fishes trapped in this practice of systemic killing (Spires, 2021: 84).
14 In a short story based in Singapore and focused on healing through gardening, the protagonist mentions in passing meeting with friends at McDonald's. While the inclusion of this multinational fast food chain may be an intentional move on the part of the author, for instance, to build character, it also signals the persistence in the imagined future of a capitalist, imperialist, and exploitative economy that clashes with solarpunk ideals as well as with the objective of the collection to offer optimistic visions of future urban spaces.
15 On the historical weaponisation of the natural/unnatural false dichotomy, see Bedford (2021: 1–16).
16 For a more thorough discussion of this topic, see the introduction to Chao et al. (2022: 1–21).
17 By siding with Val Plumwood's philosophical theory of eco-animalism, which supports the fundamental edibility of the human (Plumwood, 2012: 77–90), the editors dismiss veganism as a 'delusion' for disrupting the predator-prey dynamic (Rupprecht et al., 2021: 7). These claims not only invalidate the indeed relational nature of veganism, but also offer a constrained view of veganism as devoted exclusively to questions of food consumption, thus disregarding its political dimension and core commitment to dismantling systems of oppression and exploitation. A convincing and superbly articulated critique of Plumwood's claim regarding ontological veganism as imperialist can be found in Struthers Montford and Taylor (2020: 129–156).
18 My deepest appreciation to Priya Sarukkai Chabria for engaging in an enriching email correspondence with me on the themes explored in her short stories and, more broadly, in the edited collection in which it is included.

19 For a substantial overview of multispecies soundscapes in (anglophone) literature, see De Bruyn (2020) and Yong (2022).
20 On the political voice of more-than-human animals, see Meijer (2019).
21 This topic is discussed in greater depth by the writer in Sengupta (2021).
22 Conceptually similar is Roanne van Voorst's speculative experiment in 'futures anthropology' on food practices free from animal exploitation (2021).
23 See www.tamikobeyer.com/books. My sincerest gratitude to Tamiko Beyer for accepting my invitation to have a conversation on her collection of poems, radical imagination, and collective liberation.
24 An emblematic manifestation of prison abolition can be found in the following lines: 'Black women and femmes chanted incantations of their ancestors and the cages of Rikers blew open. San Quentin dissolved. Framingham shattered. All the prisons evaporated' (Beyer, 2021: 56).
25 Examples of this violent history in the United States are, for instance, lynching trees and liberty trees.
26 Humanity must not be intended as a monolithic agent of violence; rather, its definition relies on Sylvia Wynter's genealogy of humanity (2003) and on Syl Ko's model of the 'human' (2017).
27 Several recent publications incorporate a vegan liberationist praxis into their methodological framework. See for instance Quinn and Westwood (2018), Giraud (2021), and Hodge et al. (2021). For a brief definition of total liberation, see Pellow (2019: 295–304). For a call to action directed to those concerned with urban nature and biodiversity to adopt more critical and repoliticised understandings of 'nature' and multispecies relations, see Arcari et al. (2021: 940–965).
28 On the concept of radical hope, see Strazds (2019).
29 Shared with the author's consent from our email correspondence.

Bibliography

Abram, David (1997), *The Spell of the Sensuous: Perception and Language in a More-Than-Human World* (New York: Random House Vintage).

Alberro, Heather (2021), 'Interspecies', in John Parham (ed.), *The Cambridge Companion to Literature and the Anthropocene* (Cambridge: Cambridge University Press), 273–288.

Almiron, Núria and Jordi Xifra (eds.) (2020), *Climate Change Denial and Public Relations: Strategic Communication and Interest Groups in Climate Inaction* (London: Routledge).

Arcari, Paula, Fiona Probyn-Rapsey and Haley Singer (2021), 'Where Species Don't Meet: Invisibilized Animals, Urban Nature and City Limits', *Environment and Planning E: Nature and Space*, 4:3, 940–965, https://doi.org/10.1177/2514848620939870

Bedford, Anna (2021), 'Transecology – (Re)Claiming the Natural, Belonging, Intimacy, and Impurity', in Douglas A. Vakoch (ed.), *Transecology: Transgender Perspectives on Environment and Nature* (London: Routledge), 1–16.

Berardi, Franco (2017), *Futurability: The Age of Impotence and the Horizon of Possibility* (New York: Verso Books).

Beyer, Tamiko (2010), 'Notes Towards a Queer::Eco::Poetics', *Doveglion Lit*, 29 November, [Online], [Accessed 28 September 2022], available from: doveglionlit.wordpress.com/2010/11/29

Beyer, Tamiko (2021), *Last Days* (Farmington, MA: Alice James Books).

brown, adrienne maree (2017), *Emergent Strategy: Shaping Change, Changing Worlds* (Chico, CA: AK Press).

Cameron, Rob (2019), 'In Search of Afro-Solarpunk, Part 2: Social Justice Is Survival Technology', *Tor.Com*, 30 October, [Online], [Accessed 28 September 2022], available from: tor.com/2019/10/30

Chabria, Priya Sarukkai (2021), 'Listen: A Memoir', in Christoph Rupprecht, Deborah Cleland, Norie Tamura, Rajat Chaudhuri and Sarena Ulibarri (eds.), *Multispecies Cities: Solarpunk Urban Futures* (Albuquerque: World Weaver Press), 11–21.

Chagani, Fayaz (2016), 'Can the Postcolonial Animal Speak?', *Society & Animals*, 24:6, 619–637, https://doi.org/10.1163/15685306-12341421

Chao, Sophie, Karin Bolender and Eben Kirkey (eds.) (2022), *The Promise of Multispecies Justice* (Durham and London: Duke University Press).

Chehelnabi, Amin (2021), 'Wandjina', in Christoph Rupprecht, Deborah Cleland, Norie Tamura, Rajat Chaudhuri and Sarena Ulibarri (eds.), *Multispecies Cities: Solarpunk Urban Futures* (Albuquerque: World Weaver Press).

Chin-Tanner, Wendy (2013), 'A Conversation with Tamiko Beyer', *Lantern Review Blog: Asian American Poetry Unbound*, 17 July, [Online], [Accessed 28 September 2022], available from: lanternreview.com/blog/2013/07/17

Chng, Joyce (2021), 'It Is the Year 2115', in Christoph Rupprecht, Deborah Cleland, Norie Tamura, Rajat Chaudhuri and Sarena Ulibarri (eds.), *Multispecies Cities: Solarpunk Urban Futures* (Albuquerque: World Weaver Press).

Colling, Sarat (2020), *Animal Resistance in the Global Capitalist Era* (East Lansing, MI: Michigan State University Press).

Cook, Julia and Herman Cuervo (2019), 'Agency, Futurity and Representation: Conceptualizing Hope in Recent Sociological Work', *The Sociological Review*, 67:5, 1102–1117, https://doi.org/10.1177/0038026119859177

De Bruyn, Ben (2020), *The Novel and the Multispecies Soundscape* (Basingstoke: Palgrave Macmillan).

de Haan, Olivier (2019), *How Crucial Is Callenbach's Deep Ecological Utopia for the 21st Century Environmental Crisis? An Ecocritical Reading of Callenbach's Ecotopia (1975)*, unpublished Masters Dissertation, Faculté de philosophie, arts et lettres, Université catholique de Louvain (Prom.: Bragard, Véronique), [Online], [Accessed 3 April 2023], available from: hdl.handle.net/2078.1/thesis:18154

Eckersley, Robyn (1992), *Environmentalism and Political Theory: Toward an Ecocentric Approach* (Albany, NY: SUNY Press).

Ergas, Christina (2021), *Surviving Collapse: Building Community Toward Radical Sustainability* (Oxford: Oxford University Press).

Felski, Rita (2008), *The Uses of Literature* (Oxford: Blackwell Publishing).

Flynn, Adam (2014), 'Solarpunk: Notes Toward a Manifesto', *Hieroglyph*, [Online], [Accessed 28 September 2022], available from: hieroglyph.asu.edu/2014/09

Frumkin, Howard (2022), 'Hope, Health, and the Climate Crisis', *The Journal of Climate Change and Health*, 5, 1–7, https://doi.org/10.1016/j.joclim.2022.100115

Ghosh, Amitav (2016), *The Great Derangement: Climate Change and the Unthinkable* (Chicago: Chicago University Press).

Giraud, Eva Haifa (2021), *Veganism: Politics, Practice, and Theory* (London: Bloomsbury).

Grzyb, Liz and Cat Sparks (eds.) (2017), *Ecopunk! Speculative Tales of Radical Futures* (Greenwood, Western Australia: Ticonderoga Publications).

Gumbs, Alexis Pauline (2020), *Undrowned: Black Feminist Lessons from Marine Mammals* (Edinburgh: AK Press).

Harvey, David (2000), *Spaces of Hope* (Berkeley: University of California Press).

Hodge, Paul, Andrew McGregor, Simon Springer, Ophélie Véron and Richard J. White (eds.) (2021), *Vegan Geographies: Spaces Beyond Violence, Ethics Beyond Speciesism* (New York: Lantern Media).

Hribal, Jason (2011), *Fear of the Animal Planet: The Hidden History of Animal Resistance* (Stirling: AK Press).
Iovino, Serenella (2015), 'The Living Diffractions of Matter and Text: Narrative Agency, Strategic Anthropomorphism, and How Interpretation Works', *Anglia*, 133:1, 69–86, https://doi.org/10.1515/ang-2015-0006
Johnson, Isaijah (2020), 'Solarpunk and the Pedagogical Value of Utopia', *Journal of Sustainable Education*, 23, [Online], [Accessed 28 September 2022], available from: susted.com/wordpress/content/solarpunk
Kelsey, Elin (ed.) (2014), 'Beyond Doom and Gloom: An Exploration through Letters', *RCC Perspectives*, 6.
Kimmerer, Robin Wall (2013), *Braiding Sweetgrass: Indigenous Wisdom, Scientific Knowledge and the Teachings of Plants* (Minneapolis: Milkweed Editions).
Ko, Aph and Syl Ko (2017), *Aphro-ism: Essays on Pop Culture, Feminism, and Black Veganism from Two Sisters* (New York: Lantern Media).
Kopnina, Helen, Haydn Washington, Bron Taylor and John J. Piccolo (2018), 'Anthropocentrism: More than Just a Misunderstood Problem', *Journal of Agricultural and Environmental Ethics*, 31:1, 109–127, https://doi.org/10.1007/s10806-018-9711-1
Lodi-Ribeiro, Gerson (ed.) (2018), *SolarPunk: Ecological and Fantastical Stories in a Sustainable World*, trans. Fábio Fernandes (Albuquerque, NM: World Weaver Press).
Malm, Andreas (2021), 'Warming', in *The Cambridge Companion to Literature and the Anthropocene* (Cambridge: Cambridge University Press), 242–257.
McFarland, Meyari (2021), 'Old Man's Sea', in Christoph Rupprecht, Deborah Cleland, Norie Tamura, Rajat Chaudhuri and Sarena Ulibarri (eds.), *Multispecies Cities: Solarpunk Urban Futures* (Albuquerque: World Weaver Press), 36–47.
McFarland, Sarah E. (2021), *Ecocollapse Fiction and Cultures of Human Extinction* (London: Bloomsbury).
Medina, José (2013), *The Epistemology of Resistance: Gender and Racial Oppression, Epistemic Injustice, and Resistant Imaginations* (Oxford: Oxford University Press).
Meijer, Eva (2019), *When Animals Speak: Toward an Interspecies Democracy* (New York: New York University Press).
Nixon, Rob (2011), *Slow Violence and the Environmentalism of the Poor* (Cambridge, MA: Harvard University Press).
Pellow, David Naguib (2019), 'Total Liberation', in Matthew Schnider-Mayerson and Brent Ryan Bellamy (eds.), *An Ecotopian Lexicon* (Minneapolis: University of Minnesota Press), 295–304.
Plumwood, Val (2012), *The Eye of the Crocodile* (Canberra: ANU Press).
Quinn, Emelia (2021), 'Vegan Studies and Queer Theory', in Laura Wright (ed.), *The Routledge Handbook of Vegan Studies* (London: Routledge), 261–271.
Quinn, Emelia and Benjamin Westwood (eds.) (2018), *Thinking Veganism in Literature and Culture: Towards a Vegan Theory* (Basingstoke: Palgrave Macmillan).
Räpple, Eva Maria (2019), *The Environmental Crisis and Art: Thoughtlessness, Responsibility, and Imagination* (New York: Lexington Books).
Ring, Dave (ed.) (2020), *Glitter + Ashes: Queer Tales of a World That Wouldn't Die* (Washington, DC: Neon Hemlock).
Rupprecht, Christoph, Deborah Cleland, Norie Tamura, Rajat Chaudhuri and Sarena Ulibarri (eds.) (2021), *Multispecies Cities: Solarpunk Urban Futures* (Albuquerque: World Weaver Press).
Sengupta, Sayantani (2021), 'A New Anthology of Speculative Fiction Counters Dystopias with Hopeful Futures', *Scroll.in*, 5, [Online], [Accessed 28 September 2022], available from: scroll.in/article/1006815

Shewry, Teresa (2015), *Hope at Sea: Possible Ecologies in Oceanic Literature* (Minneapolis: University of Minnesota Press).
Shotwell, Alexis (2016), *Against Purity: Living Ethically in Compromised Times* (Minneapolis and London: University of Minnesota Press).
Singh, Vandana (2008), 'The Creatures We Don't See: Thoughts on the Animal Other', *Jeff Vandermeer*, 10 October, [Online], [Accessed 28 September 2022], available from: jeffvandermeer.com/2008/10/10
Singh, Vandana (2014), 'Alternative Visions: Some Musings on Diversity in SF', *Antariksh Yatra: Journeys in Space, Time, and Imagination*, 27 May, [Online], [Accessed 28 September 2022], available from: vandanasingh.wordpress.com/2014/05/27
Singh, Vandana (2021), 'A Speculative Manifesto', *Antariksh Yatra: Journeys of Space, Time and the Imagination*, 20 October, [Online], [Accessed 28 September 2022], available from: vandanasingh.wordpress.com/2021/10/20
Sjunneson Henry, Elsa, Dominik Parisien, Nicolette Barischoff, S. Qiouyi Lu and Judith Torr (eds.) (2018), 'Disabled People Destroy Science Fiction', *Uncanny Magazine*, 24, [Online], [Accessed 28 September 2022], available from: uncannymagazine.com/issues
Spires, D.A. Xiaolin (2021), 'The Exuberant Vitality of Hatchling Habitats', in Christoph Rupprecht, Deborah Cleland, Norie Tamura, Rajat Chaudhuri and Sarena Ulibarri (eds.), *Multispecies Cities: Solarpunk Urban Futures* (Albuquerque: World Weaver Press).
Strazds, Laila (2019), 'Radical Hope: Transforming Sustainability', *The Journal of Sustainability Education*, 21, [Online], [Accessed 28 September 2022], available from: susted.com/wordpress/wp-content
Struthers Montford, Kelly and Chloë Taylor (2020), 'Beyond Edibility: Towards a Nonspeciesist, Decolonial Food Ontology', in Kelly Montford Struthers and Chloë Taylor (eds.), *Colonialism and Animality. Anti-Colonial Perspectives in Critical Animal Studies* (London: Routledge), 129–156.
Tanaka, Natsumi (2021), 'A Life with Cibi', in Christoph Rupprecht, Deborah Cleland, Norie Tamura, Rajat Chaudhuri and Sarena Ulibarri (eds.), *Multispecies Cities: Solarpunk Urban Futures* (Albuquerque: World Weaver Press).
Ulibarri, Sarena (ed.) (2018), *Glass and Gardens: Solarpunk Summers* (Albuquerque, NM: World Weaver Press).
van Voorst, Roanne (2021), *Once Upon a Time We Ate Animals: The Future of Food* (New York: Harper Collins).
Vogelaar, Alison, Brack W. Hale and Alexandra Peat (eds.) (2018), *The Discourses of Environmental Collapse: Imagining the End* (London: Routledge).
Wagner, Phoebe and Brontë Wieland (eds.) (2017), *Sunvault: Stories of Solarpunk and Eco-Speculation* (Nashville: Upper Rubber Boot Books).
White, Kyle (2017), 'Indigenous Climate Change Studies: Indigenizing Futures, Decolonizing the Anthropocene', *English Language Notes*, 55:1, 153–162.
Wray, Britt (2022), *Generation Dread: Finding Purpose in an Age of Climate Crisis* (Toronto: Knopf Canada).
Wynter, Sylvia (2003), 'Unsettling the Coloniality of Being/Power/Truth/Freedom: Towards the Human, After Man, Its Overrepresentation – An Argument', *CR: The New Centennial Review*, 3:3, 257–337, https://doi.org/10.1353/ncr.2004.0015
Yong, Ed (2022), *An Immense World: How Animal Senses Reveal the Hidden Realms Around Us* (New York: Penguin Random House).

9 Nature and masculinity in Barbara Kingsolver's *Prodigal Summer*

Writing an ethical shift in environmental perception

Chloé Bour--Lang

As we are increasingly facing unprecedented crises, whether social or sanitary, the relationship between human beings and their environment is propelled to the centre of ethical and philosophical discussions. Literature has undoubtedly been at the forefront of this exploration, from fantasies of wilderness to post-apocalyptic scenarios featuring desolate landscapes. Paralleling the rising concerns regarding socio-political issues relative to climate change, literature has seen an emergence of new genres which address ecological matters. Yet, when diving into the world of environmental fiction, one is readily met with dark speculative narratives and other dystopian stories. Catastrophist discourse can therefore emerge as a dominant mode of reacting to the impending consequences of climate change, greatly contributing to an atmosphere seeped in distress and moroseness, and, eventually, feeding into a general sentiment of hopelessness. Perhaps even worse, as Elizabeth Tavella articulately posits in Chapter 8 of this volume, the paralysing effects of this 'failure of the imagination' can potentially fuel an ideology of denial and of neoliberal fatalism.

However, there has recently been a growing interest in literary works that focus on the revitalising potential of the ecological crisis, offering counter-discourses that shift the readers' attention to the proposal of revolutionary paradigm shifts instead. Amongst the works that move away from the sensationalist potential of major ecocatastrophes, contemporary American author Barbara Kingsolver chooses to tell the stories of everyday people and their attempts at negotiating the consequences of centuries of environmental neglect and destruction. In her collection of essays *Small Wonder* (2002), she thus declares: 'The feeling I dread most is fear but not despair,' before offering insight into her ability to find reasons to keep going in spite of catastrophic and depressing events (2002: 15). Her works of fiction, from short stories to novels, all reflect this to a certain extent, showcasing characters who overcome personal as well as collective traumas. Mingling foci on personal loss with more generalised and systemic concerns, Kingsolver consistently calls attention to social and environmental issues, her

DOI: 10.4324/9781032726953-10

distinctive style infused with her activist roots and her academic training in biology.[1] In her fictional stories, environmental science thus becomes a filter through which Kingsolver's characters (and, by extension, her readers) may experience the world and may learn to shape a vision that will, according to her, correct the ideological shortcomings which stem from a lack of proper scientific knowledge (2002: 212–213). Thus, in her 2000 novel *Prodigal Summer*, the protagonists, underscored by reconnection and revitalisation, look for means to move away from the anxiety of a devastated world towards new possibilities. Interweaving three narratives set in contemporary southern Appalachia, the novel follows Deanna Wolfe, Lusa Landowski, and Garnett Walker in the rural setting of fictional Zebulon Valley as their personal issues come to intersect with those of their surroundings. The characters' stories hence centre on their relation to each other, but also to the natural environment they inhabit and with which they interact. Their personal struggles – tied to grief and isolation – are entangled with environmental issues, and as they find solutions for the latter, they begin walking the path of their own improvement and healing.

Although Kingsolver's fiction proposes an encouraging and heartening addition to ecofiction, with its emphasis on the possibility for ideological changes, and consists of a notable contribution to the contemporary American environmental imagination and its political framing, her work remains relatively unknown to the European public.[2] With this chapter, I thus hope to contribute to the discussion regarding the relevance of her work in a world in need of narratives of resilience, but also to participate in showing the potential of such stories to a wider audience. More precisely, my intent is to show how Kingsolver demonstrates a potential for environmental imagination which resists defeatist narrativisation of climate change. I therefore believe that exploring the reinvigorating potential of *Prodigal Summer* (2000), offers the possibility of diving into a different and vivifying theoretical frame, namely the rapidly growing field of ecofeminism, which philosopher Karen Warren has defined as follows:

> An ecofeminist ethic is both a critique of male domination of both women and nature and an attempt to frame an ethic free of male-gender bias about women and nature. It not only recognizes the multiple voices of women, located differently by race, class, age, [and] ethnic considerations, it centralizes those voices. Ecofeminism builds on the multiple perspectives of those whose perspectives are typically omitted or undervalued in dominant discourses, for example Chipko women, in developing a global perspective on the role of male domination in the exploitation of women and nature. An ecofeminist perspective is thereby . . . structurally pluralistic, inclusivist, and contextualist, emphasizing through concrete examples the crucial role context plays in understanding sexist and naturist practice.
>
> (1988: 151)

Originating simultaneously from different activist circles in several countries around the 1970s, ecofeminist philosophy admits the interconnectedness of climate change with many contemporary issues, either as a root cause or as an aggravating agent, a position that is central to Kingsolver's writing and that establishes her as an author recognised as ecofeminist. If this framing has previously been explored in relation to the author's production, most analyses have focused on female characters[3] and only marginally touched upon her treatment of masculinity.[4] In this chapter, I therefore aim to contribute to the analysis of Kingsolver's perspective on gender by centring on her male characters and on their masculinity as informing their relationship with nature. More specifically, my framework will be informed by the subcategory of vegetarian-ecofeminism, which draws attention to animals and animality, and, in this regard, will mostly be based on works by Carol J. Adams, Josephine Donovan, or Greta Gaard. These ecofeminists wish to 'remedy a perceived problem in feminist theory, animal advocacy, and environmentalism, namely, a lack of attention to the intersecting structures of power that reinforce the "othering" of women and animals, and contribute to the increasing destruction of the environment' (Adams and Gruen, 2014: 1). In this regard, vegetarian-ecofeminism allows for cross-examination of the oppression and domination that women, nature, and non-human animals alike sustain while analysing these shared experiences as mutually reinforcing. Using Kingsolver's representation of the relationship between human and non-human animals as a doorway into her ecofeminist approach to ethics, this chapter explores the author's portrayal of hunting practices as symptomatic of a masculinist destructive ideology. My discussion will then lead to the analysis of the rewriting of the figure of the huntress in *Prodigal Summer*, with a specific focus on the transformative potential of tracking methods as they become a way for practices consistent with the ethics of care.

Through the undertones of realism in *Prodigal Summer*, the author acknowledges the anguish that comes with the Anthropocene, using a grim, death-saturated atmosphere as a backdrop and a starting point to the three intertwined stories. In 'Predators,' Deanna Wolfe, a lone forest ranger, grapples with her own place in the world of humans as she tries to defend the land under her protection from invasive and irresponsible hunters. 'Moth Love' follows newly widowed Lusa Landowski and her attempt to save her late husband's failing farm while she tries to find her place as an outsider in a seemingly hostile family. The third plot, 'Old Chestnuts,' allows old Garnett Walker to begin to overcome his grief over his wife's passing and to escape his dated and destructive agricultural beliefs and resulting practices. As the title of *Prodigal Summer* echoes that of *Silent Spring* (1962), it shapes the reader's encounter with the novel, for it establishes it as a response or a follow-up to Rachel Carson's influential book.[5] The threatening atmosphere of Kingsolver's novel is undoubtedly indebted to *Silent Spring*, as it describes a silenced and damaged nature which no longer teems with life because of the deleterious effects of pesticides. While drawing on Carson's criticism of anthropogenic environmental damage, Kingsolver expands her reflection by offering a novel that questions and denounces the cultural biases that enabled such a devastating outcome. More specifically, the author paints a fictional backdrop that is strewn with allusions to Carson and her essay, going as far as naming Deanna's step-sister Rachel. Kingsolver's expansion on Carson's work is also perceptible through her creation of a literary text that incorporates her ideas into the fictional lives of ordinary Americans, which are still directly impacted by indiscriminate use of pesticides some forty years after the publication of *Silent Spring*. Therefore, not only does Kingsolver connect the inhabitants of Zebulon Valley to the fate of the natural world, but she also draws a link between social and ideological issues, which she presents as analogous to the ones detailed in *Silent Spring*. By doing so, she further absorbs Carson's essay into her own wider ethical and philosophical framing in which ecological and social issues are interconnected.

By focusing on this framing for this chapter, my intention is to foreground the processes of domination at play in *Prodigal Summer*, starting with the analysis of the destruction of nature as provoked by masculine behaviours. Here, I would like to draw attention to the distinction between masculinity as a socialised way of behaving in and interacting with the world, and being biologically male, since my critical framework owes to materialist feminism in admitting that gender is a social construct.[6] As a consequence, I recognise that women can internalise a masculinist ideology and that they might engage in behaviours considered to be masculine.[7] However, in the case of *Prodigal Summer*, I would argue that masculine traits are mostly, if not exclusively, embodied by cisgender male characters, and, as a consequence, this chapter will concentrate on masculinity as embodied by said characters.

Observing the masculine murderous dominating mindset

Without taking away from their complexity and verisimilitude, the characters of *Prodigal Summer* can be regarded as an expression of Kingsolver's criticism of socially constructed gendered practices, chiefly in regard to relation to nature. The men of the novel almost systematically display stereotypically masculine behaviours which are oppressive to both women and the natural elements that surround them. In her 1985 inquiry into the roots of patriarchy, *Beyond Power: On Women, Men and Morals*, Marilyn French affirms that 'Our culture, which worships above all else the power to kill, has reached the point of wishing to annihilate all that is "feminine" in our world' (1985: 523). Similarly, Ellyn Kaschak states that the 'construction of a separate and individual self directly reflects the masculinist predilection to make invisible the interconnections between people and all living things' (1992: 150). By representing the oppression of women and nature as intertwined, and therefore examining how male characters interact with both, *Prodigal Summer* becomes a notable literary exploration of the gendered politics of climate change.

In the novel, it is firstly through the practice of chemically aided agriculture that the looming sense of ecological threat is brought up.[8] Following a rather traditional subdivision of tasks and a gendered separation of domestic and public spaces, men are expected to be in charge of agricultural responsibilities in Zebulon County. Their methods follow those of conventional agriculture, based on the heavy use of agrochemicals. One of the protagonists, Garnett Walker, represents the approach to cultivation that the novel deems senseless and noxious. The feud between him and his neighbour, Nannie Rawley, is sparked precisely by Garnett Walker's chemical inclinations. While the old man persists in spraying an ever-increasing amount of pesticides on his crops, among which are the infamous Sevin and Roundup, Mrs. Rawley actively engages in protesting against the use of chemicals while growing her apples organically. Garnett Walker's agricultural vision is shaped not only by the underlying social expectations of his Appalachian mode of living, but also by more intricate and subtle factors that result from his American culture and his newfound Christian fundamentalism.[9] Indeed, Kingsolver created Garnett Walker as a born-again Christian who turned to religion as a means of dealing with the loss of his beloved wife, Ellen, making his inflexible faith a defining feature of his character. To explain and justify his gardening practices, he returns repeatedly to the Bible, which he quotes at length. As a result, his superficial and restrictive interpretation of religious texts means that the consequences of using toxic products are inconsequential to him. He believes that human beings – and primarily men – have been created to be in charge of the natural world and its non-human animals, who must be viewed as 'gifts to [God's] favored children [to be used] for [their] own purposes, even if this occasionally causes this one or that one to go extinct after a while' (2000: 188).

In Garnett Walker's understanding, killing animals up to extinction and potentially creating ecocatastrophe do not pose any ethical problems. According to his religious beliefs, which prevent him from realising that he is partly guilty of the ecologically bound failure of his own plantations, he is doing God's work. A few chapters of *Prodigal Summer* are dedicated to the epistolary exchanges between Mr. Walker and Mrs. Rawley, exposing the man's apparent misunderstanding of his own religion, and his dangerously biased interpretation of the idea of stewardship. In this regard, it appears that Garnett Walker's lack of critical outlook and his word-for-word application of the Bible are on the verge of biblical literalism.[10] The exchange of letters between the two characters thus creates a detailed debate, allowing Kingsolver to dismantle what she presents as a damaging discourse. Through Nannie Rawley's voice, she proposes that her compatriots distance themselves from a non-critical reading of religious texts, and that they use contemporary scientific knowledge to shed light on God's word.

The absurdity and arrogance of Garnett Walker's oppressive take on what his role should be are additionally illustrated by his vain attempt at recreating the American chestnut, an iconic American tree that was decimated by an exogenous form of blight in the first half of the twentieth century. His experimental efforts aimed at recreating life and at helping this important species subsist are laudable, but are constantly undermined by his restrictive ideological groundings. Garnett Walker's repeated failures thus foreshadow the overall sterility of his own literal and metaphorical gardens; the narration indicates that he is to blame for his own shortcomings. The vanity of his enterprise thus further reveals Kingsolver's didactic agenda: an ideology that relies on distrusting and controlling nature may damage it and make it unviable altogether. In other words, the author offers a targeted criticism of one character's vain attempts at manipulating life to expose the absurdity of trying to save the environment without deeper ethical changes. Creating life with agricultural and conceptual tools that are lethal is delusory. Furthermore, the chestnut trees and their blighted fate connote a bad omen for Garnett who, once again, is oblivious to the ongoing irony of his situation. Indeed, in the very own history of the destruction of the American chestnut lies a truth that he continues to overlook: ecotopes have a fragile balance that can easily be disturbed by exogenous components. By not acknowledging this web of interconnectedness, Garnett Walker dooms himself to failure. In this regard, Garnett's approach is reminiscent of restorationism, which, in the context of ecology and from an ecofeminist perspective, raises a few issues. Eric Katz, in words that perfectly summarise the ethical concerns pertaining to Garnett's practices, explains that 'ecological restoration is a form of domination, another instance of humanity's attempt to control the natural world' (1996: 224). Focusing on trying to recreate an ersatz chestnut tree, the old man completely embodies the 'self-deception' that Katz sees in ecological restoration, for it 'is an expression of a misguided faith in human omnipotence, the human power to control and shape the natural world' (1996: 224).

The same is true for his social life, as he actively participates in the exclusion he suffers from. Indeed, because of his inflexible principles, Garnett is estranged from his community while he surely (yet unconsciously) undermines his neighbour's attempt at befriending him. Despite his unpleasant moods and ecological disrespect, Nannie Rawley tries to foster a positive relationship with the old man. Yet, Garnett is entrenched in his own beliefs, which fester in the tension stemming from their disagreement on their rightful place in the natural world. Because of Nannie's organic methods, his neighbour suspects she is guilty of witchcraft and believes she should be put back in her place. Garnett's accusation of sorcery not only is reminiscent of a truly dated religious perception of women, but also participates in foregrounding the interconnectedness of his views on women and nature. Emphasising this interconnectedness, Carolyn Merchant's words could adequately sum up Garnett's convictions: 'The witch, symbol of the violence of nature, raised storms, caused illness, destroyed crops, obstructed generation, and killed infants. Disorderly woman, like chaotic nature, needed to be controlled' (1980: 127). To a certain extent, even Nannie's presence in her own garden is perceived as sacrilegious and threatening, as she evokes both a witch and Eve, classical examples of sin. Garnett therefore believes he should control Nannie for her own sake and that of his own chestnut breeding project. Both his conception of nature and that of women thus inform his perception of his surroundings, directly impacting his daily life and degrading its quality.

Kingsolver's criticism of patriarchal ideology and its grim consequences is further embodied by the character of Eddie Bondo, a young hunter. The reader first encounters him as he steps uninvited onto the protected lands of forest ranger Deanna Wolfe, in circumstances that anticipate his consistently ambiguous behaviour. Eddie Bondo's intentions are quickly uncovered by the ranger: he is attracted by the prospect of killing animals in Appalachia, on a land where it is forbidden to do so. He indeed bears multiple characteristics of a predator, mirroring the animals that he seeks to hunt and that Deanna wishes to protect. The tension between the intentions of the two characters is redolent of that which exists between Garnett and Nannie, in a similar gendered pairing where the man pursues his individualistic objectives while the woman embodies a form of knowledge or natural wisdom which may be regarded as more progressive thinking. Bondo's hunting practices as well as his way of behaving towards his lover both fall under the predatory lens he appears to look through, and which is amplified by the animalisation of his character. Indeed, hunting is the very reason for his presence in Appalachia, but he also finds interest in engaging in another form of game, which is a hunt for a sexual partner.

The very first paragraphs of the novel chillingly foreshadow Bondo's sudden arrival by prompting us to observe Deanna through the eyes of a stalker. Watching through the hunter's menacing gaze, the readers discover Deanna in her element, at home and free in the wilderness of the forest. Yet a man trespasses the frontier of this protected space and lurks in the shadows 'with a gun . . . hiding inside a copse of leafy beech trees,' 'watching her' (2000: 3). Upon Bondo's intrusion, it is suggested that only another animal – a predator at that, as the title of the storyline suggests – could have crept up on her unnoticed. If Deanna is assuredly a lover of animals and chiefly of predators, her control over her own space and over herself is repeatedly challenged by the young man's sense of entitlement and lack of self-consciousness. There, Deanna's identity is being renegotiated: if her first name positions her character as a rewriting of the Roman goddess of hunting, Diana, and her last name, Wolfe, as a woman akin to wolves, the power they normally connote is being undermined. Since Bondo is presented as a predator who happens to particularly enjoy hunting other predators, Deanna becomes (his) prey, dispossessed of her caring authority over other animals. Additionally, following Marti Kheel's analysis of the game of the hunt (1996), since Deanna refuses to fully partake in the seduction process, and therefore resists Bondo, she is immediately positioned as hunted. Like a prey animal is at risk of losing their physical integrity, Deanna risks the loss of another kind of integrity. Giving herself to the enemy would compromise her own ethics by validating Bondo's undiscriminating predatory behaviours. For Deanna, this would consist in a betrayal of her inmost beliefs; for her, 'to kill a natural predator is a sin' (2000: 181). Notwithstanding any unease experienced by the reader at the beginning of Bondo and Deanna's relationship, the interaction between these characters eventually evolves to appear (mostly) consensual, as the young man gradually softens.[11] In many ways, predation and seduction are entangled in their plotline, as both he and Deanna negotiate their status. Following a patriarchal dynamic, the identities of the hunter and the hunted are gender coded.

Although the reader will learn to appreciate Bondo as his behaviour changes, his dangerous outlook on both the female protagonist and the natural world will never fully dissipate, despite Deanna's repeated attempts at educating him. In turn, he will appear armed, manifest sexist behaviours such as urinating in front of his new lover and disappearing after impregnating her. This attitude is consistently presented as acquired and as aggravated by Bondo's young age and lack of critical skills as well as ascribed to his attachment to his culture. Not only can Bondo's calling be observed as proceeding from an idealised vision of man, informed by the myth of 'man-the-hunter' (Haraway, 1991: 86), but it also emulates the archetypal American hunter, a founding figure of the New World. In his extensive study *Hunting and the American Imagination*, Daniel Justin Hermann explains:

> the hunter lived closest to wild nature; through his roaming and rambling, he comprehended the geography and topography of North America; he understood the wilderness and its wild fauna. The backwoods hunter's resemblance to the Indian made him seem indigenous, a man whose character and conduct were products of his wilderness environment, a man whose tie to the land was more profound than that of any civilized individual.
>
> (2001: 63)

Like Hermann's hunter, Bondo evolves within the wild nature emulated by Deanna's protected grounds as a pseudo-indigenous hunter, appearing as if he was part of this environment, comfortable, and free from doubt. Yet, by entering a territory that is not his, for it belongs to the animals who live there and is under the care of Deanna, Bondo, unaware of the unspoken rules that predated his arrival, violates the sanctity of the ranger's lands. By entering this space and doing as he pleases, he does not acknowledge the disruption he causes to others' lives. The hunting motif thus acts as a magnifier of his entitled and careless attitude, which Deanna denounces:

> They have those hunts all over . . . There's one going on right now in Arizona, the Predator Hunt Extreme, with a ten-thousand-dollar prize for whoever shoots the most. . . . Can you feature the damage those men will do to the state of Arizona in just one weekend, the plague of mice and grasshoppers they'll cause?
>
> (2000: 324)

She thus finds Bondo at fault for not thinking in terms of ecosystems, and consequently for overlooking the ramifications of his actions. Because Bondo hunts not out of need but instead for pleasure, Deanna disapproves of his practice. Indeed, the young man positions himself above other predators, choosing them as his victims and therefore breaking a form of natural balance put forth continually by the ranger; in *Prodigal Summer*, predators appear to be the most valuable animals of all. For Bondo, taking the place of a wolf, a coyote, or a bobcat would anchor his position as dominant and as the animal at the top of the pyramid. In a way, the hunter's attempt at taking the place of these predatory animals is perceived by Deanna as an affront to the sacred biological order and as a sign of hubris.

Hence, just like Garnett Walker, Eddie Bondo is a product of a culture of domination over nature, in which men decide the fate of the rest of creation and, taking upon themselves a Godlike role of creation and destruction, are free to use it as they please. Furthermore, both male characters display certain behaviours that perpetuate social violence against women, as they prevent them from doing as they please in their own spaces. They invade their garden and guarded territories, either with carcinogenic and lethal chemicals or physically, although these spaces are clearly delimited and marked as either organic or protected. In both cases, the intrusions come with the threat of death; while in a relationship with Bondo, Deanna runs the risk of having to mourn the coyote family she longs for. With Garnett, Nannie could lose her bread-winning orchard along with its entire ecosystem. The strategies of domination, if complex and sometimes subtle, are multiple and greatly impact the lives of these women who see different layers of their integrity threatened by the carelessness and arrogance of the men that come into their lives.

Both male characters also sustain a certain ideal of masculinity and American identity, embodying what Hermann perceives as the key archetypes of the gardener and the hunter (2001: 45). These archetypes are not only similar in that they both foreground a desire to control nature, but it can also be argued that they are both built in opposition to femininity and feminine values such as care and attention, for they specifically rely on domination and control. In a way, Kingsolver gives room to these archetypes to be tried and proved wrong, hence suggesting that we move past them and their principles which sustain socio-ecological crises. Indeed, according to Kingsolver as well as to ecofeminist critics, masculine behaviours such as Garnett's and Bondo's are responsible for devitalising the world, reducing its ecological abundance and, in turn, filling it with the ghosts of animals already extinct or about to become extinct. In *Prodigal Summer*, a few literal ghosts appear and disappear, giving the environment an uncanny quality as it becomes a memento of the impact of environmentally unethical behaviours. Coyotes thus become 'small golden ghosts of the vanished red wolf, returning' (2000: 16), luna moths are 'ethereal green ghost[s] of the upper forests' (2000: 36), while both Lusa and Deanna come to the realisation that they live 'among ghosts' (2000: 62, 78).

Kingsolver's novel can thus be read as a detailed criticism of contemporary American culture, of its patriarchal and ecocidal groundings, and as a political fiction that relies on scientific discourse to educate its readership. Through the voices of educated and oppressed women, male characters and readers alike are compelled to embark on a journey of cultural re-evaluation.

Making room for care and attention

The construction of the characters in *Prodigal Summer* thus puts forth Kingsolver's awareness of gender dynamics, specifically in rural areas. The plot itself benefits from this perspective, since her female protagonists are subjected to the masculine behaviours of the men in their lives, despite being headstrong and having undeniable agency. *Prodigal Summer* is also a story of women dealing with firmly established social expectations relative to their roles as women, and of their attempt at finding their place in spaces that are usually regarded as masculine. In the novel, Lusa, Nannie, and Deanna all engage in activities such as husbandry, agriculture, and forestry, which are normally carried out by men in Zebulon Valley, that is, in areas outside of the domestic.[12]

As the novel begins, the third-person narrator invites the readers to look at Deanna from this dominating masculine standpoint, briefly presenting her as an intruder in a territory which should be one of masculinity:

> If someone in this forest had been watching her – a man with a gun, for instance, hiding inside a copse of leafy beech trees – he would have noticed how quickly she moved up the path and how direly she scowled at the ground ahead of her feet. He would have judged her an angry woman on the trail of something hateful. He would have been wrong.
> (2000: 3)

Through this make-believe internal focalisation, the perspective of hegemonic masculinity is momentarily in charge of the representation of this setting. This (presumably) imaginary armed man is thereby in control of the environment around him, as he sees without being seen in return. In these few lines, the readers get to observe Deanna through the eyes of a man, an armed man at that, only to be invited to dismantle his imagined thought process. By co-opting the male gaze, the narrator is able to signal the rejection of this dominant position, labelling it as inaccurate.[13] The decision to make this someone a 'he' where a 'they' could have been expected, and to not keep the neutrality of 'someone,' indeed forces the reader to adopt this gendered point of view. Coupled with the conditional form 'if someone had/he would have,' the opening paragraph becomes an invitation to participate in a make-believe situation where readers are to actively engage in the creation of the story as they wear the filtered spectacles the narrator hands them.

By shifting the point of view, announcing that the male gaze is 'wrong,' the narrator asks that the readers depart from the hegemonic masculine vision, thereby positioning their voice as non-masculine or possibly feminine. Kingsolver's posture thus appears to be revealed through this reflection, as she repositions her voice as well as her readers' gaze within a feminist framework, where strongheaded women are not 'angry' and 'hateful' just because they do not answer to societal expectations of gender roles. The shift of perspective and of ideology is immediate. It also conditions the readers' perception of the upcoming stories and characters, in that it prepares them to pay attention to how men look at women as well as to how their actions impact their lives, further drawing attention to the ways in which masculine characters shape feminine experiences.

One of the prominent pervasive testimonies of the implications of masculinity for women is woven through the storylines of all the female characters in *Prodigal Summer*: Lusa, Deanna, and Nannie alike share the experience of loss as induced by behaviours resulting from a masculinist ideology. While Deanna and Nannie both fear for the lives of the beings forming the ecosystems they participate in, they also grieve the disappearance of a family member. It is indeed suggested that the death of Rachel, Nannie's daughter and Deanna's stepsister, is a product of environmental issues. It is through Garnett Walker's reductive and offensive thought process that the readers come to learn of Rachel's tragic story, as he reflects that he 'suspected a connection between that long-ago birth of a deformed child and [Nannie's] terror of chemicals' (2000: 138). The girl was born with a disability and, according to her mother, 'spent fifteen years' dying as a result (2000: 284). While there is no certainty that her health issues are linked to pesticides, Garnett's reflection infers that there is a connection to be made. As for the third female protagonist, Lusa Landowski, she is immediately met with the loss of her husband in an accident resulting from a dire economic situation, which in turn is a consequence of the devitalising tobacco culture. Between the lines, the narrative reveals that cultivating tobacco has brought death on at least three different levels: it degrades the land because it is a monocrop; it kills the individuals that consume it; and, as it is not a sustainable crop, it eventually threatens the lives of the farmers who rely on it to make a living. Lusa's husband died in an accident that would not have happened if it were not for 'tobacco's failure,' the narrator affirms through internal focalisation. For Lusa, it is 'the drop in price supports that had pressed him to take part-time work driving grain deliveries for Southern States' (2000: 50). This tragic event shapes Lusa's entire story, completely transforming her life. As for Garnett, he himself laboriously discovers the connection between his fondness for pesticides and his wife's death; however, it is only through the critical gaze of Nannie, who is a woman, that this discovery is made possible:

> 'That god-awful Sevin you've been spraying on your trees every blooming day of the week! . . . I have to breathe it. If I get lung cancer, it will be on your conscience.' Her hail of words stopped; their gazes briefly met and then fell to the grass around each other's feet. Ellen had died of lung cancer, metastasized to the brain . . . It dawned on him with a deeper dread that it might be possibly true.
>
> (2000: 275)

The observation, made through an ecofeminist framework, that most (if not all) of the deaths in *Prodigal Summer* result from actions that are modelled by patriarchal ideology calls for a delving into ecofeminism's own story. Initiated by women who precisely linked social and environmental concerns, the movement exposed that if, as a general rule, women were not the ones making the decisions leading to ecological disasters, they however were their primary victims. Retracing the history of ecofeminism,

Merchant recalls that 'the majority of activists in the grassroots movement against toxics are women . . . Many became involved when they experienced miscarriages or their children suffered birth defects or contracted leukemia or other forms of cancer' (Merchant, 1996: 12). The narrative perspective through which these losses are approached in *Prodigal Summer* thus systematically draws attention to masculine guilt and feminine suffering, but also to masculine unconscious resignation and feminine resilience. As we have seen, if Garnett is incapable of realising that overusing pesticides possibly played a role in his wife's death, it is quite the opposite for the women in *Prodigal Summer*, who are determined to find solutions. Where Deanna is actively participating in the preservation and revitalisation of wilderness, Nannie both takes care of her patch of land and educates Garnett, and Lusa is busy with finding solutions to save her late husband's farm. By doing so, Lusa takes on a man's role in the eyes of the locals.

Kingsolver's emphasis on women's potential and agency calls attention to a vivifying, feminine point of view as a response to the deleterious effects of patriarchy. The behaviour of some female characters directly subverts that of masculine characters, for they mirror them just enough to directly answer to them, while also showing that these men have the capacity to grow and learn for the better. Deanna, for instance, tracks just as much as Bondo does, but uses the same methods to radically different ends. The young man is indeed surprised to find Deanna without a weapon, which allows her to make him (as well as the readers) realise that killing is not the only goal of tracking (2000: 7). While Bondo's approach is presented as conservative and irresponsible, the woman's reclaiming of the practice opens the possibility for a caring approach. When Deanna observes the tracks, spies on animals, smells stumps, and tries to make herself invisible to her prey, she uses hunting techniques. Yet, it is not to make sure they do not flee before she has a chance to kill them, but rather that she may avoid disturbing them while making sure they and their ecosystem are healthy. Hence, according to French philosopher Baptiste Morizot, tracking becomes a way of paying attention to her environment, and thus caring for it. By observing the remains and tracks of animals, Deanna takes care of them, eventually using her tracking skills to monitor the environment's overall well-being: 'Si l'on désarticule . . . le pistage de l'acte de prédation, il devient désormais une certaine forme de l'attention' [If we dissociate . . . tracking from the act of predation, it becomes a form of attentiveness] (Morizot, 2018: 142, my translation).

By transforming hunting practices, Kingsolver recognises the importance of this motif in the construction of the American identity, following Hermann's analysis of the archetype of the hunter (2001). She thus acknowledges this fundamental heritage, while offering a subversive proposal that still honours it, branching away from its patriarchal roots without completely erasing this structural narrative. Rather, as she strips hunting of its violent character, she offers what can be read as an ecofeminist rewriting, creating an American narrative that does not lead to ecological disasters. Masculine behaviours are therefore re-evaluated and replaced by attributes that are historically and culturally perceived as more feminine, opening a space for a genuine paradigm shift. Kingsolver displaces the archetype of the ideal American – the American native, as Hermann calls him (2001: 7) – towards feminine figures who are caring, attentive, and nurturing. In this regard, contrary to Bondo and his fellow masculine and death-inclined trackers, Deanna can then be described as practising the ethics of care, as understood by Carol Gilligan:

> An ethics of care directs our attention to the need for responsiveness in relationships (paying attention, listening, responding) and to the costs of losing connection with oneself or with others. Its logic is inductive, contextual, psychological, rather than deductive or mathematical.
>
> (Gilligan, 2018)

Central to ecofeminist philosophy and practices, the ethics of care better enable the understanding of the evolution of human beings' relationship to nature, which Kingsolver's characters undergo as they learn to care for the environment and to listen to it. For instance, Nannie's organic cultivation is anchored in her will to care and protect, as Garnett notices that 'She was the sworn friend and protector of all creatures great and small, right down to the ticks, fleas, and corn maggots, evidently' (2000: 88). The women in *Prodigal Summer* are connected to each other by this indiscriminate consideration for animals, including some insects that are otherwise despised, antagonised, or depreciated. If Nannie appreciates insects as detested as ticks, fleas, and maggots, Lusa is an entomologist with a particular love for moths and their acute sense of smell. Both protagonists are united in the way they are characterised, with their attributes and practices being evocative of witches. Lusa is redheaded and of Jewish heritage, and Nannie is clearly perceived as a witch by Garnett; moreover, both are marginalised by their way of interacting with nature.[14] Lusa's interest in moths makes her particularly aware of the role of her own sense of smell in the way she experiences life, thus amplifying one of her means for attention to her surroundings and further reconnecting her to animality. Lusa's perceptive abilities, as well as her respect for moths, create a textual link between Lusa and Deanna, who also knows to be attentive and care for these animals. She repeatedly notices the presence of moths, actively observes them, and interprets both their biological and symbolic signals:

> She watched it crawl slowly up the screen on furry yellow legs. It suddenly twitched, opening its wings to reveal the dark eyes on its underwings meant to startle predators, and then it flew off to some safer hideout. Deanna felt the same impulse to bolt – to flee this risky mate gleaned from her forest.
>
> (2000: 31)

Most notably, Deanna particularly cares for one of the most hated non-human animals in the United States, the coyote, whom she tries to shelter from the harmful eyes of human predators.[15] In the novel, the love and care Deanna and other women give stem from a conjunction of their ability to adopt an animalised point of view and their significant scientific knowledge. Deanna's way of being in relation to the world and of inhabiting it is undeniably informed by her affection for coyotes, to the point that the relationship she builds with them becomes one of kinship. In some ways, Kingsolver portrays this woman as a coyote, constantly blurring and challenging the frontiers between human and non-human animals. She achieves this with the repeated slippage between Deanna as prey and the coyotes as prey, Deanna's ways of moving in her environment, and her very surname that contributes to the motif of being situated at the intersection between coyote and woman. Lusa's character is written in a comparable manner, with moths as kindred beings. She understands herself in terms of animal faculties, decoding her own instincts and feelings through the framework of animality. The only justification she finds of her love for her husband, for they appear so different from each other, thus stems from her animalised perception; if she is attracted to Cole, it is because of pheromones. Lusa's memories entangle moth love (the title of her storyline) and human love, creating a hybridised couple that flutter between human animal and moth. The contours of their bodies shift as the narrator takes on Lusa's voice:

> Cole, the man who buried his face in every fold of her skin to inhale her scent. He could only love sex more if he had antennae the shape of feathers, like a moth, for combing the air around her, and elaborately branched coremata he could evert from his abdomen for the purpose of calling back to her with his own scent.
>
> (2000: 40)

The women in *Prodigal Summer* understand themselves as an integral part of the animal realm.

Yet, these characters not only observe non-human animals including insects from a place of kinship and recognition, but also share a sound scientific training. Deanna and Lusa graduated in wildlife biology and in entomology respectively, and Nannie displays scientific inclination in her approach to gardening and caring for animals. The three women are thus capable of identifying with non-human animals because of their knowledge, and also because of their ability to think in terms of systems. Both faculties are undoubtedly enhanced by their higher scientific education, which they share with Kingsolver. Although their background in science could separate them from the natural world as a result of a dynamic where culture sets one free from nature, Kingsolver actually suggests that understanding non-human animals creates a way for her female characters to identify with animals. Both Lusa and Deanna show a greater understanding of their emotions and feelings because they analyse themselves as they would with coyotes and moths, thereby further inscribing themselves in the ecosystem of these animals, which they occupy both physically and intellectually. This means that Lusa, Nannie, and Deanna's practice of care stems as much from their scientific knowledge as from a more emotional and primal connection to nature, a fact that creates a feminised space for scientific discourse.

Despite this clear scientific and rational anchoring, Kingsolver's female characters are in touch with other aspects of their perceptive faculties, among which is the eerie perception of ghosts. Indeed, one of the peculiarities of this novel resides in the liminality of its genre, with a strong realistic grounding scattered with little pockets of fantastic occurrences with a gothic undercurrent. Ghosts are regularly mentioned in the novel, whether on a literal or metaphorical level, for the reader does not know with certainty whether they actually exist. The fact that they are visible and intelligible to women reveals yet another layer of the ability of the female characters to navigate sensorial and intellectual spaces which are normally deemed inaccessible. Bearers of knowledge and seers of the imperceptible, these women are integrated into their ecosystems and create multiple complex bridges between worlds that have been separated by an ideology of duality. Through reconnecting abilities, Kingsolver's women are literary guides towards the new ethics of care.

Conclusions: acknowledging the ideological shortcomings of masculinity for a renewed ecological approach

By entangling the devastating effects of the hegemonic patriarchal culture of the contemporary United States and the revitalising potential of a caring ethical anchoring, Kingsolver offers her readers a novel that can be seen as a starting point for an exploration of environmental ethics. Drawing away from catastrophic literature, she acknowledges the dystopian potential of climate change, but also transcends its potentially paralysing effects. She uses the apparent sterility of the post-Carson world to create a space for the marginalised voices of women in which she sees transformative potential while adopting a framework that is unmistakably reminiscent of ecofeminist thought. The trajectories of her female characters are inextricably linked to that of their natural environment, and, by changing their outlooks and behaviours in relation to their surroundings, these characters set their lives in a motion that is presented as virtuous. Kingsolver's characters create the opportunity for detailed criticism and argumentation that rely on scientific discourse through the voices of educated – yet marginalised – women, who do not have to emulate masculine characteristics to become protagonists in their stories where they initiate paradigmatic change. By creating male characters who engage in stereotypically nefarious behaviours, Kingsolver shows how hegemonic masculinity is at least partially to blame for the socio-ecological crises which continue to shape the twenty-first century. By doing so, she introduces the possibility of reclaiming traditionally feminine values and invites men to participate, suggesting they side with women in the rediscovery of fruitful and life-bearing principles. As a consequence, Kingsolver foregrounds the idea that climate change is socially induced, and that solutions will accordingly arise from addressing the interconnectedness between environmental and social issues. In other words, the author calls for a recognition of the oppression of social groups as indissociable from environmental issues, thus contributing to the expansion of environmental imagination and its articulation, whilst bringing ecofeminist ideas to the diverse public that constitutes her readership.

Notes

1 When Kingsolver was a student, she was involved in the Sanctuary movement for Central American refugees' protection and in protests against nuclear sites. Later, she dipped her toes in journalism when she published her first book, *Holding the Line: Women in the Great Arizona Mine Strike of 1983*, telling the story of an eighteen-month-long strike in Arizona mines.

2 For a more detailed discussion on the concept of environmental imagination, see Buell (1995).

3 See for example Jacobson (2005), Van Tassel (2008), or Bénédicte Meillon's body of work on Kingsolver's fiction.

4 In her doctoral thesis dedicated to Kingsolver's short stories, Bénédicte Meillon briefly mentioned that Kingsolver's characterisation of male protagonists moved away from conventional representations of masculinity. In the case of *Prodigal Summer*, I argue that male characters rather undergo a process which challenges their embodiment of a more conventional masculinity.

5 Carson's book was instrumental in bringing attention to the potentially catastrophic consequences of pesticide use, from the destruction of ecosystems to their carcinogenic effects on human beings. Some ecofeminists consider her groundbreaking book to be the foundational work for ecofeminist thought, for her observation of environmental issues as connected with other concerns, including health-related ones. Carson is perceived as one of the first to offer a framing of interconnectedness in response to 'the wanton human domination of the natural world,' a perspective that is now shared by ecofeminists (Diamond and Orenstein, 1990: ix).

6 See Merchant (1980), Butler (1990), and Connell (1995).
7 Following the terminology used in gender studies and masculinity studies, masculinism here refers to 'the ideology and justification of male dominance' (Pease, 2021: 538).
8 The protagonists' relationship to nature has already been explored in, among other studies, Wenz (2003), Van Tassel (2008), and Leder (2009).
9 Appalachian landscapes have a long history of being exploited for their natural resources. If this appropriation of the environment has sparked an ecological consciousness for some, other Appalachians, however, engaged in disastrous practices such as mountaintop removal, tobacco culture, or mining (Sovine Reid, 1979: 242; Cory, 2019: 1–5). These activities were at the core of regional economic growth and, as a consequence, were instrumental in the employment of locals. Garnett Walker's character embodies a certain ambivalence of Appalachians in that he both represents a wounded Appalachian who grieves part of his identity with the loss of the American chestnut, and the Appalachian who partakes in exploiting his natural environment for a lack of better knowledge.
10 Biblical literalism is the belief that, unless specified in the text, the Bible should be interpreted literally.
11 Despite the impossible-to-ignore implications that come with talking about non-consensual relationships, mostly in regard to physical abuse, I still believe that there is an ongoing unease on the side of Deanna that indicates a lack of agreement on a spiritual and ethical level. Bondo gradually makes his way into her home, her body, her principles, transforming her but also temporarily weakening her groundings. This ambiguous subtext, I believe, calls for this questioning regarding Deanna's consent.
12 For an analysis of the gendered division of the domestic space, see Jacobson (2005).
13 The term was introduced by British film theorist Laura Mulvey (1975) to designate how women are most often represented as perceived through the eyes of men, which in turn shapes women's representation in popular culture.
14 In her essay *In Defense of Witches: The Legacy of the Witch Hunts and Why Women Are Still on Trial*, Mona Chollet draws this comparison between the representation of witches and of Jews: 'The demonization of women as witches had much in common with anti-Semitism. Terms such as witches' "sabbath" and their "synagogue" were used; like Jews, witches were suspected of conspiring to destroy Christianity and both groups were depicted with hooked noses' (2022: 7).
15 In his book *Coyote America*, Dan Flores retraces the history of coyotes in the United States. While these mammals 'may now have more fans . . . than ever before,' they still suffer from an ongoing negative reputation altogether. Flores explains that in a 1980s Yale study, coyotes were ranked 'dead last in public appeal – behind rattlesnakes, skunks, vultures, rats, and cockroaches,' and that they were still, as of 2016, the 'most persecuted large mammal in American history"' (2016: 15).

Bibliography

Adams, Carol J. and Lori Gruen (eds.) (2014), *Ecofeminism: Feminist Intersections with Other Animals and the Earth* (New York: Bloomsbury).

Buell, Lawrence (1995), *The Environmental Imagination: Thoreau, Nature Writing, and the Formation of American Culture* (Cambridge: Belknap Press of Harvard University Press).

Butler, Judith (1990), *Gender Trouble: Feminism and the Subversion of Identity* (London: Routledge).
Carson, Rachel (1962), *Silent Spring* (Boston: Houghton Mifflin).
Chollet, Mona (2022), *In Defense of Witches: The Legacy of the Witch Hunts and Why Women Are Still on Trial* (New York: St. Martin's Press).
Connell, R.W. (1995), *Masculinities* (Berkeley: University of California Press).
Cory, Jessica (2019), 'Introduction', in Jessica Cory (ed.), *Mountains Piled Upon Mountains: Appalachian Nature Writing in the Anthropocene* (Morgantown: West Virginia University Press).
Diamond, Irene and Gloria Feman Orenstein (1990), *Reweaving the World: The Emergence of Ecofeminism* (San Francisco: Sierra Club Books).
Donovan, Josephine and Carol J. Adams (eds.) (2007), *The Feminist Care Tradition in Animal Ethics* (New York: Columbia University Press).
Flores, Dan (2016), *Coyote America: A Natural and Supernatural History* (New York: Basic Books).
French, Marilyn (1985), *Beyond Power: On Women, Men and Morals* (New York: Summit).
Gaard, Greta (2015), 'Ecofeminism and Climate Change', *Women's Studies International Forum*, 49 (March–April), 20–33, https://doi.org/10.1016/j.wsif.2015.02.004
Haraway, Donna J. (1991), *Simians, Cyborgs and Women: The Reinvention of Nature* (New York: Routledge).
Hermann, Daniel Justin (2001), *Hunting and the American Imagination* (London: Smithsonian Institution Press).
Jacobson, Kristin J. (2005), 'The Neodomestic American Novel: The Politics of Home in Barbara Kingsolver's *The Poisonwood Bible*', *Tulsa Studies in Women's Literature*, 24:1 (Spring), 105–127, https://doi.org/10.2307/20455213
Kaschak, Ellyn (1992), *Endangered Lives: A New Psychology of Women's Experience* (New York: Basic Books).
Katz, Eric (1996), 'The Problem of Ecological Restoration', *Environmental Ethics*, 18:2 (Summer), 222–224, https://doi.org/10.5840/enviroethics199618236
Kheel, Marti (1996), 'The Killing Game: An Ecofeminist Critique of Hunting', *Journal of the Philosophy of Sport*, XXIII, 30–44, https://doi.org/10.1080/00948705.1996.9714529
Kingsolver, Barbara (1989), *Holding the Line: Women in the Great Arizona Mine Strike of 1983* (Ithaca, NY: Cornell University Press).
Kingsolver, Barbara (2000), *Prodigal Summer* (New York: HarperCollins).
Kingsolver, Barbara (2002), *Small Wonder* (New York: HarperCollins).
Leder, Priscilla (2009), 'Contingency, Cultivation, and Choice: The Garden Ethic in Barbara Kingsolver's *Prodigal Summer*', *Interdisciplinary Studies in Literature and Environment*, 16:2 (Spring), 227–243, https://doi.org/10.1093/isle/isp034
Meillon, Bénédicte (2005), 'La nouvelle-oxymore de Barbara Kingsolver: la révélation des écritures et l'écriture des révélations' (Doctoral Thesis, Université de Toulouse-le-Mirail).
Merchant, Carolyn (1980), *The Death of Nature: Women, Ecology and the Scientific Revolution* (San Francisco: Harper and Row).
Merchant, Carolyn (1996), *Earthcare: Women and the Environment* (New York: Routledge).
Morizot, Baptiste (2018), *Sur la piste animale* (Arles: Acte Sud).
Mulvey, Laura (1975), 'Visual Pleasure and Narrative Cinema', *Screen*, 16:3 (Autumn), 6–19.

Pease, Bob (2021), 'From Ecomasculinity to Profeminist Environmentalism: Recreating Men's Relationship with Nature', in Paul M. Pule and Martin Hultman (eds.), *Men, Masculinities, and Earth: Contending with the (m)Anthropocene* (Basingstoke: Palgrave Macmillan), 537–558.

Sovine Reid, Melanie (1979), 'On the Study of Religion in Appalachia: A Review/Essay', *Appalachian Journal*, 6:3 (Spring), 239–244.

Van Tassel, Kristin (2008), 'Ecofeminism and New Agrarianism: The Female Farmer in Barbara Kingsolver's *Prodigal Summer* and Charles Frazier's *Cold Mountain*', *Interdisciplinary Studies in Literature and Environment*, 15:2 (Summer), 83–102, https://doi.org/10.1093/isle/15.2.83

Warren, Karen J. (1988), 'Toward an Ecofeminist Ethics', *Studies in the Humanities*, 15:2, 140–156.

Webteam (2018), 'Carol Gilligan', *Ethics of Care*, 8 July, [Online], [Accessed 20 March 2022], available from: https://ethicsofcare.org/carol-gilligan/

Wenz, Peter S. (2003), 'Leopold's Novel: The Land Ethic in Barbara Kingsolver's *Prodigal Summer*', *Ethics and the Environment*, 8:2 (Autumn), 106–125, https://doi.org/10.2979/ete.2003.8.2.106

10 Backward looking is necessarily organic

Female artists revisit traumatic pasts and reimagine present and future alliances[1]

Katarzyna Bojarska

It has for a long time been assumed – though this position has consistently been revised in recent years – that history is liberation from nature, a time after prehistory, the overtaking of this time and its regulation by a series of conquests, revolutions, and other events that are distinguished, above all, by the scale of violence (Chakrabarty, 2009). Researchers and scholars of the Anthropocene point to the necessity of combining the natural and the cultural, what we conceptualise as human (universal) history and natural history, society and the environment (Bonneuil and Fressoz, 2016). Despite the huge investment in the 'never again' and the political, cultural, and artistic activities that followed World War II in the form of awareness-rising campaigns, films, exhibitions, and publications, humans have not learned from history a better life or better relationships. Wars have not been prevented, imperial colonial politics persist, and so do dominance and racism. At the time of writing, I can clearly see – in Bucha, Irpin, Hostomel, and other Ukrainian towns and cities – that the experience and the critical historical reception of genocide – the Holocaust predominantly – has not stopped people from committing genocidal acts. Racial violence, the enforcement of white supremacy, colonialism, economic exploitation, and slavery do not belong exclusively to the past, but rather continue to be pressing, if not overwhelming, problems of the present. The question arises, then, how we make history visible to ourselves. How do we visualise a historicity that allows for a truly critical approach to the legacy of environmental violence, of the harm done to the 'organic' by human practices of imperialism, capitalism, colonialism, racism, and patriarchy, a historicity that opens the possibility for change?

In this chapter I will be taking a closer look at both individual and collaborative projects by choreographer and performer Agata Siniarska, and works in multiple media and forms by visual artist Karolina Grzywnowicz (Figure 10.1). I propose seeing their creative and critical practices as forms of engagement in what Isabelle Stengers calls 'changing the

DOI: 10.4324/9781032726953-11

Figure 10.1 Katarzyna Wolińska during a performance of *Second Nature* by Agata Siniarska and Karolina Grzywnowicz, Erbstücke Festival in Dresden, Germany, 26 April 2019.

Photo: Michał Ramus.

story' (Haraway, 2016; Stengers, 2009). What I mean by this can best be explained in reference to Donna Haraway's words:

> human exceptionalism and bounded individualism, those old saws of Western philosophy and political economics, become unthinkable . . . when organisms plus environments can hardly be remembered for the same reasons that even Western-indebted people can no longer figure themselves as individuals and societies of individuals in human-only histories.
>
> (Haraway, 2016)

If these histories are no longer possible, what are the other stories? Haraway is inquiring about relevant forms of revolt and pointedly asks her crucial question: 'How to matter and not just want to matter?' (Haraway, 2016), a question pertaining to theoreticians, critics, and artists. The two artists discussed in this chapter ask precisely this in their feminist artistic practices. These might not be grand paradigm shifts, but rather closer to what anthropologist and filmmaker Elizabeth Povinelli has called 'an idea that matters to some part of the world – that helps that part of the world matter forth' (Povinelli, 2021: 2).

In the following analysis, I am inspired, above all, by Siniarska's and Grzywnowicz's artworks and their discerning commentary, by Judith Butler's reflections on relationality and dependence presented in *The Force of Nonviolence*, and by Catherine Malabou's essay *The Brain of History, or the Mentality of the Anthropocene*. The latter provides an interesting shift in thinking about human responsibility for nature from consciousness to brain – a site of plasticity and metabolism as well as addiction – and suggests that we treat ecology as a new libidinal economy. Further, Sussan Schuppli's idea of 'material witness' and her reflection on memory's numerous interdependencies provides important points of reference. Last but not least, a recent book by Elizabeth Povinelli, *Between Gaia and Ground* (2021), with its anthropological decolonial feminist perspective, has allowed me to see the artistic interventions discussed here in the framework of ancestral catastrophe.

Thinking about the current state of climate and environment, as well as of the global political situation, Povinelli distinguishes between a coming catastrophe and an ancestral catastrophe, the first of which is figured as an event to come, an eruption that in an unspecified future will mark the end and establish a new beginning of human and natural history: 'The coming climatic catastrophe builds on, intensifies, and collapses into these concepts of birth and death, event, and finitude. The ancestral catastrophe does not' (Povinelli, 2021: 2). First and foremost, as Povinelli points out, they operate in different kinds of temporalities. Ancestral catastrophe locates itself in the long history of colonialism, and the 'contemporary climatic, environmental, and social collapse' have to be inscribed into this narrative:

> Ancestral catastrophes are past and present; they keep arriving out of the ground of colonialism and racism rather than emerging over the horizon of liberal progress. Ancestral catastrophes ground environmental damage in the colonial sphere rather than in the biosphere . . . in errancies rather than in ends; in waywardness rather than in war; in manoeuvres, endurance, and stubbornness rather than in domination or resistance, despair, or hope.
>
> (Povinelli, 2021: 3)

From this standpoint, the entanglement of human existence must be seen as something entirely different from what it has been, and so too must (human) history. The concept of ancestral catastrophe, as I understand it, allows one to discuss memory and postmemory of the traumatic violence of World War II and the Nazi genocide of European Jews, in a framework that includes thinking about and acting against damage, dominance, and exploitation in a more-than-human world and through interspecies connections.[2]

How, then, to address the dark legacy of World War II and its ramifications affecting landscapes both mental and literal, and simultaneously be in one's own time perceiving past catastrophe and present (as it develops) as meaningfully and unavoidably entangled, and as such doing damage to human, but also non-human (or, to use Schuppli's term, 'material'), witnesses. The artistic projects I discuss in this chapter pose a very important question of how to address the past, where to look for accounts, how to deal with witnesses, and, last but not least, how to approach the changing of a story, so that it may deal with our current state of historical experience and with us as historical subjects. As Catherine Malabou has observed:

> It is striking to note that today's victims of socio-political traumas demonstrate the same profile as victims of natural catastrophes (tsunamis, earthquakes, floods) or grave accidents (serious domestic accidents, explosions, fires). We have entered a new age of political violence, in which politics is defined by renouncing hope of endowing violence with political meaning. This effacement of meaning is the new face of the social – bearing witness to an emergent, globalized psychic pathology that is identical in all cases and all contexts.
>
> (Malabou, 2019b: 152)

Malabou adds in another part of the same conversation that 'the ecological crisis puts us precisely in a schizophrenic situation, having to choose, without being able to choose, between ourselves and the environment' (Malabou, 2019b: 154). How, then, are we to establish meaning where it is denied by the sheer impact and force of violence? How are we to make connections, or rather, regain relationality?

If our historicity should be defined first and foremost by our concrete location in time and in relation to our ancestors and successors, the key question seems to be what defines that time and who and what we count as a community in time. Standing at the threshold of this new community, we need to abandon the worn-out rituals of liberal individuals looking back at the past violence, struck with horror and unable to respond. Butler offers the possibility, or rather the necessity, to think of this backward-looking subject differently. Above all, it is a being always already in a relationship with other beings, and aware of their own dependence and interconnectedness, which does not mean failure of self-liberation or of self-reliance, nor is it a reason for shame (Butler, 2020). In my opinion, this 'radical equality' needs to be framed not only within the ecological and feminist theoretical approach, but also, as I will try to show, with the artistic one.

According to Butler, individualism normalises aggression, while violence is reduced to individual acts. In an interview, Butler emphasises that by placing the issue of 'non-violence' at the centre of her reflections, she is in fact attempting to speak of 'social obligations' and 'social relationality,' which for the purposes of these reflections can be understood as historical obligation and historical relationality (Gessen, 2020). One might ask, paraphrasing Butler, what we owe to those with whom we have lived on Earth. Why should we owe humans or other living creatures care? What role could care play in an experience of our own historicity? Why should we engage in non-violent relationships, and can we do so without radically changing our position in the natural world? (Gessen, 2020). Butler thus insists that violence inflicted upon others (violence that is not episodic but structural) is a blow to this fundamental bond of coexistence and interdependence, and it is this bond and dependence that are primarily to be protected.

Thinking about interdependence also allows us to see historical consciousness and responsibility differently. Malabou argues that the key to this shift in perspective (or 'changing the story'), rooted in a recognition of the profound interaction of history and nature, is the brain and its processes. Conceived of as a refuge not of rationality but of plasticity, the brain, she writes, may be our greatest ally and guide, for it is responsible for both our biological and cultural adaptations: 'Human practices alter or affect brain-body chemistry, and in return, brain-body chemistry alters or affects human practices' (Malabou, 2019a: 176). The brain is the environment in which metabolism occurs; it operates under the logic of addiction. Addiction, in turn organises our attachments to the phenomena of the world we live in. From this perspective, the question of historical responsibility seems to be the following: how can we feel truly responsible for what we have done to Earth if our acts are dictated by the compulsivity of addiction and it seems impossible to be break free? 'Only new addictions,' writes Malabou, 'can help break old ones. Ecology has to be a new libidinal economy' (Malabou, 2019a: 179).[3]

What remains is the question of binding politics and history, and, in the next step, thinking about art and its (eco)political efficacy. Ecology and mentality, powerlessness and interdependence, do not necessarily lead to depoliticisation of either history or art. On the contrary, as many critics and curators have recently pointed out, they may be terms for a new political vocabulary that takes into account not only the need to salvage the memory of past victims, but to salvage us as subjects of lethal historical entanglement while this is still possible.

As Jill Bennett postulated in the 'notebook' commissioned for the exhibition dOCUMENTA 13, *Living in the Anthropocene* (2012), feminist and ecological aesthetics may trigger a truly transdisciplinary revolution:

> Ecological thought is changing the way in which our practices might operate in the future. Thinking ecologically means attuning, perceiving,

> and doing what we know how to do differently, in different spaces, dimensions, relationships. This is what it is to be in the midst of a paradigm shift, to be actively living in the Anthropocene.
>
> (Bennett, 2012: 17)

Such transdisciplinarity, be it in the artistic or in academic field, is open to alliances previously unimagined and solidarities that will allow various cross-examinations of traumatic pasts and legacies of extreme violence, as well as struggles for equality and emancipation. Such transdisciplinarity might allow us to elaborate a better theory of living together and examine what made us exclude others and include ourselves into past projects, too often falsely labelled as universal.

Second Nature

It is hard to grasp the 'real' identity of performance artist Agata Siniarska, as she introduces herself as a 'fictional character with many talents and a criminal past' (Siniarska, n.d. 'About'), or as a 'cyborgiczna forma o zmiennym stopniu produkcji, wielu talentach . . . świadomie działająca rozdział po rozdziale w przestrzeniach choreograficznych sekwencji' [a cyborg-like form with varying degrees of production, multiple talents, . . . consciously operating chapter by chapter within the spaces of choreographed sequences] (Komuna Warszawa, n.d.). For one, she is a highly educated and versatile artist who, as we learn from the website of Muzeum Sztuki in Łódź, 'is interested in building up knowledge that examines diverse media, protocols, and self-developed strategies without imposing any hierarchies. She is also interested in all sorts of knowledge aimed to engage, not to explain.'[4] Her projects are feminist and emancipatory, concentrating on the questions of female identity, bodily autonomy, sexual pleasure, and love, and rebelling against established norms and protocols of behaviour. Siniarska's later works, created during the second decade of the twenty-first century, acquire a more ecofeminist and ecological character. She turns towards the intertwined narratives of the 'degradation of nature and the exploitation of women' (Obarska, 2019), and her performances increasingly expand the presence and agency of non-human protagonists. In 2017, at Studio Theatre in Warsaw, Siniarska presented her solo performance entitled *Ślepowidzenie* [*Blindsight*] where, as she claimed, air, temperature, bacteria, aliens, and other species were participating in her work and were co-responsible for its production. She aimed to explore the entangled webs of implication, where she does not exist as a separate entity: 'Jestem porowata, nie mogę być pewna wszelkich granic . . . Jestem częścią ogromnego kolektywu. Nic do mnie w pełni nie należy' [I am porous, I cannot be sure of any boundaries . . . I am part of a vast collective. Nothing fully belongs to me]. Instead, she is 'ukryta w przestrzeni publicznej prezentacji, skradziona lub nieświadomie zagarnięta' [hidden in the public space of presentation, stolen or unconsciously seized] (Czwarty 'Powrót (do) przyszłości').[5] In a lecture performance entitled *Hyperdances* (2016), Siniarska followed and expanded on American choreographer Yvonne Reiner's concept of dance and movement, stating:

Hyperdance can be uncountability of molecules.
Hyperdance can be tectonic movements.
Hyperdance could be the biosphere.
Hyperdance could be the sum total of all the nuclear materials on Earth.
Hyperdance could be the very long-lasting product of direct human manufacture, such as Styrofoam or plastic bags or non-human product: the sum noise of the waves.

(Siniarska, n.d. 'In the Beginning Was a Copy')

Siniarska departs from Reiner's theory and practice of dance and arrives at the following realisation of what the present moment demands, the present moment of dance and the political, social now:

. . . no longer street/stage/museum, but the whole Earth!
Not only human bodies, but the entanglements of human and non-human vibrating matter.
No collective action, but an infinite multiplicity of tangles.
Not Dolly Parton but Dolly the sheep.

(Siniarska, n.d. 'In the Beginning Was a Copy')

Dance, or art more generally, needs not only to leave art institutions, but to reconceptualise the ordinary and the local (street), moving towards a bigger context, the planetary. It also needs to expand beyond the human body (and its relationship with other bodies), beyond the collective (understood as group of people), and into the realm of clones and boundless, untamed matter (Piasecka and Siniarska, 2021).

With her project *The Soft Act of Killing* (2019), Siniarska offers a 'solo performance for 50,000 dead cells, 10 billion bacteria, 30,000 roentgen radiation per hour, 23 litters of air, 7.8 richter magnitudes and the audience' (Susch Museum). And she pursues and develops the inquiry she set for herself in her earlier works:

> This lecture performance is a learning process of situating myself and my artistic practice in the so called 'compromised times.' The world is a terrible mess. It is toxic, irradiated, and full of injustice. I aim not to cause suffering, destruction, death, but simply by living, buying things,

> throwing them away, I have a terrible effect on the ecosystem. I am a detective that discovers to be a criminal and so is my dance practice. What is the role of my artistic practice in the present time? Does this world need my dance? Can I practice responsibility towards the world through my practice? Is my dance dangerous? Can I learn how to stay with (Haraway) the trouble of living and dying together on a damaged Earth through my dance? Can dance provide the means to building more liveable futures?[6]
>
> (Susch Museum)

First, Siniarska offers a distressing evaluation of the current human and more-than-human condition, and of her own position as someone who is implicated in, if not responsible for, injustice and damage done to the ecosystem. Then, the artist persistently inquires about the meaning and aim of her choreographic practice, and about the conditions of possibility of its agency, or even life-saving potential, in the face of what she recognises precisely as an 'ancestral catastrophe.' Inspired by Haraway, Siniarska asks if dancing can let her 'stay with the trouble,' working out solutions that can be transformative and shareable. By working with her body and its movement, she orients herself towards other bodies and forms of movement, to the audience but also further, towards other organisms, elements, sounds, etc. With her rather unspectacular moves, gestures, and activities, as simple and banal as breathing, she is trying to find a position of response, responsibility, and repair vis-à-vis ancestral catastrophe. It is as if she is trying to create new attachments, and then possibly new addictions, such as those discussed and theorised by Malabou.

Undoubtedly, the (expanded) audience is a crucial reference for the artist. She performs not only, or not merely, in front of people, but rather in relation to all matter surrounding her, the elements of the world she is also made of. Despite meticulously constructed choreography, she seeks interaction and unexpected restructuring both on the level of bodily experience as well as on an intellectual and affective one. In this regard, any performance becomes a collective work and effort:

> W momencie spotkania z publicznością, performans wymyka się mojej wiedzy, działam w przestrzeni jednej z jego wersji, a wersji jest tyle, ile w danym momencie materii, tej ludzkiej i nie-ludzkiej w przestrzeni spotkania. Każda z tych wersji wytwarza inną wiedzę.
>
> [At the moment of meeting the audience, the performance escapes my knowledge, I act in the space of one of its versions, and there are as many versions, as much matter, human and non-human, as there is in the space of the encounter. Each of these versions produces different knowledge.]
>
> (Choreografia w sieci, n.d.)

The instability of meaning resonates with the porosity of the body and its surroundings. Knowledge is produced spontaneously at these collective encounters, and it is transmitted, or transmits itself, more like a virus than a structured thought.

In 2019, Siniarska was invited by the curators of the series *Choreographic Territories: New Paths of the Avant-Garde* to reinterpret the work of the modern dance choreographer Pola Nireńska (1910–1992).[7] The project was meant to encourage a young generation of dancers to reach for the legacy of nomadic Polish avant-garde women choreographers and dancers such as Bronisława Niżyńska, Yanka Rudzka, Pola Nireńska, and Irena Prusicka. In the interwar period, modern dance was an international language of emancipation from rigid conventions and social roles, but also the violent politics of the time (see Szymajda, 2017).[8]

Born into a middle-class, rather traditional Jewish family of manufacturers in Warsaw, Pola Nireńska had to struggle against her traditional background in order to pursue her dream of becoming a dancer. She was allowed to study at the Wigman School in Dresden, and arrived there in 1929. Nireńska graduated in 1932 and toured the United States with Wigman's company. Upon the Nazis' rise to power and antisemitic practices at the Wigman school, Nireńska left Germany and moved back to Warsaw. Nireńska become alarmed by the rising antisemitism in Europe, and decided to leave the continent. Travelling a lot, she became increasingly aware of, and threatened by, growing racist violence. In 1935 Nireńska emigrated to England where she found refuge throughout World War II, continued practising as a dancer and choreographer, and stayed until her emigration to the United States in 1949. Except for her parents, who in 1939 emigrated to Israel, the rest of her family perished in the Holocaust. After initial hardships in the United States, Nireńska settled in Washington, D.C., where in 1960 she opened the Pola Nirenska Dance Company. Five years later she married Jan Karski, a Polish resistance fighter and diplomat during World War II and a courier to the Polish government-in-exile and to Poland's Western Allies between 1940 and 1943, who reported about Nazi genocide, the destruction of the Warsaw Ghetto, and extermination camps in Poland. Nireńska kept working, mostly as a choreographer. In 1980 Karski testified before Congress in support of funding for the United States Holocaust Memorial Museum, and was one of the crucial protagonists of Claude Lanzmann's 1985 film, *Shoah*. It is believed that these events made Nireńska revisit the traumatic past of European Jews. In the 1980s, she committed herself to work on *In Memory of Those I Loved . . . Who Are No More* (later titled *Holocaust Tetralogy*). During this work, Nireńska experienced a nervous breakdown which took her out of work for nearly two years. The *Holocaust Tetralogy* included three group dances and one solo: *Life* (1983), *Dirge* (1982), *Shout* (1986), and *The Train* (1988), the latter of which made use of images of a train car and gas chamber, with female dancers crowded anxious and grieving (Faber, 2019; Mozingo, 2008). As Karen A. Mozingo puts it, 'The four sections of the tetralogy depict the destruction of the family as their space is narrowed, their path forced, and their narrative twisted toward a deathly closure' (Mozingo, 2008: 137–138). Nireńska committed suicide in 1992.

Siniarska executed her project, *Second Nature*, whose point of departure was Nireńska's life and work, and in particular her final autobiographical *Tetralogy* project, in collaboration with dancer Katarzyna Wolińska and visual artist Karolina Grzywnowicz (Figure 10.2).[9] I propose reading this collaborative piece as a new development in Siniarska's creative reflection on the role of her dance practice vis-à-vis past and present catastrophes. To a certain extent, two processes were conducted autonomously: Grzywnowicz did her artistic research on the installation by following the route already mapped out in her earlier study of the history of plants, exemplified by the project *Weeds* (2015c), to which I will return later. This installation acted not so much as a stage set as an 'open form,' a reflective and sensual environment activated by the bodies, including Wolinska's own body,

Figure 10.2 Katarzyna Wolińska during a performance of *Second Nature* by Agata Siniarska and Karolina Grzywnowicz, Erbstücke Festival in Dresden, Germany, 26 April 2019.

Photo: Michał Ramus.

surrounding it and moving within it. Grzywnowicz's archival research was matched by a peculiar archaeology of movement, reaching deep into the traumatic choreographies of the traumatised bodies from the photographs and verbal testimonies of those who survived.

Siniarska was not interested in a reconstruction of Nireńska's *Tetralogy of the Holocaust*, nor in its reinterpretation, but rather in an encounter with that history through her own history and that of her fellow artists.[10] As I have already noted, the artist seems to be locating herself within the ancestral catastrophe, as described by Povinelli, rather than after what we could call historical catastrophe and before climate catastrophe; she linked her own historicity (the sense of living with catastrophe and violence on a planetary scale) with the bodily relations of human and non-human worlds both past and present. On the choreographic level, Siniarska struggled with a body lying flat, a body that is asleep or already a corpse. As she herself put it: 'jeżeli położę ciało, to to ciało jest już bardzo dramatyczne. Leżące ciało ma tak niesamowity ciężar, że trudno jest go przełamać' [If I lay the body down, that body is already very dramatic (Figure 10.3). The lying body has such an incredible weight that

Figure 10.3 Katarzyna Wolińska during a performance of *Second Nature* by Agata Siniarska and Karolina Grzywnowicz, Erbstücke Festival in Dresden, Germany, 26 April 2019.

Photo: Michał Ramus.

it's hard to break through] (Piasecka and Siniarska, 2021). Modern dance choreographers were interested in stressing the gravity of the dancer's body, in grounding it – unlike ballet where the bodies, the feet, barely touch the ground. The dramaturgy of Nireńska's *Tetralogy* led downwards to the grave, the stillness and muteness of the perished lives and bodies. Siniarska decided to work with the gestural dictionary of modern dance, borrowing figures and expressive gestures for her choreography so that they become a meaningful marker of the history of dance. She also wanted to work with the body brought back from underground, from nonexistence, and bearing its traces.

The premiere of *Second Nature* took place at the HELLERAU European Centre of Art in Dresden (Hellerau, 2019), the very place where Pola Nireńska had studied with German dancer and choreographer Mary Wigman, and where she experienced both the rise of Nazism and artistic and sexual emancipation. For Grzywnowicz, Siniarska, and Wolińska, the encounter with Nireńska's story on site meant an opportunity to get closer to a person's traumatic memory, her biography, but also to a distant dialect of modern dance diction: intense stylisation, emotional expressiveness, and expression of liminal states. The three artists entered this relationship without the desire to understand or appropriate Nireńska's life and her life story, and without the need to understand or translate it. This is evidenced by the convincing contention of critic Zuzanna Berendt:

> W tańcu Wolińskiej elementy, które można identyfikować z techniką i stylem tańca modern pojawiały się wyraziście odróżniając się od tego, co można by nazwać zwyczajnym trybem ruchu. Kiedy tancerka neutralną ekspresję twarzy zastępuje otwarciem ust, przywodzącym na myśl ikonografie związane z lamentacją, błyskawicznie zmienia się możliwy do odebrania przez widza emocjonalny przekaz jej ruchu. Siniarska pokazuje w jaki sposób oddziałują na publiczność strategie tańca wyrazistego i udowadnia, że działają one w sposób skuteczny. Jednocześnie, umieszczając je w bliskości środków skromnych i minimalnych, wywołuje subtelny efekt obcości, skłaniający do refleksji nad tym jakiego rodzaju emocje budzą w nas reprezentacje Zagłady. Nagła zmiana ekspresji w niespodziewany i zachwycający sposób wpływa na postrzeganie przestrzeni współdzielonej przez widzów i tancerkę. Rozpacz, jaką można odczytać z twarzy, błyskawicznie naznacza otoczenie w jakim się znajduje jako krajobraz związany z traumatycznym przeżyciem utraty.
>
> [In Wolińska's dance, the elements that can be identified with the technique and style of modern dance appeared distinctly different from what can be called the ordinary mode of movement. When the dancer replaces the neutral expression of her face with the opening of her mouth,

> reminiscent of iconographies associated with lamentation, the emotional message of her movement as it can be received by the viewer changes instantly. Siniarska shows how expressive dance strategies affect the audience and proves that they are effective. At the same time, by placing them in the proximity of modest and minimal means of expression, she creates a subtle effect of defamiliarisation, prompting us to reflect on what kind of emotions representations of the Holocaust evoke in us. The sudden change of expression affects the perception of the space shared by the audience and the dancer in an unexpected and delightful way. The despair that can be read on her face instantly marks her surroundings as a landscape associated with the traumatic experience of loss.]
>
> (Berendt)[11]

This loss invoked by Berendt is omnipresent as the artists oscillate between past despair (Nireńska's despair expressed by the gestures of modern dance) and present despair (in the face of the past and present violence, destruction, and trauma), between the need for closure and the necessity of maintaining an open narrative in which anyone can participate in their own right.

Aware that they were working with choreography as testimony, Siniarska and Wolińska asked themselves first and foremost what Nireńska's body was like: what kind of art and life practices and discipline had it followed? The artists worked with fragmented information and did not pretend otherwise. They did not succumb to archive fever; on the contrary, the remnants of Nireńska's personal history and oeuvre made it possible to create a consciously hybrid and anachronistic work. As Wolinska said in an interview, 'We are trying to breathe life into a form that is partly dead and, in order to come alive, needs a body dancing here and now' (Obarska). This dancing body returns today to the issue of political violence, which is as alive as it can be and which today also concerns the bodies of women, migrants, and members of the LGBT community, as well as nature. As Siniarska noted, 'The question of the role of choreographic language in the face of political or economic violence seems very contemporary to me' (Obarska). The contemporaneity of the encounter between choreography and political violence stems from a particular historical sensibility which rejects linearity in favour of analogy and treats dance first and foremost as a language for producing meanings, building relationships with the world, and creating a community of experience.

Let us now turn towards Grzywnowicz's installation, which provided the setting and an integral element of *Second Nature* – a quasi-garden consisting solely of plants that were both politicised by the Nazis as part of their *Blut und Boden* policy, and were complicit in the performance of Nazi violence. The artist worked with the concept of a garden as tamed nature, but also with the idea of the *Naturgarten*, dating back to the first half of the twentieth century and enthusiastically adopted by the Nazis (Mauch, 2004:

85–86). According to this predominantly racist concept, gardens were supposed to consist only of plants native to the site, while all non-native species needed to be removed.[12] For Grzywnowicz, this mode of thinking, through classification, selection, and extermination, provided a link to the biography of Pola Nireńska, who as a foreigner in Germany, had to leave Mary Wigman's dance team in the 1930s.[13] For the Nazis, the *Naturgarten* provided a metaphor of an ideal society with no outsiders (Figure 10.4).

Grzywnowicz is conscious of a crucial role the environment plays in contemporary studies of past violence, as embodied by the environmental history of the Holocaust and other genocides (Katz, 2001; Cole, 2020; Małczyński et al., 2020). She writes:

> Going beyond the anthropocentric perspective allows us to see nature as an actor, witness and co-participant of history, and landscape to be read as a living archive in which traces of past events are recorded. . . . Adopting an environmental perspective, taking into account such aspects as location, landscape, introduces an expanded group of witnesses to the stage.
>
> (Grzywnowicz, 2019b)

So, what we encounter in the installation part of *Second Nature* is a congregation of possible witnesses. While at first glance it seems to be not more than a nicely arranged set of plants, intended to provide a background for the performance, a closer scrutiny reveals the underlying logic of Grzywnowicz's installation and its integral part in the foreground of the project. All the plants used are species historically abused by the Nazis as 'camouflage, raw material and an involuntary ally,' the artist explains, and continues:

> lupins, which were used to conceal places of violence (extermination camps like Treblinka, Bełżec, or mass graves in forests throughout Poland), pines that were planted at the site of the demolished Sobibór camp, or poplars and rowan which were used to cover fragments of the Auschwitz camp. Furthermore, plants that were in Rudolf Höss's garden in front of his villa in Auschwitz have been used in the installation. The Commander was proud that his wife, Hedwig, had a 'flower paradise' there.
>
> (Grzywnowicz, 2019a)[14]

The hidden layer is not so hidden in Grzywnowicz's installation; the stories of specific plants can be found on information plates. Additionally, in her description of the research-based project, she critically addresses the politics of nature at memorial sites and museums:

> [it] is based on the division of the natural world into 'higher' organisms (trees), which are cared for because of their historical value, and 'lower' organisms, considered as weeds and intruders and blunted because of

Figure 10.4 Karolina Grzywnowicz's installation and its detail. Erbstücke Festival in Dresden, Germany, 26 April 2019.

Photo: Michał Ramus.

> their effect on the appearance of the camp. Since 1994 in Auschwitz-Birkenau [Museum], the latter have been regularly exposed to Roundup, which is a 'total herbicide.' This means that it destroys all herbaceous plants in the area.
>
> (Grzywnowicz, 2019a)[15]

By scrutinising the practices of contemporary museums on sites of extermination, the artist points to long-lasting processes and procedures which are signs of the continuous ancestral catastrophe that Povinelli discusses: classification, exploitation, and extermination for the sake of an aesthetic outcome. The installation is about the logic of the camp-museum's memory politics and system of 'caring' for a site visited annually by hundreds of thousands of people. As such, Grzywnowicz appears to be caring both for past and present alike. One can be easily fooled by the pleasant look of the garden the artist has designed and arranged for *Second Nature*, which makes the knowledge that follows surprising and all the more distressing. Inscribing Nireńska's biographical experience in the broader context of the environmental history of the Nazi genocide, Siniarska and Grzywnowicz's work draws attention to the persistence of past violence in the surrounding landscape of eco-witnesses, as well as to the persistence of violent practices against nature and contemporary exploitation of the environment.

Weeds

In the introductory chapter to her book entitled *Material Witness*, artist, researcher, and member of the group Forensic Architecture,[16] Susan Schuppli defines her title in the following way:

> Material witnesses are nonhuman entities and machinic ecologies that archive their complex interactions with the world, producing ontological transformations and informatic dispositions that can be forensically decoded and reassembled back into a history. Material witnesses operate as double agents: harbouring direct evidence of events as well as providing circumstantial evidence of the interlocutory methods and epistemic frameworks whereby such matter comes to be consequential. Material witness is, in effect, a Möbius-like concept that continually twists between divulging 'evidence of the event' and exposing the 'event of evidence.'
>
> (Schuppli, 2020: 3)

Schuppli is interested in how the relationship between witness/evidence is played out in the context of recorded violence, at the sites of violence and in the vast unknown, which can only become legible once we 'change the story,' first by tuning in to new procedures and questioning old protocols and epistemic approaches. When approaching this vast territory, Schuppli touches upon a fundamental paradox and quotes Karen Barad:

> How did language come to be more trustworthy than matter? Why are language and culture granted their own agency and historicity, while

matter is figured as passive and immutable or at best inherits a potential for change derivatively from language and culture? How does one even go about inquiring after the material conditions that have led us to such a brute reversal of naturalist beliefs when materiality itself is always already figured within a linguistic domain as its condition of possibility?

(Schuppli, 2020: 10–11)

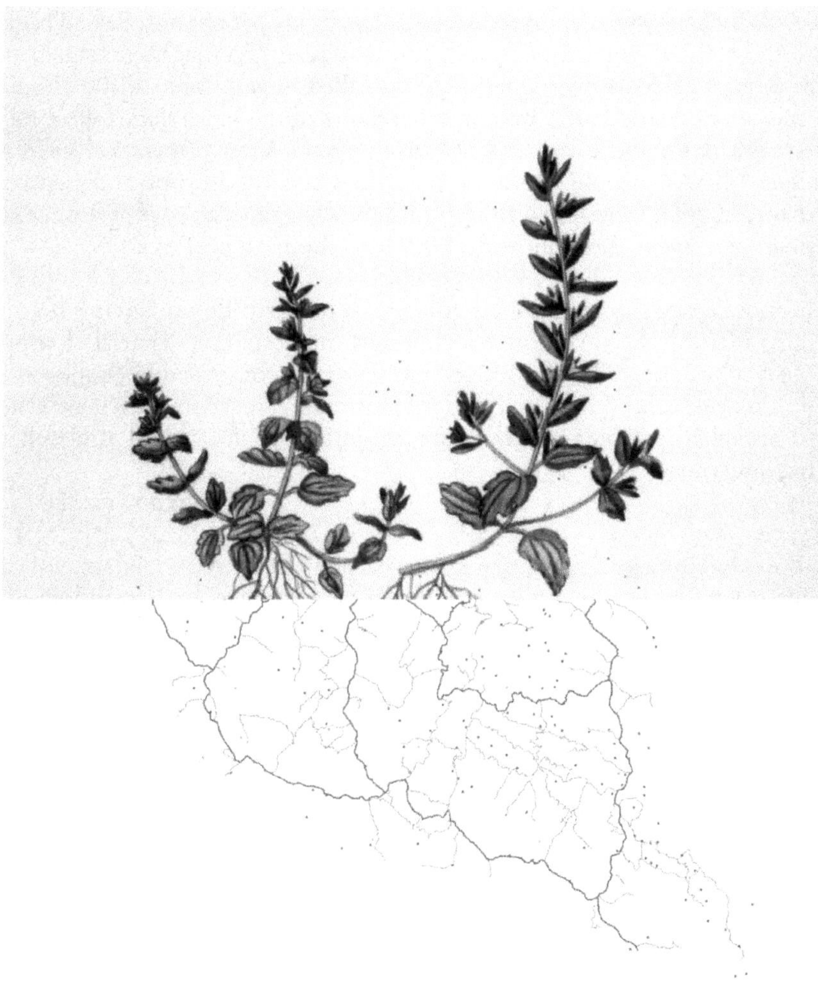

Figure 10.5 Karolina Grzywnowicz, *Weeds* (2015).

In her research-based artwork, Grzywnowicz treats her informants, weeds, as a reliable source of knowledge about past violence. She investigates the depopulated and seemingly mute landscape of south-eastern Poland, the region of Bieszczady. Once densely inhabited, the sites she investigates are now overgrown by vegetation, a sign once immediately decoded as an epitome of neglect and oblivion. Yet in her research, Grzywnowicz looked beyond the cultural muteness or historiographical invisibility of the plants that have taken over spaces previously inhabited by the people who fell prey to post–World War II mass resettlements. Following historical sources, we know that around 620,000 civilians were deported from the Bieszczady Mountains, Beskid Niski, Beskid Sądecki, Pogórze Przemyskie, and Roztocze, and that their settlements were destroyed. Moreover, during the so-called 'repatriation of the Ukrainian minority,' nearly half a million inhabitants of the borderlands were relocated eastward to the Soviet Union. During 'Akcja Wisła' [Operation Vistula], another 140,000 people, this time minorities such as Boykos and Lemkos, were forced out and resettled to territories newly gained as part of the Polish-German agreement (Kochanowski, 2001). As the artist observes:

> In most cases the abandoned villages were burnt down leaving no visible trace of the farmhouses, churches and buildings that had previously stood there. The buildings that were not consumed by flames were deconstructed and used as building materials to harden the roads. Once so prevalent, villages are gone forever, and the multicultural atmosphere has drawn to a close.
>
> (Grzywnowicz, 2015a)

So, how does one know or learn how to look in order to see, in spite of 'no visible traces,' to establish new regimes of visibility and understanding of a natural environment saturated with historical violence? Grzywnowicz shares the details of her research trips and onsite investigations: penetrating areas where she uncovered layers of subsequent processes of natural succession, coming across single fruit trees which grow surprisingly inside an arrangement of plants that looks like a forest, or remnants of orchards which are an obvious sign of human lives that were part of this ecosystem. These are rather evident oddities in the landscape that can trigger further inquiry. Also, as meticulously documented by the artist, there are dozens of species of plants whose 'testimonies' can be unearthed: 'gooseberries, redcurrants, hazelnut, walnut, and linden trees, accompanied by the ornamental plants that used to be cultivated in home gardens (such as rudbeckia lacinata, daffodils, irises, asters), the plants that are connected with certain rituals (periwinkle, evergreen boxwood, *viburnum opulus*),' as well as 'stinging nettle, couch grass, or alpine dock,' all of which reveal traces of human lives and their brutal ends on this territory (Grzywnowicz, 2015b). The artist's research also follows the lives of humans and plants, their various entangled narratives. As Paweł Mościcki points out in an essay accompanying her project:

> the lesser periwinkle (*Vinca minor*) symbolised immortality, which is why it was used to decorate Easter baskets. For the same reason it accompanied funeral ceremonies and was planted on graves. Unmarried women were buried in a periwinkle garland. At the same time, it was an indispensable element of wedding ceremonies: the bride wore a periwinkle garland and prior to the wedding, houses were decorated with periwinkle wreaths.
> (Mościcki, 2015)

Grzywnowicz is interested not only in cataloguing her findings, but also in finding their meaning, their history, and their deeply rooted presence in the present moment. Obviously, as Schuppli notes:

> the mere fact that materials capture and archive eventful processes within their substratum, or harbour information as metadata, does not convert such entities into de facto material witnesses capable of testifying before the tribunals of history. Matter becomes a material witness only when the complex histories entangled within objects are unfolded, transformed into legible formats, and offered up for public consideration and debate.
> (Schuppli, 2020: 16)

Figure 10.6 Karolina Grzywnowicz, *Weeds*, at the exhibition Floraphilia. Plants as archive, Akademie der Künste der Welt, Cologne 15.09.2018–18.11.2018.

What does it take, then, to make matter, matter? Weeds, the plants discarded from a garden, that grow in a superfluous way, unnecessary, become, for Grzywnowicz, a means of communicating not only with the past, but also with the present of her own oblivion and her memory of beings both human and non-human. She enters a dense network of connections and relations, and her intervention can be considered a violent one, and as such not entirely unproblematic (Zajączkowska, 2019).[17] Resettlement of part of a meadow into a gallery space can be treated as re-enactment of historical violence, this time displaced onto plants deprived of their habitat and their right to resist. I find this an important element of the project, as it reveals numerous problems related to living not in the aftermath, but rather in the *long durée* of violence, the ancestral catastrophe which leaves little room for righteousness. Paradoxically, in an analogy to Jacek Małczyński's argument regarding Łukasz Surowiec's *Berlin-Birkenau* project (2012), *Weeds* can be also interpreted as pointing to humans' destructive power and disregard for the natural world's life, memory, and creativity. The question one could ask in this context is whether there are forms of human interaction with nature that remain innocent and ethical, if the regimes of ethics and responsibility are human *per se*. The artist's aim, it seems, is not so much care and protection, as it is raising consciousness and teaching the difficult art of seeing and memory. Grzywnowicz offers both new questions and new answers about past violence, and the present moment of revisiting this past and coming to terms with it.

The artwork *Weeds* has been presented on several occasions and at various venues (Figure 10.7). It is a multimedia project that includes an installation made of a section of a meadow (a plant colony) relocated by the artist:

> I cut it in squares: half by half-meter . . . and 30 centimetres in depth – that's how plants root. I drove with it at night so it wouldn't get steamed up in the car. In the gallery, it grows on a support and drainage system similar to the one used to make roof gardens. We also set up a system of lights to mimic sunlight. The meadow is accompanied by a map with marked locations of former villages to reflect the scale of the phenomenon.
>
> (Sulej and Grzywnowicz, 2015)

Looking at it in a gallery space can be both a confusing and an instructive experience: a piece of landscape to a certain extent always looks nice, comforting, or may bring about nostalgic feelings; yet upon realising what (who) these plants actually are, and what they are supposed to do to us, dissonance and discomfort and introduced. On the one hand, we do not want to know, or we already know anyway, what this will add to the burden of a dark heritage. Yet, it is not about the past and what we know of

Figure 10.7 Karolina Grzywnowicz, *Weeds*, at the Zachęta National Gallery of Art, Project Space, Warsaw, 4.09.2015–1.11.2015.

Photo: Marek Krzyżanek.

it, but about what we do not know about our present, about the blindness we willingly (socially) perform, and the surprise that there is still more to know, more to our being in history than we have hitherto acknowledged. This is precisely what Grzywnowicz seems to be aiming at: not so much the correction of our epistemic habits, but rather our practices of relationality and commemoration.

After being exhibited at the Project Space of the Zachęta National Gallery of Art in Warsaw, the installation was moved to a public space in the Powiśle district of the city, in the vicinity of the Museum of Modern Art, where it remains to this day, 'evidence of the event' and the 'event of evidence,' as Schuppli would have it (Schuppli, 2020: 3). The transformation of matter continues, and so do our historical sensibilities. To understand *Weeds* in all its complexity, one needs to locate the artist's gesture as – to use Schuppli's words again – 'undertaken with the specific aim of radicalizing the political relationships that such matter maintains to acts of witnessing and forms of testimony' (Schuppli, 2020: 28). Both acts of witnessing and forms of testimony need to be constantly rethought in relation to matter and media, both of which surround human beings and partake in their history, not as mere recorders, but also active and responsive agents whose role should not be ignored.

Coming after, as is the case with Grzywnowicz and artists of her generation, necessitates not so much new but rather adequate forms of response that need to relate themselves to institutions and disciplines, genres and affects, all of which have imposed authority on the knowledge of the past. To 'radicalize what matters as evidence' (Schuppli, 2020: 19) is also to radicalise what counts as autobiography, interpretation, reinterpretation, restaging, etc.; it is an intervention made in the present about the status of the past and forms of historicity, that is, about the approach to that past. Depending on how we see the connection and interdependence between the past and the present, whether we wish to 'change the story' for the sake of tomorrow, will doubtless influence what we chose as our concepts of futurity, creativity, and relationality, but also survival.

Conclusions

Both Grzywnowicz and Siniarska perform a kind of ecological backward-looking, thus practising an ecological mentality, as theorised by Malabou. Armed with imagination and investigative skills, as well as with academic research to support them, they allow us to see more, to want to see more, and to look where we thought there was nothing to see (or where we were not expecting to look). Arriving at the sites of past violence or revisiting the histories of those who suffered this violence, they introduce new approaches. There are ethics of investigative care at work in their projects. Finally, one could also say their projects deconstruct the understanding of life on the sites of death, and do so within ancestral catastrophe. They point to a continuity of life, but in different forms: different forms of life and different forms of continuity.

Siniarska continues her research, now in the framework of *Museum of the Anthropocene*. Grzywnowicz keeps investigating lives in the aftermath of violence, most recently in the Palestinian camps of Aida, Dheisheh, Fawwar, al-Arroub, Balata, Jalazone, and Qalandiya, where she recorded lullabies performed by female inmates. The lullabies are archives as acts of resistance against Israel's brutal 'necropolitics' (Mbembe, 2003). In the projects discussed in this chapter, I have hoped to show how certain historical sensibilities are shaped by means of artistic interventions, how practices of life and protocols of knowledge are shifted, and how territories for transdisciplinarity open to recognise and counter an ongoing destruction.

Notes

1 Research leading to this publication was funded by National Science Centre, Poland, grant number: 2021/41/B/HS2/01437. I would like to thank Daniel Malone for proofreading the chapter.

2 Eric Katz discussed the relationship between genocide and ecocide with recourse to the concept of domination (Katz, 2001, 2015). I would like to express my gratitude to the editors of this volume for directing me to these sources.
3 These thoughts proved useful also for the comparative critical analysis of Joanna Rajkowska's exhibition entitled *Rhizopolis*, and *Tu Muranów* [Here, Muranow] made with the participation of Artur Żmijewski. See Bojarska (2021).
4 See also the brochure accompanying Siniarska's project *Prototypes 04. (Land) Slide*, curated by Katarzyna Słoboda at MS2, 2020.
5 In 'On the Movement of Plants and the (Im)Possibility of Plant-and-Human Choreographies.' Magdalena Zamorska discusses in detail her research collaboration with Siniarska, which specifically attempts to work out choreographic techniques for a moving-together of human and plant, and of the embodied practice *becoming-plant* (Zamorska, 2020).
6 The artist's words have been transcribed verbatim.
7 The series, begun in 2018 as part of celebrations of the 100th anniversary of Poland regaining independence, was sponsored by the Adam Mickiewicz Institute. *Second Nature* opened the second season of the series.
8 See my discussion of the possibilities and limitations of a biographical study, exemplified by Weronika Kostyrko's monumental biography of Nireńska, with reference to *Second Nature* (Bojarska, 2020).
9 Installation: Karolina Grzywnowicz; choreography and performance: Agata Siniarska and Katarzyna Wolińska; dramaturgy: Mikołaj Szymanówka; production: Stowarzyszenie Twórców 'Cztery wymiary to dla nas za mało.' Premiere: 26 April 2019.
10 In 2018, POLIN Museum of the History of Polish Jews hosted a performance entitled *The Other Side*, which integrated a dance narrative about Nireńska's artistic biography into the museum's spectacular architecture. Two years earlier, *Lament* (a part of *Tetralogy of the Holocaust*) was reconstructed in Washington, D.C., at the invitation of Paul Gordon Emerson.
11 I am grateful to the author and to Karolina Grzywnowicz for kindly sharing this unpublished text with me. See also Berendt (2020: 145–156).
12 For a further analysis of the Nazis' racist politics and attitude to nature, see Wilson (2012), Blood (2010), and Brüggemeier et al. (2005). For an interesting example of a research-exhibition, see also *Race and Forest*, prepared by the interdisciplinary research team INTERPRT, which concentrates on intersections between international humanitarian law and legal protection of the environment. The project was presented at Biennale Warszawa in spring and summer 2020. See the online documentation at https://biennalewarszawa.pl/en/rasa-i-las-2/. See also the collective's website at www.interprt.org/ (Accessed 12 May 2022).
13 It is unclear why Nireńska had to leave Wigman's dance company and Germany. Her biographer, Weronika Kostyrko, provides two different explanations: either it was because of growing antisemitism or because she was Wigman's lover, which under Nazi rule could end the latter's career (Kostyrko, 2019).
14 The artist's words have been transcribed verbatim.
15 'Blunted' is the word originally used by the artist. 'Eradicated' would surely be a better choice here.
16 Forensic Architecture (FA) is a research agency working with architectural techniques and technologies to investigate state and corporate terror and the violations of human rights. The group is led by architect Eyal Weizman. They are based at Goldsmiths, University of London. See 'Forensic Architecture.'

17 A poet, writer, cultural critic, and scholar of botany, Urszula Zajączkowska has discussed some of the contemporary artworks that make use of plants and their problematic approach to the natural world and violence, including *Weeds* by Grzywnowicz, a project she considers rather anthropocentric and violent in that it disrespects the lives of plants and subjugates them to the logic of art production.

Bibliography

Barad, Karen (2007), *Meeting the Universe Halfway: Quantum Physics and the Entanglement of Matter and Meaning* (Durham: Duke University Press).

Bennett, Jill (2012), *Living in the Anthropocene* (Ostfildern: Hatje Cantz).

Berendt, Zuzanna (2020), 'Mnogie natury spektaklu. O konieczności przekraczania estetyki performatywności w badaniach nad nowym tańcem', *Kultura Współczesna. Teoria. Interpretacje. Praktyka*, 4:111, 145–156, https://doi.org/10.26112/kw.2020.111.11

Berendt, Zuzanna (unpublished), 'Doświadczenie w sieci relacji: Druga natura Karoliny Grzywnowicz i Agaty Siniarskiej' [Experience in the Network of Relationships: Karolina Grzywnowicz and Agata Siniarska's 'Second Nature'].

Blood, Philip W. (2010), 'Securing Hitler's Lebensraum: The Luftwaffe and the Białowieża Forest, 1942–1944', *Holocaust and Genocide Studies*, 22:4, 247–272, https://doi.org/10.1093/hgs/dcq024

Bojarska, Katarzyna (2020), 'Nireńska: o kruchości archiwum i ciele tańczącym tu i teraz' [Nireńska: On the Fragility of the Archive and a Body Dancing Here and Now], *Pamiętnik Teatralny*, 69:3, 209–224, https://doi.org/10.36744/pt.508

Bojarska, Katarzyna (2021), '"Nie istnieje nic innego niż zależność." Refleksja o historyczności,' *View. Theories and Practices of Visual Culture*, issue 29, [Online], [Accessed 15 May 2022], available from: www.pismowidok.org/pl/archiwum/2021/29-obrazy-i-wyobrazenia-rasy-historie/nie-istnieje-nic-innego-niz-zaleznosc

Bonneuil, Christophe and Jean-Baptiste Fressoz (2016), *The Shock of the Anthropocene*, trans. David Fernbach (London: Verso).

Brüggemeier, Franz-Josef, Mark Cioc and Thomas Zeller (eds.) (2005), *How Green Were the Nazis? Nature, Environment and Nation in the Third Reich* (Athens, OH: Ohio University Press).

Butler, Judith (2020), *The Force of Nonviolence: An Ethico-Political Bind* (London: Verso).

Chakrabarty, Dipesh (2009), 'The Climate of History: Four Theses', *Critical Inquiry*, 35:2, 197–222, https://doi.org/10.1086/596640

Choreografia w sieci. Rozmowa z Agatą Siniarską (n.d.), [Online], [Accessed 15 May 2022], available from: https://choreografiawsieci.pl/agata-siniarska/

Cole, Tim (2020), 'Expanding (Environmental) Histories of the Holocaust', *Journal of Genocide Research*, 22:2, 273–279, https://doi.org/10.1080/14623528.2020.1726653

Czwarty 'Powrót (do) przyszłości,' *Calisia.pl*, [Online], [Accessed 15 May 2022], available from: https://calisia.pl/czwarty-powrot-do-przyszlosci,10438

Faber, Rima (2019), 'Ghosts of the Past: The Creation of Pola Nirenska's *Holocaust Tetralogy*', *Dance Today*, 36 (September), 34–38.

'Forensic Architecture', [Online], [Accessed 18 May 2022], available from: https://forensic-architecture.org/

Gessen, Masha (2020), 'Judith Butler Wants Us to Reshape Our Rage', *New Yorker*, 9 February, [Online], [Accessed 15 May 2022], available from: www.newyorker.com/culture/the-new-yorker-interview/judith-butler-wants-us-to-reshape-our-rage
Grzywnowicz, Karolina (2015a), 'Weeds-project', [Online], [Accessed 15 May 2022], available from: https://karolinagrzywnowicz.com/project-3/
Grzywnowicz, Karolina (2015b), 'Weeds-plants', [Online], [Accessed 15 May 2022], available from: http://chwasty.com/rosliny/?ln=en_EN
Grzywnowicz, Karolina (2015c), 'Weeds', [Online], [Accessed 15 May 2022], available from: http://chwasty.com
Grzywnowicz, Karolina (2019a), 'Second Nature-text', [Online], [Accessed 15 May 2022], available from: https://karolinagrzywnowicz.com/text-6/
Grzywnowicz, Karolina (2019b), 'Second Nature', [Online], [Accessed 15 May 2022], available from: https://karolinagrzywnowicz.com/second-nature/
Haraway, Donna J. (1986), 'Primatology Is Politics by Other Means', in Ruth Bleir (ed.), *Feminist Approaches to Science* (New York: Pergamon Press).
Haraway, Donna J. (2016), 'Tentacular Thinking: Anthropocene, Capitalocene, Chthulucene', *e-Flux Journal*, 75, September, [Online], [Accessed 15 May 2022], available from: www.e-flux.com/journal/75/67125/tentacular-thinking-anthropocene-capitalocene-chthulucene/
Hellerau (2019), 'Interview with Agata Siniarska', [Online], [Accessed 15 May 2022], available from: www.hellerau.org/en/agata-siniarska/
Katz, Eric (2001), 'Nature's Healing Power, the Holocaust, and the Environmental Crisis', in Martin D. Yaffe (ed.), *Judaism and Environmental Ethics: A Reader* (Lanham: Lexington Books), 309–320.
Katz, Eric (2015), *Anne Frank's Tree: Nature's Confrontation with Technology, Domination, and the Holocaust* (Winwick: The White Horse Press).
Kochanowski, Jerzy (2001), 'Gathering Poles into Poland: Forced Migration from Poland's Former Eastern Territories', in Philipp Ther and Ana Siljak (eds.), *Redrawing Nations: Ethnic Cleansing in East-Central Europe, 1944–1948* (Lanham: Rowman & Littlefield), 135–154.
Komuna Warszawa (n.d.), [Online], [Accessed 15 May 2022], available from: https://komuna.warszawa.pl/spektakle/agata-siniarska/
Kostyrko, Weronika (2019), *Tancerka i Zagłada: Historia Poli Nireńskiej* (Warszawa: Czerwone i Czarne).
Malabou, Catherine (2019a), 'The Brain of History, or, the Mentality of the Anthropocene', in Magdalena Ziółkowska (ed.), *Plasticity of the Planet: On Environmental Challenge for Art and Its Institutions* (Milan: Mousse), 167–184.
Malabou, Catherine (2019b), 'The Plasticity of the World: Philosophy, Neuroscience, and Feminism for the Future. Catherine Malabou in Conversation with Ewa Majewska', in Magdalena Ziółkowska (ed.), *Plasticity of the Planet: On Environmental Challenge for Art and Its Institutions* (Milan: Mousse), 147–166.
Małczyński, Jacek, et al. (2020), 'The Environmental History of the Holocaust', *Journal of Genocide Research*, 22:2, 183–196, https://doi.org/10.1080/14623528.2020.1715533
Mauch, Christof (ed.) (2004), *Nature in German History* (New York: Berghan Books).
Mbembe, Achille (2003), 'Necropolitics', trans. Libby Meintjes, *Public Culture*, 15:1, 11–40, https://doi.org/10.1215/08992363-15-1-11
Mościcki, Paweł (2015), 'Reading the Weeds', [Online], [Accessed 15 May 2022], available from: https://karolinagrzywnowicz.com/text-5/
Mozingo, Karen A. (2008), 'Crossing the Borders of German and American Modernism: Exile and Transnationalism in the Dance Works of Valeska Gert, Lotte Goslar and Pola Nirenska' (PhD Dissertation, Graduate School of The Ohio

State University), [Online], [Accessed 15 May 2022], available from: http://efaidnbmnnnibpcajpcglclefindmkaj/https://etd.ohiolink.edu/apexprod/rws_etd/send_file/send?accession=osu1214361531&disposition=inline

Muzeum Susch (n.d.), 'Agata Siniarska: The Soft Act of Killing', [Online], [Accessed 15 May 2022], available from: www.muzeumsusch.ch/en/1198/Agata-Siniarska-THE-SOFT-ACT-OF-KILLING

Obarska, Marcelina (2019), 'Rośliny i przemoc: artystki o spektaklu *Druga natura*', *Culture.pl*, [Online], [Accessed 15 May 2022], available from: https://culture.pl/pl/artykul/jakie-jest-cialo-tanca-modern-artystki-o-spektaklu-inspirowanym-biografia-poli-nirenskiej

Piasecka, Magdalena and Agata Siniarska (2021), 'Solo o relacji', *Teatr*, 6, [Online], [Accessed 15 May 2022], available from: https://teatr-pismo.pl/16012-solo-o-relacji/

Povinelli, Elizabeth A. (2021), *Between Gaia and Ground: Four Axioms of Existence and the Ancestral Catastrophe of Late Liberalism* (Durham: Duke University Press).

Race and Forest (2020), [Online], [Accessed 15 May 2022], available from: https://biennalewarszawa.pl/en/rasa-i-las-2/

Schuppli, Susan (2020), *Material Witness: Media, Forensics, Evidence* (Cambridge, MA: MIT Press).

Schuppli, Susan (n.d.a), [Online], [Accessed 15 May 2022], available from: https://susanschuppli.com/

Siniarska, Agata (n.d.b), 'About', [Online], [Accessed 15 May 2022], available from: http://cargocollective.com/agatasiniarska/about

Siniarska, Agata (n.d.c), 'In the Beginning Was a Copy: Lecture Performance', [Online], [Accessed 15 May 2022], available from: http://cargocollective.com/agatasiniarska/LECTURE-PERFORMANCES

Siniarska, Agata (n.d.d), 'Landslip', [Online], [Accessed 15 May 2022], available from: https://msl.org.pl/prototypes-04-agata-siniarska-landslip/

Stengers, Isabelle (2009), *Au temps des catastrophes: Résister à la barbarie qui vient* (Paris: Découverte).

Sulej, Karolina and Karolina Grzywnowicz (2015), 'Akcja "Wisła" była wyrywaniem ludzi z ich ziemi jak chwastów' [Operation 'Vistula' Was the Uprooting of People from Their Land Like Weeds], *Wysokie Obcasy*, 19 May, [Online], [Accessed 15 May 2022], available from: www.wysokieobcasy.pl/wysokie-obcasy/7,53662,18828125,karolina-grzywnowicz-akcja-wisla-byla-wyrywaniem-ludzi-z.html

Szymajda, Joanna (ed.) (2017), *Polskie artystki awangardy tanecznej: Historie i rekonstrukcje* (Warszawa: Instytut Muzyki i Tańca, Instytut Adama Mickiewicza).

Wilson, Jeffrey K. (2012), *The German Forest: Nature, Identity, and a Contestation of a National Symbol 1871–1914* (Toronto: University of Toronto Press).

Zajączkowska, Urszula (2019), 'Martwa natura' [Still Life/Dead Nature], *Dwutygodnik*, 5:257, [Online], [Accessed 15 May 2022], available from: www.dwutygodnik.com/artykul/8299-martwa-natura.html

Zamorska, Magdalena (2020–2021), 'O ruchu roślin i (nie)możliwości roślinno-ludzkich choreografii', *Sensus Historiae*, 37, 53–63.

11 'Ruined and wrecked!'

Annie Proulx confronts the ecocatastrophe

Hannah Jocelyn

'All human knowledge and understanding rises from stories,' and 'Fiction reflects our perception of ourselves in our time, our general human and individual position in the universe,' writes Annie Proulx in an essay entitled 'Dangerous Ground' (2009: 8–9), part of an interdisciplinary book about regionalism in the humanities. 'Our time' is one of ecological peril. In Proulx's fiction, three affective tropes that evince that peril regularly arise: characters in motion, homes in disrepair, and ecosystems in danger. The former two themes are not necessarily caused by the latter, but they augment the sensation of solastalgia – the distress that develops when one's home is ecologically imperilled. All three tropes, woven through every novel, help the reader to feel the destruction humankind has wrought and to get a sense of what is to come: a shattered, uprooted existence. To use Proulx's words again, her protagonists are up 'against the idea of mass, whether landscape or a crushing social situation or powerful circumstances' (Bolick, 1997). By 'mass' she means some kind of overwhelming, external force – in this case the ever-escalating processes of extraction, exploitation, and degradation upon which our society is designed to depend. Her method is subtle but encompassing, a practice she calls 'deep-landscape literature,' in which she endeavours to capture the entire ecosystem of a place, providing her reader with a profound understanding of its history and anatomy. This newfound understanding, I can only imagine, is intended to spark in the reader an accompanying environmentalism. In other words, when one realises through reading about years of pollution and years of perseverance against that pollution, one should be moved to care for the place under threat. In the case of Proulx's work, the places include Newfoundland, New England, Wyoming, and Texas. This chapter will explore the thematic throughlines featured in Proulx's novels *Postcards* (1992), *The Shipping News* (1999b), *That Old Ace in the Hole* (2009), and *Barkskins* (2016). In each novel there is always at least one character on the constant move, a home falling apart, and the environment (plant, animal, mineral) is under threat from human mistreatment. With these themes, Proulx elicits

DOI: 10.4324/9781032726953-12

the reader's solastalgia and thereby creates the conditions for eco-activism. Her deep-landscape literature shows us what is at stake, and what we might do to combat the impending ecocatastrophe.

Proulx was born in Norwich, Connecticut, in 1935. She went to high school in Portland, Maine, and started college at Colby, but finished at the University of Vermont in 1969. In 1971, she moved to Montreal, Quebec, for post-graduate education, and after completing her MA and passing her doctoral oral exams in 1975, she left to live in Canaan, Vermont (on the U.S.-Canada border). She lived in Vermont for almost thirty years, and then moved to Newfoundland. She wrote an article, 'House Leaning on Wind,' for the October 1997 issue of *Architectural Digest* about her home 'of unusual character and curious construction,' 'only accessible by sea,' on the 'Great Northern Peninsula.' She describes the house as 'gaunt and nearly a wreck' with a 'roof full of holes.' In the article, deep-landscape ideas already begin to appear:

> I felt a super natural alertness to the landscape, an anxiety that it would all disappear if I slept, an eager curiosity about the people who lived here and the fateful hand of history that had seized the island's shoulders so many times.

In 1996, she left Newfoundland for Wyoming, where she bought a house by the Medicine Bow National Forest, at an altitude of 8100 feet, and spent a decade going back and forth between Canada and the United States. Not too long after, she built another house in Wyoming, and moved again, at least temporarily, into the home she called Bird Cloud, which she hoped would be her final house – but it was not to be. She moved to the Pacific Northwest, living on five acres of red-cedar woodland in the Snoqualmie Valley, just outside of Seattle, Washington. It turned out that she was allergic to the cedars, so she moved again in 2022, this time to New Hampshire (Weaver, 2022). Proulx has had a lifetime of building homes and leaving homes behind her; perhaps this is why she is able to so vividly depict similar instability in her characters.

Proulx started her career as a journalist, writing mostly for food magazines such as *Gourmet* under the name E. A. Proulx. In 1988, she published her first book, a collection of short stories called *Heart Songs, and Other Stories*. Four years later, she published her first novel, *Postcards*, followed shortly after by *The Shipping News*, which won a Pulitzer Prize and a National Book Award. She is probably best known for her short story 'Brokeback Mountain,' which was originally published in *The New Yorker* in 1997 and then appeared in the collection *Close Range: Wyoming Stories*, published in 1999. In all, she is the author of eleven books, and each one contains a close connection to the landscapes in which they are set.

That Proulx's work has ecological underpinnings is not a new contention. In 2009, Alex Hunt, an English literature scholar, edited an entire collection devoted to the ways Proulx's writing engages with regional 'weather, landscape, history, and local economy' to offer a 'critique of postmodern culture' (2009: 1), entitled *The Geographical Imagination of Annie Proulx*. I obviously agree with the consensus of that collection:

> Proulx's work seems to be moving toward an ethics of place based on environmentally sustainable ways of living and laboring on the land and on the values of local community cooperation. Developing a sense of place, Proulx's work seems increasingly to attest, leads to environmental ethos.
>
> (Hunt, 2009: 5)

But what I am arguing here is not merely that her work both illustrates and ignites a relationship with place – so much has already been proven. Proulx herself has stated outright, 'Place and history are central to the fiction I write, both in the broad, general sense and in detailed particulars' (Proulx, 1999a; see also Rood, 2001). I am arguing, instead, that she uses these three particular, repeating themes in order to engender feelings of solastalgia in her reader, feelings that can and should be turned into action: plant a garden, protest, divest land, join a social-justice organisation, call a politician, be neighbourly, boycott polluters, write propaganda, participate in mutual aid. Proulx has said that:

> [a] novel should take us, as readers, to a vantage point from which we can confront our human condition, where we can glimpse something of what we are. A novel should somehow enlarge our capacity to see ourselves as living entities in the jammed and complex contemporary world.
>
> (Proulx, 1999a)

What we do once we have seen, how productive we make that confrontation, is up to us.

English literature scholar Mark Asquith put it well when he said: 'Welcome to Proulx country – although readily locatable on a map, it is a landscape made fantastic by the immensity of its history and its oppressive influence upon those who scratch a living from its soil' (2009: 16). It is the deep-landscape method that lends Proulx's work that immensity. Because of its profound connection to place and its implicit call to action, deep-landscape literature invites the reader to participate in literary environmentalism. The terms 'ecolit,' 'environmental literature,' and 'literary environmentalism' (and others) are often used interchangeably, but it is important to distinguish between them. To me, the difference between traditional ecolit – often associated with such white male writers as Henry David Thoreau, Ernest Callenbach, and Richard Powers – and literary environmentalism is in the way the latter makes room for the reader. Ecolit is the creation of so-called expert knowledge about the wild or so-called 'natural' world and the way narrative work reflects and responds to it. In contrast, literary environmentalism is about collective participation and action. It means reading with the intention of surfacing the writer's underlying ecological stance. Ecolit is a cultural artefact, a product, whereas literary environmentalism is an act, a way of reading, critiquing, becoming involved with the author to achieve her agenda. A reader must decide from the outset to read in an environmentalist frame of mind, to choose to see what is implicit in such tropes as those Proulx emphasises. In *Slow Violence and the Environmentalism of the Poor* (2011), Rob Nixon, a leading academic in the environmental humanities, suggests that writers who can help envisage ecological destruction are activists. They help us to

> apprehend threats imaginatively that remain imperceptible to the senses, either because they are geographically remote, too vast or minute in scale, or are played out across a time span that exceeds the instance of observation or even the physiological life of the human observer.
> (2011: 15)

The writer makes visible what might have felt bygone, and when we can see that past damage and its present threat, we are motivated to act.

Proulx's work provides the imagery to recognise what is a deep-rooted and abiding threat. As she explains, it is the writer's work to make history feel present, to recover what has been lost: 'It is a writer's thought that nothing we see has meaning unless we understand how it was in the past – the element of time, the slow accretion of change, the seeming immutability of rural land forms' (2009: 18). Here she is echoing the influential landscape theorist Donald Meinig, who writes that 'Life must be lived amidst that which was made before. Every landscape is an accumulation. The past endures' (1979: 44). Deep-landscape literature surfaces that past. 'Deep landscape' is the term Proulx used in her essay 'Dangerous Ground' to describe the kind of writing that makes setting so powerfully all-encompassing that it becomes a shaper and instigator of the story. This is literature in which 'landscape seemed fixed and immutable, ordering the lives of the characters who moved within it, forcing the events of the story' (2009: 7). As Alex Hunt notes in the introduction to *The Geographical Imagination of Annie Proulx*, 'she foregrounds the geography – land forms, environmental facts, weather, and so on – and shows human individuals and communities as subject to the conditions of that geography'

(2009: 3–4). In a deep-landscape novel, 'the story that unfolds can only happen because of *where* it happens' (2009: 7, original italics). And the *where* is precise, imbued with the history and materiality of that particular place. It is regionally specific. Even before she named it deep-landscape literature, Proulx wrote with this approach: 'From place come the characters, the way things happen, the story itself' (Proulx, 1999a). 'Everything of the natural place may swell into the story,' Proulx writes, 'the underlying geology, insects, temperature extremes, shoreline, seasonal shifts of daylight' (2009: 14). The landscape begins deep in the soil, back through layers in time. Deep-landscape literature allows readers to understand how we have arrived at this inflection point in history, how the crisis we are experiencing and the catastrophe we are anticipating is part of a long saga of abuse of Earth and its human and nonhuman inhabitants. As Proulx once said in an interview in 1999, 'The present is always pasted on layers of the past (Proulx, 1999a).

Trying to define 'landscape' has been a source of academic debate. As Meinig notes in his essay 'The Beholding Eye,' there are many explanations of landscape, and variations in observing place – for example, describing landscape as *nature* implies that it is untouched, and its representation will disregard human intervention, whereas using a term like *ecosystem* suggests a scientific approach, broken into parts and processes for analysis (1979: 44). The words a writer chooses in her description signifies how she reads the place. For Proulx to choose to capture a place's entire past and present, its psychic and physical condition, is an immense undertaking. Here are her words of definition:

> Landscape is what is out there, what was out there, and, for the futurologists, what may be out there some day. . . . Landscape is geography, geology, archaeology, astrophysics, agronomy, agriculture, the violent character of the atmosphere, climate, black squirrels and wild oats, folded rock, bulldozers; it is jet trails and barbwire, government land, dry stream beds; it is politics, desert wildfire, introduced species, abandoned vehicles, roads, ghost towns, nuclear test grounds, swamps, a bakery shop, mine tailings, bridges, dead dogs. Landscape is rural, urban, suburban, semirural, small town, village; it is outports and bedroom communities; it is a remote ranch. Landscape is neither pure nor static. We may speak of the rural landscape as an independent (though shrinking) entity, but it is only possible to isolate rural countryside in the mind. For everything is linked.
>
> (2009: 10)

Describing landscape this way runs the risk of neutralising the term: if it is everything, then it is nothing. But what I believe she means to do here is to contend that all these constituent characteristics make a place singular. Every place will contain a distinct combination of tellurian actions and ecological reactions, creating a unique landscape: 'Landscape is the sum of accumulated changes wrought by the inhabitants on their marks on the land' (2009: 14). And yet, Proulx notes, a place does not stand alone; it is intertwined with the places around it and beyond, as every landscape is affected by the larger scope of land. Her idea of landscape is defined by its connections: to its past, to its present, to its potential, to its neighbour, to its nation, to its hemisphere, to the world. In the epigraph of her 2022 publication, *Fen, Bog & Swamp: A Short History of Peatland Destruction and Its Role in the Climate Crisis*, Proulx describes her enchantment with making visible invisible connections. She writes,

> It is a little like a foggy summer morning when we can see beaded spiderwebs strung between stalks and stems, between tree and ground, twig and leaf. As the sun heats the earth the droplets evaporate, and the illusion that the entire world is held together by fine spider threads evaporates with it.
>
> (2022)

These sentences evoke both the connectivity and its tenuousness. I do not think it is too much of a reach to equate the warming of the day she describes with the global heating of the planet, and the way that as it gets hotter, our grasp on the linkages dissolves.

Although Proulx does not reference it directly, deep landscape echoes (in name and in approach) a philosophy called deep ecology, created in the first wave of environmental criticism in the early 1970s by a Norwegian philosopher named Arne Naess. Deep ecology – defined in opposition to shallow ecology, which focuses primarily on mitigating pollution and resource depletion – is contingent on self-realisation, on altering our attitudes towards non-human entities such as animals, forests, even whole bioregions. The aim is to realise that each member of the ecosphere has inherent and equal value, and that natural diversity and symbiosis are crucial to the survival of the whole – without anthropocentric emphasis on humankind (Naess, 1973: 4). Humans have no claim, beyond satisfying basic vital needs, to earthly resources. Deep landscape, which almost translates the philosophy of deep ecology into a method of writing and of reading, resonates beyond literature, meaning it might be possible to take its lessons from the page into the world – using active reading to become activists.

The deep-ecology attitude is a stance long proffered by Indigenous storytellers and scholars, often called Traditional Ecological Knowledge (TEK) (LaDuke, 1994). Like deep ecology, TEK emphasises the 'web of relationships between humans, animals, plants, natural forces, spirits, and land forms in a particular locality' (Battiste and Henderson, 2000: 44). But beyond a descriptive philosophy, TEK is also 'prescriptive – that is, it provides an account of how people *ought to act* in relationship with nature' (McGregor, 2018: 109, emphasis in original). Ronald Trosper, a scholar of Indigenous economics, has outlined four tenets of TEK that are consistent among all Indigenous peoples, regardless of variations among Native nations. The first is community, a responsibility to all beings; the second is connectedness, an understanding of how we all are; the third is seventh-generation thinking, a consideration of the long-term consequences of every action; and the fourth is humility, the knowledge that we are small and 'the natural world is powerful and will be able to cause trouble if not treated properly' (Trosper, 1995: 67). Similar lessons reverberate in Proulx's deep-landscape literature. As noted, characters in her work are small compared to a large and powerful world; they are '*of* the landscape, their lives ruled by place' (Proulx, 2009: 14, emphasis in original). And with her writing, Proulx encourages care for all human and non-human lives and draws connections between eras bygone and incoming, species of plant and animal, landscapes near and far.

This understanding of the interconnection of all living and non-living things, past, present, and future, may have been informed by Proulx's earliest memory, one of dappled light shimmering through forest trees. She has described having learned 'that there was much more to a forest than trees,' coming to sense 'the extraordinary symbiotic connections between soil, land shape, weather, senescence, decay, propagation, seeds and roots, insects, birds, reptiles, and mammals, including humans' (Treisman, 2013). Proulx is deftly able to translate this expansive symbiosis from the real world to her work. One line, from *Barkskins*, beautifully captures this connection:

> as he cut, the wildness of the world receded, the vast invisible web of filaments that connected human life to animals, trees to flesh and bones to grass shivered as each tree fell and one by one the web strands snapped. (2016: 12)

In all her work, Proulx captures the tenuousness of the web – and humanity's efforts to destroy it.

In his study of ecocriticism for the Anthropocene, Ryan Hediger articulates the connection between the climate emergency and our sense of stability: 'As the climate changes, places change and some places literally disappear from under our feet. Such cases dramatically render homesickness' (2019: 285). Another way of thinking of this homesickness is as solastalgia, a term created in the early aughts by philosopher Glenn Albrecht. The histories being surfaced in deep-landscape writing – interwoven histories of settler colonialism, industrialisation, unchecked capitalism – elicit a feeling of solastalgia. The term is adapted from *nostalgia* (the urge to reminisce or return home) to express 'the homesickness you have at home' (Albrecht, 2019: 27). As Albrecht writes,

> I define 'solastalgia' as the pain or distress caused by the ongoing loss of solace and the sense of desolation connected to the present state of one's home and territory. It is the existential and lived experience of negative environmental change, manifest as an attack on one's sense of place.
>
> (Albrecht, 2019: 38)

The word *solace* describes comfort, the alleviation of anguish; the word *desolation* describes 'deprivation of comfort' (Albrecht, 2019: 38), a state of destruction and isolation. The experience of having the former replaced by the latter engenders solastalgia. Albrecht argues that now is the age of solastalgia, brought on by the climate catastrophe. As the danger intensifies, and our home (personal, regional, national, global) is increasingly imperilled, we will all feel solastalgia. Though she does not name it solastalgia, Proulx put it well when she wrote in *Fen, Bog & Swamp*, 'If your delight is in contemplating landscapes and wild places the sweetness will be laced with ever-sharpening pain' (Proulx, 2022: 6).

Proulx is offering a preview of that feeling, an alert aimed at showing what we have already lost. The three main themes in her writing are refracted through the lens of solastalgia. The loss of solace can be seen in the first theme, the proliferation of uninhabitable homes; houses crumble, and characters are discomforted for it. The desolation appears in the second theme, itinerant characters – despondent and often solitary wanderers. And the third theme is the ultimate source of solastalgia: the threat of environmental devastation. What follows is an explication of each theme as it appears in Proulx's fiction.

Characters in motion

Although not always displaced by an environmental emergency, the characters in Proulx's novels embody the homelessness and resulting homesickness that Hediger describes, communicating the feeling of solastalgia we should all expect to experience, and soon, if we continue in our complacency. The extreme but inevitable outcome of global heating, rising sea levels, and crumbling infrastructure will be increasingly fewer liveable places and increasingly more climate refugees. People will take to the roads. The characters in Proulx's novels are not necessarily displaced by ecological disaster, but they are on the move. The characters' movement allows the reader to feel what it might be like to lose one's place. They often find themselves in unfamiliar places, persistently transient, and often so from the very beginning of the book.

In an interview about her first novel, *Postcards*, published in 1992, Proulx explained that the book is

> concerned with what happens when a region has only one economic base and it goes under – the breakup and scattering of families, the subdivision of land, the outflow of old residents or the new position they adopt as service providers to the rich moving in.
>
> (Proulx, 1999a)

It is about our overdependence on an untenable industry, about small family farms losing out to mass production – and about the resulting displacement. *Postcards* is set mostly in the American South in the aftermath of the Second World War. This is a time of 'big economic transitions,' notes environmental humanities scholar Wes Berry, 'including the escalation of fossil fuel use and the corresponding decline of rural communities' (2009: 171). The opening words are: 'Even before he got up he knew he was on his way' (1992: 1). So begins the initial example of one of the themes Proulx returns to again and again, a theme that is emblematic of deep-landscape literature: a protagonist on the move. The 'he' in question is Loyal Blood, and he has just murdered his girlfriend, Billy. Loyal leaves the family farm and does not stop moving for the duration of his life. Journeying through the American West, always unfamiliar with his surroundings, Loyal is never settled: 'his way had been that of an exile for a long time' (1992: 217). Eventually Loyal buys a 'bony square of dirt' to try to make a home, but even then he is so uncomfortable that he calls it 'the place' (1992: 218), suggesting the unspecific and disconnected sensation he feels about it. There he builds 'an uneasy house' (1992: 218), which promptly burns down within the first year of construction. After this disaster, Loyal turns his truck into a mobile home, and stays 'on the slow move' (1992: 230) through Texas, New Mexico, Utah, Minnesota, and Florida. With the mobile home, he perpetuates his remoteness, his rootlessness. Soon, even his truck is stolen. Loyal dies walking on the side of a road, unsure of where he is or in what direction he is travelling, 'turned around. Completely turned around' (1992: 279). Transient and untethered, lost until his last living moment, Loyal embodies the disorientation and desolation of solastalgia. Though Loyal may have been launched into his unsettled life by his violent action, his inability to find a place to stay allows readers to get a sense of what houselessness might feel like. If we continue to count on gas and oil, continue to toxify the land and air we share with all life, continue to let the sea rise and be filled with swill, we will find ourselves roadside, moribund, and missing home.

The theme of transience continues into Proulx's next and perhaps best-known novel, *The Shipping News* (1999). Again, the first words of the book alert us to the constant motion of its characters:

> Here is an account of a few years in the life of Quoyle, born in Brooklyn and raised in a shuffle of dreary upstate towns. . . . At thirty-six, bereft, brimming with grief and thwarted love, Quoyle steered away to Newfoundland, the rock that had generated his ancestors, a place he had never been nor thought to go.
>
> (1999b: 1)

After Quoyle arrives in Newfoundland, he and his children must repeatedly move lodgings, from car to hotel to short-term rental, because none are fit to occupy – too crowded or too cold. And even when they finally settle in the family house, located on a windy cliff, the place epitomises impermanence, feeling tenuous until it disappears completely, leaving Quoyle and the kids houseless once again. They are not rendered unhoused by the present threat to the environment that underlies the novel, but they are deprived of solace, their home destroyed. In this way, the family embodies solastalgia.

In Proulx's following novel, *That Old Ace in the Hole* (2009), her protagonist, Bob Dollar, is also houseless and homesick. The book begins:

> In late March Bob Dollar . . . drove east along Texas State Highway 15 in the panhandle, down from Denver the day before, over the Raton Pass and through the dead volcano country of northeast New Mexico to the Oklahoma pistol barrel, then a wrong turn north and wasted hours before he regained the way . . . and listened to the songs about staying home, going home, being home and the errors of leaving home.
>
> (2002: 1)

Here again is a transient man, unsure of his surroundings, grappling with the problem of how to make a home; it is Proulx offering the reader a premonition of solastalgia. Dollar bounces between a bunkhouse, which is described as 'spartan' with 'no electric' and 'no sheets, towels, telephone, or water' (2002: 67), and a rented room. It is not just his search for a place to stay that keeps Dollar on the move, though; it is also his job. Employed in Denver by the Global Pork Rind to scout locations for hog farms, he spends much of the novel 'driving north and south, east and west, exploring back roads with a stack of maps on the seat beside him, looking for rundown properties' (2002: 198). In the end, after all that movement, and despite knowing that 'ruined places could not be restored' and that 'the place would turn into a massive hog farm, millions of hog bunkers and scummy lagoons spread across the old plain, waiting for what would come next' (2002: 358), he decides to stay. His fragile existence in the panhandle is perpetually laced with peril; he is homesick.

The geographical and temporal scopes of Proulx's novel *Barkskins* (2016) are the most far-reaching of any of her work. The story takes place in the years between 1693 and 2013 and includes a vast lattice of characters who travel long distances. They move between Canada and the United States, and around China, Europe, Brazil, and New Zealand. The characters are descended from two family lineages, beginning with René Sel and Charles Duquet, both of whom had arrived, in 1693, to the French colonies in North America. They are the indentured servants of a vicious man named Monsieur Trépagny. René marries a Mi'kmaq woman, and their descendants are the vehicle for illustrating the massacre of Indigenous people by white settlers, called *wenuj* by the Mi'kmaq people. A tendril of the *wenuj* wreckage is Indigenous dispossession, resulting in René's progeny being displaced for the entirety of the novel. They must travel for logging work, and what homes they have are burned or stolen by settlers who 'doubled and redoubled in number' (2016: 183) and who also decimate their resources. This era of colonialism is recognised by such environmental historians as William Cronon to be the advent of what became our current climate crisis. The movement of the Sel family is representative of the kind of transience that this ongoing crisis will necessitate: desperate, penurious. Charles becomes one of those settlers whose actions and travel ruin resources. He journeys to China, via Amsterdam, and establishes a three-corner trade route: western lumber and furs are traded for Chinese spices, opium, and silk, which are then traded for tea and coffee from the Dutch. He marries the daughter of a shipping captain so as to secure the route and adopts three sons to ensure heirs for Duke & Sons, his timber company. His children and theirs become exceptionally rich by destroying the landscape and by using international movement and transnational connections to do so. As the Duquet family uses intercontinental attachments to send North American lumber around the world, thus expanding their wealth, their effect on the global climate worsens – the company uses excessive resources in their bid to spread capitalist interests around the world. René's family's destitution and Charles's family's success are both predicated on movement. The former is representative of the experience of homesickness, the latter of the mechanisms that engender homesickness. Both allow the reader to preview the solastalgia we can expect to experience in our own lives, on this ecologically catastrophic trajectory.

Homes in disrepair

Concomitant with the homesickness engendered by transience is the homesickness caused by deteriorating houses, property, or land. This is Proulx's second theme. For the reader to feel the homesickness of solastalgia, Proulx presents protagonists whose homes are unstable or disintegrating. She is offering the very specific feeling of losing something as you are still living in it – something you depended on for comfort and protection. The reason for the disrepair is not always directly related to environmental damage, but the loss is similarly emotional and material – as ours would be, in the event of catastrophic climate change. In *Postcards*, the farm from which Loyal runs is slowly worn down, until it must be split 'into three parcels': 'The farm became a shape like a pair of trousers, each leg a division of the property' (1992: 135). One section is turned into a mobile home lot, those permanently impermanent structures; the other section is sold to a man who is overwhelmed by what he has acquired:

> Dr. Franklin Saul Witkin, forty-seven, stoop-shouldered, urban in habitat but haunted from childhood by fantasies of wilderness, sat on the stone wall staring into the chaos of his purchased landscape. There was too much to look at. Knotted branches. The urgent but senseless angular pointing of tree limbs. Grass the color of wafers. Trees lifting soundless explosions of chrome and saffron. Mountains scribbled maroon, riven by mica-shot cliffs. The yelling light. He looked up and the sky filled with swarming points. If he walked into the woods, land tilted, trees thronged like gnats, the air turned sallow and he was lost.
>
> (1992: 155)

The untamed immensity of the place disorients him, and he loses his bearings. Proulx anthropomorphises the landscape, empowering it; it is urgent, lifting, yelling. The tilted land knocks Dr. Witkin off his feet. All the new owners are unstable: the mobile homes suggest brief residence, and the lost man is merely a seasonal inhabitant. This impermanence demonstrates the tenuousness of the property in its new iteration. Even the original 'sway-backed farmhouse' falls into decrepitude, becoming nothing more than a 'tumbled heap of beams and stone' (1992: 155). In a far corner of the land, Jewell leaves herself a shed to occupy, 'a slatternly lean of paintless clapboards, the porch slipping away like melting butterscotch' (1992: 152). These are homes in disrepair. The dilapidation is the material result of an untenable industry, and is a portent for the outcome of continuing to rely on destructive methods in farming, which is desolation.

Quoyle's home in *The Shipping News* proves similarly slatternly and unsustainable. When he arrives, the house is in shambles, 'half ruined, isolated, the walls and doors of it pumiced by stony lives of dead generations' (1999b: 47). He and his family must live in a hotel until they can make the house semi-habitable, and even after they move in, it is in a perpetual state of construction. Lashed to the Newfoundland rock with cables that creak and pull, constantly threatening to snap in a great storm, the home is never complete: 'no matter what they did to the house . . . it kept its gaunt look, never altered from that first looming vision behind the scrim of fog' (1999b: 185). The dilapidation and impermanence of the house are supplemented by its lore. Legend has it that the entire structure was dragged over winter ice from an island off the coast:

> the Quoyles only been on the Point there a hundred years or so. Went there in the 1880s or 1890s, dragging that green house miles and miles across the ice, fifty men, a crowd of Quoyles and their cunny kin pulling on the ropes.
> (1999b: 162)

Quoyle begins to believe that this passage across the frozen water forever tainted the house, 'the idea fixed in him that the journey had twisted the house out of true, wrenched the timbers into a rare geometry' (1999: 185). Permanently distorted by its never-ending assembly, the house eventually goes 'missing' entirely:

> The great rock stood naked. Bolts fast in the stone, a loop of cable curled like a hawser. And nothing else. For the house of the Quoyles was gone, lifted by the wind, tumbled down the rock and into the sea in a wake of glass and snow crystals.
> (1999: 231)

The family is once again homeless and homesick.

In *That Old Ace in the Hole*, it is not just Dollar who occupies an unstable residence: all of the Woolybucket homes are in danger, surrounded by hog farms that emit a 'ferocious, a palpable, heavy ammoniac stink that burned the eyes and throat' (2002: 319) and sitting atop a rapidly depleting water source. Hog farm corporations constantly send scouts, among them Dollar, to cajole people from their homes so that they can demolish the houses and run abusive hog farms on the land. As one woman remarks: 'Do you think it's right that some mean-hearted corporation can buy up panhandle land and force out the local people? I don't know what we are goin a do' (2002: 319). The farm that her family was able to maintain for years is now in the slow process of ruin, uninhabitable because two elements required to live – air and water – have become toxic. This is a familiar, real-world kind of dispossession, in which low-income residents are the targets of pollution and poisonous by-products. Proulx gestures towards a long history of alterations to the Earth:

> In his mind's eye he saw the panhandle earth immemorially used and tumbled by probing grass roots, the cutting hooves of bison, scratchings of ancient turkeys, horses shod and unshod pounding along, the cut of iron-rimmed wheels, the slicing plow and pulverizing harrow, drumming hail, the vast scuffings of trailed cattle herds, the gouge of drill bits and scrape of bulldozers, inundations of chemicals. What was left was a kind of worn, neutral stuff, a brownish dust possessing only utility. This ghost ground, ephemeral yet enduring, was what it came down to.
>
> (2002: 359)

Before industry, there was flourishing fauna. Then came the wheels, the plough, the drills, the bulldozers, the chemicals. Mankind has reduced the land to its utility, valuing it only for what it can produce, wearing it out, rendering it ghostly, leaving merely dust – no place to make a home, for human or non-human lives.

This kind of pollution and demolition, spurred by capitalist greed, is the source of ruined homes in *Barkskins* as well. In this novel of immense scope, it is primarily the Indigenous people and non-human lives (plant and animal) who lose their place. The culprit is the colonial project of deforestation, again immediately evident on the first page of the book: 'René swung the short-handled trade tool, felt the jolt of the tree's resistance, swung again, embarking on his life's work of clearing the forest of New France' (2016: 9). Encapsulated in this line is the settlers' relentless attack, and the landscape's futile fight against it. We read, in the course of seven hundred pages covering over three hundred years, about the terrible effect of settler colonialism and capitalist ambition on what was for many more centuries a stable ecosystem. It used to be that the Mi'kmaw, whom Proulx focuses on, were at home in the land: 'Their own way of life left no scars on the land, they said, denuded no forests. They glided through the waterways and in seconds the wake of their passage vanished in the stream flow and the forests remained as they had been silent and endless' (2016: 38). With the arrival of white men, that stability is destroyed. Proulx writes,

> The smoke-thickened years passed, and Crown corvée work gangs widened the west trail to a road. More settlers came into the forest. Every morning the sound of distant and near chopping annoyed woodpeckers who imagined rivals, then, feeling outnumbered, fled to wilder parts. The trees groaned and fell, men planted maize between the stumps. The deer and moose retreated, the wolves followed them north. In its own way the forest was swallowing René Sel, its destroyer.
>
> (2016: 57)

The animals and people who used to live in and sustain the forest are forced out of their homes as the settlers ravage the landscape. They withdraw towards the last remaining wilderness, which is also doomed. As one Mi'kmaw man tells his community,

> We are sharing our land with the *Wenujs* and they take more and more. You see how their beasts destroy our food, how their boats and nets take our fish. They bring plants that vanquish our plants. Most do not mean to hurt us, but they are many and we are few. I believe they will become as a great wave sweeping over us. . . . All these woods once ours . . . and we went anywhere we wished without hindrance. That time has passed. But I wish to tell you that if we Mi'kmaw people are to survive we must constantly hold to the thought of Mi'kmaw ways in our minds. We will live in two worlds. We must keep our Mi'kmaw world – where we, the plants, animals and birds are all persons together who help each other – fresh in our thoughts and lives. We must renew and revere the vision in our minds so it can stand against this outside force that encroaches. Otherwise we could not bear it.
>
> (2016: 181)

Soon enough, the Mi'kmaw man laments, 'I have *no place*' (2016: 276, emphasis in original). What was once their home has been annihilated, as has been a way of life that values the symbiosis of all human and non-human lives, and that would protect what has been violated. In its place are the conjoined destroyers, capitalism and colonialism, whose perpetrators value the environment only for the commodities it can provide.

Ecosystems in danger

Enfolded within the first two themes – characters' movement and the loss of their homes – is the third, the ecological emergency. Extractive industries, contaminating corporations, virulent viruses, or ravenous capitalists and their technology threaten to ruin the earth. Woven through all the novels analysed here is impending environmental danger and destruction. With deep-landscape writing, Proulx imbues every setting with a sense of the long history and ongoing mechanisms of extraction, contamination, and exploitation, and then shows how that history comes to bear on the lives of her protagonists. Each novel addresses a different aspect of environmental mistreatment: in *Postcards*, it is large-scale agricultural industry; in *The Shipping News*, it is the oil industry; in *That Old Ace in the Hole*, it is the hog-farm industry; and in *Barkskins*, it is settler-colonialism and the logging industry.

Postcards is set, as Wes Berry notes, in the aftermath of World War II, during

> big economic transitions . . . including the escalation of fossil fuel use and the corresponding decline of rural communities. Free-market capitalism drove the transformation of American agriculture from the Jeffersonian ideal – an abundance of self-sufficient family farms – to the Earl Butzian ideal of agribusiness monoculture, a type of planting and harvesting enabled by large machinery powered by inexpensive products of oil refining, like diesel.
>
> (2006: 171)

Proulx shows exactly this transition, and how it has wrought ecological destruction: small family farms losing out to mass production, illegal trade in migrant labour and in ecological artefacts (like ancient skeletons and black bear gallbladders), the increased use of pesticides, the rise in dependency on oil, intensifying human abuse of what were once wild environments. Loyal goes on the run because of his crime, but the land he left withers away because of criminal negligence and governmental greed. When he takes off, it is his abandonment of the farm, more than the abandonment of his family, that upsets his father, Mink, who knows that without Loyal's help running it, the place is doomed: 'But one man couldn't bring that farm up again. It had too much against it' (1992: 21). It is a small family farm which requires a family to run. Loyal's mother comments that, in the old days (the Jeffersonian days), 'your boys would go into farmin' somewhere nearby' and that when she was a girl 'there were so many aunts and uncles, cousins, in-laws, second cousins. All of 'em livin' right around here. They'd be here now, that kind of big fam'ly if it was them times' (1992: 112). Without community, they cannot compete with the big holdings. Life on the farm becomes unbearable for the Bloods: 'Poorer every year, the work harder, the prices higher, the chances of pulling out of it fewer and fewer. It was so different now. He couldn't get his bearings' (1992: 103). Eventually, Mink attempts to burn the barn and the nine cows within it (without a thought for their lives or the terror and pain they might experience in a fire), to collect two thousand dollars in insurance money. His arson is quickly discovered, and he is arrested. In prison, he hangs himself, leaving his wife with no choice but to sell.

Loyal goes to work in a mine, where conditions are no better than on the farm:

> 'Yellowcake speculation. Talk was of production levels. The rock rats disappeared. It was all big business now, deep mines, acid leaching, chemical extraction, company prospectors, poison wastes and tailings, sand slurry choking the streams, big fish kills and mountains of dead and reeking tailings.
>
> (1992: 147)

The extractive mindset is the same, whether it plays out on farms or in the mines: take as much as possible from the land, until it can give no more. This kind of mindset and the damaging change it brings has fallout in town, too. As Loyal's mother observes:

> She saw the landscape changing. . . . Everything was changing. Brush growing up. She was critical when the road crews cut overhanging limbs from the maples. Tears streamed when they cut the trees themselves to widen the highway, hardtop now all the way to the Post Road. The village grew unaccountably, men sawed down the yellowing elms, tore up stumps with great corkscrew machines. The street spread like unpenned water to the edges of the buildings. Metal roofs glittered. At the dump, heaps of broken slates invited potshots from rat shooters, then send the bullets ricocheting back. The town sold off the timber in the watershed above the valley and for two years endured the nasal moaning of chain saws. The clear-cut left the hills as bare as the side of a scraped hog.
>
> (1992: 128)

In *Postcards*, then, we are getting a glimmer of the deforestation Proulx will diligently detail in *Barkskins*. The town's so-called progress has meant losing tree cover, paving over habitat, mounting piles of refuse, and noise pollution. Proulx is also careful, in her deep-landscape way, to show that it is not just enormous, faceless industries that affect a landscape. Each of us, in however small a way, makes change wherever we step. In this instance she is writing about Loyal:

> His blood, urine, feces and semen, the tears, strands of hair, vomit, flakes of skin, his infant and childhood teeth, the clippings of finger and toenails, all the effluvia of his body were in that soil, part of that place. The work of his hands had changed the shape of the land, the weirs in the steep ditch beside the lane, the ditch itself, the smooth fields were echoes of himself in the landscape, for the laborer's vision and strength persists after the labor is done. The air was changed with his exhalations. The deer he'd shot, the trapped fox, had died because of his intentions and commissions, and their absence in the landscape was his alteration.
>
> (1992: 77)

But the message, that we must be aware and be wary of our impact, is universal. Every microscopic material or movement has a consequence.

In *The Shipping News*, the oil industry and its inevitable waste, contaminating the water and land, is the enemy of generations of lone fishermen and of Newfoundland culture. A reserve is discovered under the Grand Banks by the coast of Killick-Claw, which boasts 'billions and billions of barrels of oil' (1999b: 198). Quoyle's friend, Billy, rants about the potential ills of this reserve, and the future its mining will bring:

> Pollution! The sea bottom strewn with clits of wires and barrels and broken metal that'll tear up any trawl. And to come? Terrible oil spills will kill off the few midget cod that's left, destroy fishery entirely, scum the landwash with black stinking ooze, ruin boats and harbors. The shipping lanes will be clogged with the oil tankers and supply boats. . . . I seen the cod and caplin go from millions of tons taken to two or three bucketsful. Seen fishing go from seasonal, inshore, small boats to the deep water year-round factory ships and draggers. Now the fish is all gone and the forests is cut down. Ruined and wrecked! No wonder there's ghosts here. It's the dead pried out of their ground by bulldozers!
>
> (1999b: 199)

Here is the environmental degradation against which we all must fight. Here is both what has happened, what is happening, and what might happen: the rubbish left behind to alter forever the ocean floor, the chemical and oil spills contaminating the water, the loss of livelihood for smaller and more sustainable fishermen, the graves desecrated by extraction. Quoyle is compelled, after hearing Billy's rant, to use his role as a journalist to save the small-scale local fishing trade. He writes in one of his articles for the paper,

> Today the most common sight on the marine horizon is the low black profile of an oil tanker. . . . Another common sight is black oil scum along miles of landwash, like the shoreline alone Cape Despond this week. Hundreds of people watched Monday morning as 14,000 metric tons of crude washed onshore from a ruptured tank. . . . Thousands of seabirds and fish struggled in the oil, fishing boats and nets were fouled. 'This is the end of this place,' said Jack Eye, 87, of Little Despond. . . . One thing is sure. There will be more oil spills, and some will be horrendous.
>
> (1999b: 201)

Quoyle is appealing to his readers, in the hope that they will see what Billy does, and then try to activate for change. The implication is that Proulx is also appealing to us, her readers, in the hope that we will see what she does. And that we will make the required change ourselves, away from extractivism and exploitation, towards stewardship and sustainability. This is literary environmentalism at work.

In *That Old Ace in the Hole*, Dollar must eventually stand up to the Global Pork Rind bosses, to defend the town against the mass production of hog products, the mass consumption of poorly generated meat, the abuse of animals, and the ruination of panhandle homes, air, land, and water. In this book, writes Berry, Proulx 'pulls us through a country of disintegrating communities, endangered species, human and animal misery, toxic wastes, and wholesale destruction of the sources of life' (2006: 170). This is a country that has a long history of exploitation, as Proulx makes clear in this tidy, deep-landscape synapsis:

> Here the Indians had lived nomadic hunting lives; traders opened routes to Santa Fe and Taos to sell calico and took peltry from the Indians in exchange for manufactured goods; army scouts came to map the terrain and tangled with the Indians . . . ranchers brought in cattle to run the free and open range and the sons of settlers became cowboys. . . . The flood of people came with railroads, small farmers . . . [and finally] oilmen and flim flam tricksters, government men. . . . Now corporate agriculturists like Global Pork Rind had moved in.
>
> (2002: 95–96)

Once Indigenous people lived with little impact on the place, then settlers cleared the land of its natural inhabitants, and each successive wave of occupiers brought its own brand of colonial destruction, up to and including the outfit Dollar represents. This relentless mistreatment of the landscape has had dangerous consequences. Not only is the place under threat from the infusion of hog farms, but its underground water source, Ogallala, is also dwindling, as is its main asset, oil, depleted from overuse:

> The water is playin out. The people built their lives on awl money expected it would last forever too. The awl is pretty much gone. And they told us the Ogallala would last forever. Now the Ogallala is finishin up.
>
> (2002: 111).

Using deep-landscape writing, Proulx draws a clear connection between a history of newcomers changing the land, wringing it of its resources, and the impoverishment and desperation the current residents are experiencing. From this we can understand how we, quite literally, reap what we sow. We can feel the solastalgia.

Barkskins offers the most sweeping example of deep-landscape literature. From this novel we are able to grasp, in ways both intimate and expansive, the enduring legacy of white settler-colonialism, the eradication of primeval forests, Indigenous dispossession and extermination, and the permanent decimation of natural ecosystems. And we are made to see how this history is not at all bygone, that it has contemporary iterations which threaten this planet and all life within it. One of the epigraphs from

Barkskins quotes from Lynn White, Jr., regarding nature, from a paper entitled 'The Historical Roots of Our Ecological Crisis' (1967):

> In Antiquity every tree, every spring, every stream, every hill had its own genius loci, its guardian spirit. These spirits were accessible to men, but were very unlike men; centaurs, fauns, and mermaids show their ambivalence. Before one cut a tree, mined a mountain, or damned a brook, it was important to placate the spirit in charge of that particular situation, and to keep it placated. By destroying pagan animism, Christianity made it possible to exploit nature in a mood of indifference to the feelings of natural objects.
>
> (White, via Proulx, 2016: i)

In an interview, Proulx has mentioned that this quotation is 'the key to the book' (Rock, 2016). The key to the book, then, is representing the social transformation or hostile takeover from traditional, in this case Indigenous, ways of knowing place to Anglo-Christian ways of destroying place. Her intention to illustrate that transformation is encapsulated in this exchange between René and Trépagny. René asks, 'Why do we cut the forest when there are so many fine clearings? Why wouldn't a man build his house in a clearing, one of those meadows that we passed when we walked here? Would it not be easier?' (2016: 17). To which Trépagny responds, scandalised,

> Easier? Yes, easier, but we are here to clear the forest, to subdue this evil wilderness. . . . Men must change this land in order to live in it. In olden times men lived like beasts. In those ancient days men had claws and long teeth, nor could they speak but only growled. . . . To be a man is to clear the forest.
>
> (2016: 17)

Derisive of a deeper connection with the natural world, indifferent to the well-being of the plants and animals, ignorant of the protective spirits, these settlers assert their dominance, and 'the stumpy ground was gouged by oxen's cloven hooves as though a ballroom of devils had clogged in the mud: the trees fell, their shadows replaced by scalding light, the mosses and ferns below them withered' (2016: 13). They leave a scorched Earth. By the end of *Barkskins*, after the relentless, unchecked logging and the loss of the old-growth forests, we are given a glimmer of what might be possible for a more sustainable future in the characters Dr. Alfred Onehube, a 'militant ecoconservationist' (2016: 689) from Manitoba, and Sapatisia Sel, the last living descendant of René, who is 'working to put forest pieces back together . . . to help the earth regain its vital diversity of tree cover'

(2016: 706). She trusts that 'the forests will help us. They are old hands at restoring themselves' (2016: 706). We can only hope.

Conclusion

Proulx does not assert outright that her work is environmental. Rather, she implies that readers should supply their own message to the reading, based on 'what was gained and what was lost' (Treisman, 2013). The only thing gained is homesickness; the loss – of healthy flora and fauna, of clean air and soil and water, of Indigenous ways of being in and knowing place – is immeasurable. As Proulx writes, 'Gone or going are the chestnut trees, elms, and old-growth forests; prairie potholes and swamps; vast flights of passenger pigeons; codfish; night skies black and without planes, satellites, or urban glow; unclimbed mountains; rivers without dams or bridges' (2009: 19). Conducting a deep-landscape reading allows us, as readers and critics, to engage in literary environmentalism, to project environmental activism onto the text, even for a writer who backs away from the eco-activist designation. It is a holistic approach to literature, including the past and future, representing ecological destruction beyond a singular calamitous act or point in time, unending or unbounded. A deep-landscape framework applied by a reader can turn work by writers who do not consider themselves environmentalists into a charge for change.

'Never has there been a time in human history when landscapes have changed and disappeared with such rapidity,' Proulx wrote (2009: 19). The situation has only worsened, and continues to worsen exponentially (see Carson, 1962; Cronon, 2011; McNeill and Engelke, 2015; Kolbert, 2015; and more). But, by marking it, deep-landscape literature intervenes to challenge this change, this disappearance. Because a story's setting must include, as Proulx writes, the 'sum of accumulated changes wrought by the inhabitants and their marks on the land' (2009: 14), it inevitably comments on the cooperating harms responsible for environmental demise. This act, Proulx notes, 'literally grounds the work, gives it a strength and a sense of truth not possible to achieve through words on the page in any other way' (2009: 14). The truth is that there is a more symbiotic way to be in place, a more sustainable way to relate to non-human lives. Deep-landscape writing gives a text its environmental vigour. In Proulx, Nixon would find another activist making noticeable protracted forms of suffering, this time through novels in which the landscape is 'fixed and immutable, ordering the lives of the characters who move within it, forcing the events of the story' (2009: 7). Her body of work serves as an example of how the land dictates the lives within it, how we will soon all carry a sense of solastalgia. Using these themes to make visible the imperceptibly slow erosion of the environment – what has happened *and* what is to come – allows Proulx to activate on its behalf now, and to encourage her readers to do the same.

Bibliography

Albrecht, Glenn (2019), *Earth Emotions: New Words for a New World* (Ithaca, NY: Cornell University Press).
Asquith, Mark (2009), *Annie Proulx's Brokeback Mountain and Postcards* (New York, NY: Bloomsbury Publishing).
Battiste, Marie and James Youngblood (Sa'ke'j) Henderson (2000), *Protecting Indigenous Knowledge and Heritage: A Global Challenge* (Chicago, IL: University of Chicago Press).
Berry, Wes (2006), 'Capitalism vs. Localism: Economies of Scale in Annie Proulx's *Postcards* and *That Old Ace in the Hole*', in Alex Hunt (ed.), *The Geographical Imagination of Annie Proulx: Rethinking Regionalism* (Washington, DC: Lexington Books), 169–182.
Bolick, Katie (1997), 'A Conversation with E. Annie Proulx', *The Atlantic Online*, [Online], [Accessed 14 June 2022], available from: www.theatlantic.com/past/docs/unbound/factfict/eapint.htm
Carson, Rachel (1962), *Silent Spring* (New York, NY: Houghton Mifflin Harcourt).
Cronon, William (2011 [1983]), *Changes in the Land: Indians, Colonists, and the Ecology of New England* (New York, NY: Farrar, Straus and Giroux).
Hediger, Ryan (2019), *Homesickness: Of Trauma and the Longing for Place in a Changing Environment* (Minneapolis: University of Minnesota Press).
Hunt, Alex (2009), 'Introduction: The Insistence of Geography in the Writing of Annie Proulx', in Alex Hunt (ed.), *The Geographical Imagination of Annie Proulx: Rethinking Regionalism* (Washington, DC: Lexington Books), 1–10.
Kolbert, Elizabeth (2015), *The Sixth Extinction: An Unnatural History* (New York, NY: Henry Holt).
LaDuke, Winona (1994), 'Traditional Ecological Knowledge and Environmental Futures', *Colorado Journal of International Environmental Literature and Policy*, 5:1, 127–148.
McGregor, Joan (2018), 'Toward a Philosophical Understanding of TEK and Ecofeminism', in Melissa K. Nelson and Dan Shilling (eds.), *Traditional Ecological Knowledge: Learning from Indigenous Practices for Environmental Sustainability* (New York, NY: Cambridge University Press).
McNeill, J.R. and Peter Engelke (2015), *The Great Acceleration: An Environmental History of the Anthropocene Since 1945* (Boston, MA: Harvard University Press).
Meinig, Donald W. (1979), 'The Beholding Eye: Ten Versions of the Same Scene', in *The Interpretation of Ordinary Landscapes: Geographical Essays* (New York, NY: Oxford University Press), 33–50.
Naess, Arne (1973), 'The Shallow and the Deep, Long-Range Ecology Movement', *Inquiry: An Interdisciplinary Journal of Philosophy*, 16:1–4, 95–100.
Nixon, Rob (2011), *Slow Violence and the Environmentalism of the Poor* (Cambridge, MA: Harvard University Press).Proulx, Annie (1992), *Postcards* (New York: Scribner).
Proulx, Annie (1997), 'House Leaning on Wind', *Architectural Digest*. https://archive.architecturaldigest.com/article/1997/10/e-annie-proulx

Proulx, Annie (1999a), 'An Interview with Annie Proulx', *The Missouri Review*, [Online], [Accessed 15 February 2023], available from: https://missourireview.com/article/an-interview-with-annie-proulx/

Proulx, Annie (1999b), *The Shipping News* (New York: Scribner).

Proulx, Annie (2002), *That Old Ace in the Hole* (New York, NY: Scribner).

Proulx, Annie (2009), 'Dangerous Ground: Landscape in American Fiction', in T.R. Mahoney and W.J. Katz (eds.), *Regionalism and the Humanities* (Lincoln, NB: University of Nebraska Press), 6–25.

Proulx, Annie (2016), *Barkskins* (New York: Scribner).

Proulx, Annie (2022), *Fen, Bog & Swamp: A Short History of Peatland Destruction and Its Role in the Climate Crisis* (New York: Scribner).

Rock, Lucy (2016), 'Annie Proulx: "I've Had a Life. I Know How Slippery Things Can Be"', *The Guardian*, [Online], [Accessed 2 July 2022], www.theguardian.com/books/2016/jun/05/annie-proulx-ive-had-a-life-i-see-how-slippery-things-can-be

Rood, Karen (2001), *Understanding Annie Proulx* (Columbia, SC: University of South Carolina Press).

Treisman, Deborah (2013), 'This Week in Fiction: Annie Proulx', *The New Yorker*, [Online], [Accessed 14 June 2022], available from: www.newyorker.com/online/blogs/books/2013/06/this-week-in-fiction-annie-proulx.html

Trosper, Ronald (1995), 'Traditional American Indian Economic Policy', *American Indian Culture and Research Journal*, 19:1, 65–95.

Weaver, Courtney (2022), 'Annie Proulx: "You've Got to Keep Moving!"', *Financial Times*, [Online], [Accessed 21 February 2023], available from: www.ft.com/content/57ec8e5e-227b-4cf4-884e-e2d50ae6427a

12 Congolese Anthropocenes, wounds of extraction, arts of resistance

Transcultural materialism in Fiston Mujila's *Tram 83* and Sammy Baloji's *The Beautiful Time*

Spring Ulmer

Montaging the human and non-human: an introduction

Ongoing histories of mineral extraction undergird Fiston Mwanza Mujila's award-winning novel *Tram 83* (2014) and Sammy Baloji's photographic series *The Beautiful Time* (2010). In *Tram 83*, Lucien, a writer, visits his old friend, a hustler named Requiem, in the 'Ville-Pays' (Mujila, 2014: 18, translated in English as 'City-State,' Mujila, 2015: 10), where everything and everyone is reduced to mined rock and its value. In the often drunken escapades that follow, Lucien and Requiem attempt to rekindle a friendship that grew cold after Lucien ran off with Requiem's wife, Jacqueline. Philosophical differences – Lucien has retained his Marxist idealism, whereas Requiem has returned from serving in the military with a jaded realism – find the two friends ever more embittered. The estranged friends remain estranged to the novel's end, even as they find themselves united in struggle against those in power. In *The Beautiful Time*, Baloji montages colour photographs of present-day mining sites together with colonial-era black-and-white photographs of miners and officials sourced from the former Union Minière du Haut Katanga mining company's archives. Intriguingly, representations of both the man-made and natural environments in *Tram 83* and *The Beautiful Time* are as detailed as depictions of the human; as a result, the non-human and human overlap and are at times indistinguishable. In other words, the setting in both *Tram 83* and *The Beautiful Time* overwhelms and plays a larger-than-scenic role. Metal structures of railway stations, railroad tracks, mining infrastructure, and mines take on a peculiar agency, at once shaping, mangling, and polluting humans and the surrounding ecosystems. Both Mujila and Baloji hail from the Katanga region of the Democratic Republic of the Congo (DRC), one

DOI: 10.4324/9781032726953-13

of the most mineral-rich regions in the world, where the unearthing of copper, cobalt, and uranium has completely altered the region and those living there. Unsurprisingly, then, *Tram 83* and *The Beautiful Time* call attention to the genocidal suffering and ecological degradation – what I will call here the Black Anthropocenic condition – wrought upon the human and the non-human by colonial and neocolonial resource theft and forced labour.

A transcultural materialist praxis, defined by Héctor Hoyos (2019: 2) as the fusing of historical materialist approaches together with the biopolitical and other-than-human understandings of new materialism, is at root a '*critique* of extractivism' and as such informs my reading of energy mineral extraction and resistance in *Tram 83* and *The Beautiful Time*. Hoyos argues that exploitation of nature and labour under economic models that revolve around 'extracting natural resources and selling them on the world market' is a strategy that disenfranchises those who do not 'reap the benefits of added value' (2019: 4–5). Such exploitation flies in the face of the traditional, precolonial African cosmological, balanced relationship between the human and the non-human, as outlined by Michael Onyebichi Eze (2017: 625–629). African cosmology, according to Eze, is a worldview in which everything is connected and found in relationship with everything else, and in which every *thing* (even dead material) radiates a life force or wave particle that interacts with other forces. Conversely, as Eze maintains, western thinking itself is mired in ways that industrialisation has imposed an ideological shift from human resources to material resources and required human dominance of nature, privileging humans over the environment. A transcultural materialist praxis must then, in essence, focus on reversing such human privileging. Reading for the non-human (the natural and man-made environment) as it connects with the human in *Tram 83* and *The Beautiful Time*, I aim to draw attention to Mujila's and Baloji's refusal to look away from the Black Anthropocenic condition, which is other than the Anthropocenic condition, in that Black Anthropocenes began with land and resource theft, and with colonialism, not with the current geological era of human dominance over climate and the environment. As Kathryn Yusoff argues, Black bodies have been forced to deliver 'the Anthropocene as a geologic event into the world,' through colonial, settler colonial, and neocolonial exploits that 'have been ending worlds for as long as they have been in existence' (2018: 82, 2018: preface xiii). Whereas Neo-Marxists, like Jason Moore and Andreas Malm, argue that our epoch is in fact the Capitalocene, a term that connects our major ecological crises specifically to the global spread of capitalism and its socio-economic-ecological injustices, Yusoff's term 'Black Anthropocenes' acknowledges the proximity of black and brown bodies to harm. This harm has been and continues to be perpetuated via black and brown bodies' 'intimacy with the inhuman [which has been] organized by historical geographies of extraction, grammars of geology, imperial global geographies, and contemporary environmental racism' (Yusoff, 2018: preface xii). For this reason, I employ the term 'Black Anthropocenes' to describe the literary and artistic preoccupations of Mujila in *Tram 83* and Baloji in *The Beautiful Time*.

'Littérature-locomotive': Mujila's literature of rape

Tram 83 is a novel of epic proportion, an experimental novel, infused with innovative jazz-inspired sensibility, in which the experimental prose jumpcuts from scene to scene as if discordantly, and throughout which refrains echo, providing musical riffs and counter-rhythms, the most common among them being: 'Au commencement était la pierre' [In the beginning was the stone], 'La gare se résumait à une construction métallique inachevée, démolie par des obus, des rails et des locomotives qui ramenaient à la mémoire la ligne de chemin de fer construite par Stanley' [the station with its unfinished metal structure, gutted by artillery, train tracks, and locomotives that called to mind the railroad built by Stanley], and 'Vous avez l'heure, citoyen?' [Do you have the time, citizen?] (Mujila, 2014: 9–10; Mujila, 2015: 1–2). These lyrical riffs steer Mujila's 'littérature-locomotive' [locomotive literature] (2014: 144, 2015: 150).

The novel opens in a crowded train station in a fictive 'Ville-Pays' [City-State] (2014: 18, 2015: 10), a small, throbbing mining capital hard not to read as Mujila's native Lubumbashi, with the first of the previously mentioned riffs, a caterwauling sampling of the Gospel of John: 'Au commencement était la pierre' [In the beginning was the stone] (2014: 9, 2015: 1). Readers learn immediately that extractive minerals are worshipped in this place and that the stone (a.k.a. capital) does all the talking. Next comes the grating cry of the railroad in the second riff: 'La gare se résumait à une construction métallique inachevée, démolie par des obus, des rails et des locomotives qui ramenaient à la mémoire la ligne de chemin de fer construite par Stanley' [The station with its unfinished metal structure, gutted by artillery, train tracks, and locomotives that called to mind the railroad built by Stanley] (Mujila, 2014: 9, 2015: 1). The Stanley being referred to here is, of course, Henry Morton Stanley, the explorer commissioned by Belgium's King Léopold II, who famously declared: 'Without the railroad, the Congo is not worth a penny' (Samatar, 2015: 1). The novel's setting, in other words, thrusts us within the ruins of colonial resource and land theft.

Mujila's prose thereafter cuts from 'une gare dont la construction métallique est inachevée' [a station whose metal structure is unfinished] (2014: 9, 2015: 1) to the novel's protagonist Requiem, who takes a moment to 'jauger la marchandise' [size up the merchandise] (2014: 10, 2015: 2) of 'Une fille, habillée comme on s'habille un vendredi soir' [A girl, dressed for a Friday night] (2014: 10, 2015: 2) who has just propositioned him with the line: 'Vous avez l'heure, citoyen?' [Do you have the time, citizen?] (Mujila, 2014: 10, 2015: 2). Her proposition signals what becomes a repeated inquiry throughout the novel, an inquiry that has everything to do with the capitalistic equating of time with money; she is asking, as it were, whether Requiem has time for her sex work. As it happens, Requiem is waiting for 'ce salaud de Lucien' [that son of a bitch Lucien] (2014: 10, 2015: 2) to arrive from the 'Arrière-Pays' [Back-Country] (2014: 22, 2015: 15), and does, in fact, have a bit of time. He gropes the girl and makes a plan to meet her later on at the nightclub 'Tram 83.'

So begins Mujila's rape narrative – and here I refer to the rape of the mined earth as well as to the rape of girls and women. 'Tu mangeras à la sueur de tes seins' [You will eat by the sweat of your breasts] (2014: 7, 2015: 0), the novel's epigraph intones, prepping the reader for a tale in which labour and bodies are confused, the non-human and the human lose their borders, and Afropolitans flock to Tram 83 where the women who work the joint pile make-up onto their hormonally altered bodies in an attempt to stop the aging process. The plunder of the Black body as extraction frontier and concomitant with the exploitation of the earth becomes ever more obvious when the Black body is gendered female and exploited, as happens in this work. Obvious echoes of Peter Abrahams's classic novel *Mine Boy* (1946), in which a young man migrates from the South African countryside to work in Johannesburg's gold mines and also must navigate a culture of urban boozing and prostitution, make Mujila's *Tram 83* all the more a timeless literary portrayal of survival in the face of Black Anthropocenes.

Stony, impenetrable visages in Baloji's timeless photographs

Like Mujila's *Tram* 83, Baloji's *The Beautiful Time* also concerns itself with Lubumbashi's rusting, corroded, train-track split, hulking, open-air mining industry. Bogumil Jewsiewicki notes in his curatorial essay for Baloji's 2010 Museum for African Art exhibition 'The Beautiful Time' that Baloji refers to his photographs of Lubumbashi's colonial-era abandoned mining structures and machinery as 'portraits' (2010: 34). The photographs could be said, indeed, to have broken from the industrial landscape photography genre and entered the pores of the populous, evaporating the 'distinction between photograph and reality,' as Susan Sontag argues photographs do in this post-Abu Ghraib era by exposing truths those in power can no longer deny; in other words, the leaked Abu Ghraib 'photographs are us,' Sontag insists, because they so blatantly expose who we are (2004).

Baloji's photographs arguably become us, as well, in the same way the toxic sludge that Baloji photographs enters the bodies of those living in the vicinity of the mining slag heaps. Baloji's photographs also refuse to relegate colonial archival images to the past. Baloji achieves this timelessness by unburying photographs from the Union Minière du Haut-Katanga archive and montaging the industrial past with the post-industrialist present. Baloji maintains that colonialism created 'a kind of amnesia in order to continue' and that the colonised produced 'a new identity while hiding part of another one that was there before' (Baloji, 2015). In his digitally montaged photographs, in which he collages colonial-era archival photographs of Black miners atop his own colour photographs of these same copper mines, Baloji more than gestures to this amnesia. The miners seem to float, ghost-like, in his montages, some dressed in prison stripes, labouring (with shovels or carrying loads on their heads) in neck-chains (Figure 12.1). The pictured miners were taken by force from Rwanda or Rhodesia, 'as prisoners, slaves in a way,' Baloji relays in a videotaped interview, drawing the viewer's attention to a miner's gaze: 'the prisoner is in front of us and is looking at us . . . There is a kind of presence, but it belongs to the past, but it is in front of us and is looking to us' (Baloji, 2015). The miners' nondescript expressions in the archival photographs call to mind Tina Campt's theorisation of Black photographed subjects' 'quotidian practices of refusal – a refusal to engage the colonial . . . gaze' or to allow that gaze 'to subsume' (2017: 59). All in all, the forms of the early twentieth-century miners when cut and pasted atop 'portraits' of the twenty-first century mines provoke. The exploited workers and mined landscapes blatantly resemble one another, if only because the workers and the mined lands both appear to be attempting to protect themselves against further exploitation by presentation of identically stony, impenetrable visages (Figure 12.2). The human and non-human refusal to engage

Figure 12.1 Sans Title 12. The montage combines past and present scenes from a copper mine in Lubumbashi, Democratic Republic of the Congo.

Figure 12.2 Sans Title 8: Baloji montages early twentieth-century miners atop present-day Lubumbashi.

the colonial gaze is revolutionary in that these workers and the nature pictured here are (at least in this onlooker's mind) interconnected by way of their resistance to being represented by their abusers as anything other than abused and exploited for capital gain.

Neither photographed era (the colonial past or the neocolonial present) showcases anything other than Black Anthropocenic survival, even as workers themselves may, as Jewsiewicki contests they do, look to the past and proclaim 'C'était la belle époque' [It was the beautiful time] (2010: 12). The only difference is that, in the early twentieth century, to be employed by Union Minière du Haut-Katanga [Mining Union of Upper-Katanga] meant that at least miners' medical care, housing, and food costs were assured, whereas in the post-industrial era when Baloji was growing up, no such social services or securities existed for copper and cobalt miners and their families, and one could argue that the situation is even worse today. In fact, Baloji refers to colonialism not as 'a thing of the past' but as 'the continuation of that system,' and furthers that 'In colonialism, you can also find the beginning of capitalism itself.' He adds: 'One of the reasons why King Leopold took Congo was to create a place for free trade. Investors were coming from everywhere . . . and it's still going on now' (Madsen and Baloji, 2019). Indeed, what started with Leopold II's genocidal reign and the extraction and exportation of rubber and ivory became, at the urging of the United States after the discovery of uranium, the colonising of the Congo. This was followed by the mining region of Katanga's secession under Moïse Tshombe, whereby the mining industry in effect became the State, and later by the post-independence assassination of Patrice Lumumba by the CIA, which occurred shortly after Lumumba announced his plan to nationalise mining. Today's unregulated cobalt and uranium fields, where artisanal miners fall ill with radiation sickness and die in mine shaft collapses, ensure that DRC remains a land wherein, as Baloji argues, mining still exists and the economy revolves around mineral resources (Baloji, 2015).

Resistance in *The Beautiful Time* and *Tram 83*

Extractive industry has, according to Eze, destabilised African communities at the same time as it has destroyed 'cultural ways of life, living traditions, and the natural environment. This shift from *human resources* to *material resources* has become the ideological platform for environmental policies in Africa,' and, Eze argues, 'bourgeoning industrial capital emphasizes state control of resources in which national governments are only interested in making deals for exploration, mining, and fracking without considering the environmental impact' (2017: 628, original italics). Achille Mbembe likewise spells out just how global capitalism eviscerates the Global South in *Out of the Dark Night: Essays on Decolonization* by unpacking how capitalism has transformed 'life itself into a commodity' by valuing 'all beings and species . . . in terms of their availability for consumption' (2019: 21–22). It is this 'logic of *extraction*' in today's new phase of globalisation, Mbembe theorises after Marx, that confirms global unemployability and the rise of surplus or superfluous populations (2019: 31, emphasis added). Ours is, Mbembe furthers, 'a new age of capital when people and things can become the objects of a sudden process of devaluation and expendability,' the determination of all value resting upon 'massive *racial subsidies* or "discounts"' (2019: 31–32, original italics). In addition, the extractive economy scales race, Mbembe asserts, 'as it attempts to extract value from – or to mark – people either as disposable or as waste,' and it is this racial subsidy, he adds, that 'is precisely what allows global capital to feel no sense of responsibility for its actions, its crimes against humanity, and the horrendous damage done' (2019: 32–33).

Black lives as commodities date back to the beginning of colonialism. As Fred Moten and Stefano Harney attest:

> what of those who were located at every point, which is to say at no point, in the production process? What of those who were not just labor but commodity, not just in production but in circulation, not just in circulation but in distribution as property, not just property but property that reproduced and realized itself? The standpoint of no standpoint, everywhere and nowhere, of never and to come, of thing and nothing.
>
> (2013: 93)

Accordingly, most all transcultural materialist critiques of African and African diasporic literatures of extraction then must, in essence, wrangle with the human and specifically the Black human as commodifiable object, given that the slave trade 'provided energy for the extraction of natural resources' and made it possible for 'fortunes . . . [to] thus [be] built on genocide and on a massive exploitation of nature' (Bassey, 2012: 10). Neocolonial relations continue the pillage, Nimmo Bassey contends, pointing to the decimation of African economies by international financial institutions that conscript African nations into debt (2012: 10–15). Structural adjustment programs, Bassey furthers, have not only 'led to privatization of state-owned enterprises' but kept African nations unable to pay off their debt, leaving them tied to 'unrestrained resource extraction so as to satisfy the unyielding demands of the creditors.' Bassey calls this financial indebtedness 'a new form of slavery, where the slave actually believes that he is free' (2012: 14–15).

Yet, in *Tram 83* and *The Beautiful Time*, the human and the non-human operate in concert as Lubumbashi's notorious slagheap grows ever higher. In these works, global capital cannot do away with memory; rather, the African cosmological precolonial exists in traces of human and non-human interrelation. Baloji hints at this when he states that 'the colonial process . . . strategically erases or hides information, producing an abstract relationship to reality,' such that when you are colonised you still have access to another preexisting worldview: 'you have these two understandings of the land, or sets of knowledge, which are in confrontation with one another' (Baloji, 2015). Just as Katrzyna Bojarska writes, in Chapter 10, of the persistence of past violences both to the human and non-human, and of how such violences become legible only after artistic reframing of the evidence, which thereby establishes new ways of seeing that counter the ongoing destruction, thanks to Baloji's and Mujila's keen investigations into and their artistic representations of extractivism, we see in their works not only the devastation of colonial and global capitalist ravaging, but the reclamation of precolonial memory and the forging of resistance to ongoing Black Anthropocenes.

Baloji, in *The Beautiful Time*, interrogates the documentation of colonial labour. The viewer is invited to contemplate the purpose of those original black-and-white photographs that were made and archived by the Belgian-owned Union Minière du Haut-Katanga. Josiah Rungano Mhute writes of colonial mining photography in Zimbabwe as mapping and packaging 'capital advancement' for 'European audiences' in the form of making a spectacle out of the vulnerability of African subjects perceived as capital labour (2001: 114–115). Baloji calls upon archival images to insist on the ongoingness of extractive exploitation and to remember the dead, reposition them, and reclaim their labour, as well as to trouble his own relationship with the medium of photography, rooted as it is in colonialist history. In an interview with Léopold Lambert, Baloji confesses:

> I asked myself many questions about photography. Given how it emerged with this colonial and extractive reality, how can I transcend or distort this? . . . How do we ultimately appropriate these colonial tools? . . . How do we integrate them to go beyond them[?]
>
> (Lambert and Baloji, 2021)

Baloji's work, when analysed in this light, is radical. His handling of the negatives shot by former colonial mine owners, his reprinting, and manipulating them is a quiet yet intense power, a way of beginning to appropriate and subvert photography. At the same time, his montages offer a critique of extractivism.

Such resistance is also found on display in Mujila's *Tram 83*, as Lucien shops his writing around only to be strong-armed at nearly every turn by those who stand between his work and its publication. His writings are deemed politically dangerous, and he is accused of 'atteinte à la sûreté de l'État, incitation méthodique et systématique à la révolte' [breaching national security and planned and systematic incitement to revolt] (2014: 63, 2015: 61). As the novel progresses, his freedom as a Black African author is routinely curtailed. African literature is even defined for him by a white European publicist:

> On en a déjà assez de la misère, de la pauvreté, de la syphilis et de la violence dans la littérature africaine . . . Le personage principal dans le roman africain est toujours célibataire, névrosé, pervers, dépressif, sans enfants, sans domicile et traîne toutes les dettes du monde.
>
> [We've already had enough of squalor, poverty, syphilis, and violence in African literature . . . The main character in the African novel is always single, neurotic, perverse, depressive, childless, homeless, and overburdened with debt.]
>
> (2014: 50, 2015: 46)

Additionally, he is called upon by a diasporic friend in France to be a representative of the Black race: 'Tous les noirs du monde entier attendent désespérément que tu achèves ce texte. Lucien, grouille-toi, ton théâtre-conte, la diaspora est au bout de sa patience. Lucien, nous les noirs de France nous attendons que ton œuvre nous réhabilite' [All the blacks the world over are waiting desperately for you to complete this text. Chop-chop with your stage-tale, Lucien! The diaspora has run out of patience. Lucien. We blacks of France are waiting for your work to redeem us] (2014: 132, 2015: 136).

Lucien's power, his resistance, lies in his refusal to play by anyone's rules. In a rare moment in which his quasi-interior monologue hijacks the narrative (and in which he refers to himself in the second person), Lucien lets loose:

> Le général t'appelle et te propose un marché: 'Écris des textes en mon honneur et j'assure et ton confort financier et ton confort matériel.' Mais toi, tu retorques: 'Je ne cède pas aux chantages, je suis un écrivain et non un griot au service du roi.' Quel raisonnement! . . . Même chose dans l'Arrière-Pays: tu as de sérieux problèmes, tu reçois des menaces du matin au soir, tu frôles la mendicité. 'Rédige des poésies patriotiques, on te laisse la vie tranquille, on te file un poste à l'université et de multiples autres privilèges.' Mais tu craches dans la soupe. Tu pourrais avoir la chance de posséder de belles carrosseries, de belles femmes, des appartements en Europe mais tu craches dans la soupe. Tu as la vie devant toi, tu craches dans la soupe.
>
> [The General calls and makes you a deal; 'Write your texts in my honor and I will guarantee both your financial and your material comfort.' And you retort: 'I don't give in to blackmail, I am a writer, not a griot in the service of the king.' What a way of thinking! Same thing in the Back-Country: you're in quite some difficulty, you're being threatened morning till night, you're on the verge of beggary. 'Compose your patriotic poetries, and we'll let you lead a quiet life. We'll give you a position at the university and a host of other privileges.' But you throw it back in their faces. You could be lucky enough to possess beautiful bodies, beautiful women, apartments in Europe, but you throw it back in their faces. Life is yours for the taking, and you throw it back in their faces.]
> (2014: 178–179, 2015: 187)

Lucien's refusal to literarily conform eventually wins him publication, though what matters ultimately is that his book's publication gets under the skin of the dissident General who oversees the Ville-Pays and profits exclusively off mineral extraction. Lucien's writings bother the dissident General so much, in fact, that he ends up putting a price on Lucien's head. Such is the power, Mujila gestures here, of literature to incite, to challenge, to radicalise. By the end of the novel, everyone has Lucien's book's final words on their lips:

> *un corps-vide*
> *un corps-chose*
> *un corps-déchet*
> *un corps-de-chien*
> *le corps-sans-tête*

d'un Général de basse-cour
trainait dans la gadoue
en état de délabrement avancé!

[*an empty-body*
a thingy-body
a trashy-body
a doggy-body
the headless-body
of a farmyard General
lying in the sludge
in a state of advanced deterioration!]
(2014: 195, 2015: 204)

In this way Lucien decapitates the head of the Ville-Pays, literarily, killing off the dissident General and thereby freeing the country and its people.

As it happens, like his character Lucien, Mujila himself has experience refusing to literarily conform as a Black African diasporic writer. He refuses to write patriotically, or in service to expectations others impose upon him. In fact, Mujila tells Robert Glasser that he invokes his own 'right of opacity, and African subjectivity':

> That means that I am African, that I am Congolese, but also that I am very much my parents' child. I am marked by my own history, my own culture. . . . I am not the spokesman for a continent, a country, a nationality or even my auntie, even though my remarks may concern the society from which I come. But one should not become blinded by one's own affiliations and identity.
>
> (2016)

Refusal to conform, then, is to refuse to be subaltern.

Enter the importance of sovereignty as that which is crucial to refusing enslavement by extractivism. After outlining extractivism as that which ultimately ensures climate chaos for the African continent, Bassey argues for 'community ownership of community resources,' so that 'people actually have full rights over their lands and territory,' which would mean that they 'have the right to free, prior, informed consent before any action or activity is executed on their land or territory' (2012: 146). What might be seen as an immediate struggle for a clean environment, climate justice, and food security, is really, according to Bassey, 'the transformation of society into one in which the people regain true sovereignty' (2012: 160).

As witnessed in *The Beautiful Time* and *Tram 83*, 'environmental crises are being reenvisioned as political, social, and cultural problems, not just scientific ones' (Heise, 2016: 25). Given that resource extraction is responsible for half the world's carbon emissions and that extraction, at root, as Yusoff attests (2018), excludes Blackness from the wealth of its own labour, poisons Black bodies and ecosystems, and, as Bassey puts it (2012), cooks the Global South, how can the narrative of the climate crisis *not* be re-envisioned? Extraction and looting of natural resources, according to Mbembe,

> goes hand in hand with brutal attempts to immobilize and spatially fix whole categories of people, or, paradoxically free them as a way of forcing them to scatter over broad areas no longer contained by the boundaries of a territorial state.
>
> (2016: 86)

Mbembe furthers that the present '*body of extraction*' is no longer 'the human being of African origin' but a human reduced to 'a thing, an object, a sellable, buyable, or possessable commodity' and that the human being of African origin is 'a *genus of subaltern humanity*, which, as a superfluous and almost excessive part for which capital has no use, seems destined for . . . expulsion' (2016: 178, original italics). Baloji and Mujila register such disposability. But precolonial whispers remain; not all has been destroyed. Baloji and Mujila are listening hard to interconnected human and non-human relationships for clues as to how to proceed. Revolt, in their works, is in the offing.

Regardless the name of the current era – Anthropocene, Capitalocene, Black Anthropocenes, of inarguable importance is Mujila's railroad cry, lamenting

> ceux que la terre avait engloutis, tous ceux que les trains avaient emportés à la suite d'un déraillement, l'amertume et les yeux rivés sur ceux qui partirent chercher leur vie de l'autre côté de l'océan et qui ne sont jamais arrivés trahis par les vagues.
>
> [those whom the earth had swallowed up, all those whom the trains had taken following a derailment, the bitterness and the eyes riveted on those who'd left to seek new lives across the ocean and who'd never got there betrayed by the waves.]
>
> (2014: 171, 2015: 180)

An era of intense climate-induced migration is upon us, and, as climate activist Vanessa Nakate writes, even though the African continent is 'responsible for only between 2 and 3 per cent of global energy-related carbon dioxide emissions,' it will 'bear almost half the costs of adapting to the consequences of climate change, and seven of ten countries most susceptible to the harshest effects of the climate crisis are in Africa' (2021: 2–3). Mujila and Baloji know what is coming because it has, in the form of Black Anthropocenes, already occurred. They are making art that rises from the refuse, from uranium's radiation, from the ashes.

Decolonising temporality: Baloji's montages and Mujila's music

Baloji's montages of men in chains digging minerals comment on the racist discipline of geology; so, too, do his 'portraits' of present-day industrial ruins and toxic effluence (Figure 12.3). Yet, it is the queer temporality of Baloji's

Figure 12.3 DSC02799: A rusted structure at a copper mine in Lubumbashi.

photomontages that most visually insist upon the abuses of capitalism, refusing to assign the archival images to the past. To better understand just how time functions in Baloji's montages, let me briefly turn to South African photographer Nomusa Makhubu's conflation of history and the present in *Self-Portrait Project* (2007/2013), a photographic series in which Makhubu overlays her self-portrait atop colonial-era photographic representations of African women. Makhubu's project aims to interrupt the viewing of persons subjected to the colonial gaze and thereby alter the meaning of the colonial photographic archive. Like Baloji who argues about colonisation's continuance by another name, by contemporising the past, Makhubu hopes viewers see that colonial apartheid structures continue to exist in South Africa. She puts it this way in an interview with Claire Counihan:

> Apartheid was simply made informal. There are no white people living in townships; when black people move to suburbs in numbers property values decline; a majority of black people depend on the public sector when there is a parasitic private sector for the needs of the wealthy, who are predominantly white. Private schools, private healthcare, private transport, and private security/policing are the reason the public sector fails.
>
> (2016: 313)

Makhubu continues, 'It's the constant representation of the black body as a body that is disposable in South Africa' (Counihan, 2016: 313). Ultimately, Makhubu's self-portraits are self-proclaimed explorations of her own 'disposable' identity: 'Having grown up in townships in the Northern part of the country,' she says, 'I lost touch with my cultural traditions. And every attempt to understand this background, turned seemingly fabricated' ('Nomusa Makhubu: Intertwined 2005–2015', 2016: para. 2). While working on the *Self-Portrait Project*, Makhubu wondered whether she was becoming just another female subject; was she simply re-inscribing the wounds upon herself retrogressively? To answer this question, she returned to the Africanist conception of circular time. She writes, 'time [to the Africanist] is a maze which one finds oneself walking in and out of. The present is constituted of the past and future' (2014: 25). What Makhubu discovered when she applied African circular notions of time to her own work was that rather than her work being 'perceived as a retrogressive "return,"' suddenly 'the re-enactment of women's experiences in colonial photography' could 'be used to turn the direction of the weapon' (2014: 25). Thusly, the power in Baloji's and Makhubu's montages resides in the way these photographic works gesture towards a timeless knowing that challenges Western notions of time and white supremacist notions of the Black body. In Baloji's *and* Makhubu's photographic works, the gaze decolonises – looks back (to the past) and forward (towards the future) simultaneously.

In a decolonial move of his own, Mujila similarly troubles Western understandings of time, infusing his language with the jazz rhythms that play in *Tram 83*: rhythms that are African polymetric and polyrhythmic in origin, contrapuntal turns that date back centuries to circular African dance patterns and African ritual ceremonies (Gioia, 2021: 1–32). Mujila's prose is blared, polyvocal, its riffs overlapping call and responses that circle back again and again. Jazz emerged in the United States out of African slaves' songs being forced into white man's monometric and monorhythmic structures (Schuller, 1986: 3–62). 'Syncopation, preceding or following the main beats,' was the only 'workable compromise,' Gunther Schuller writes (1986: 15). Like Baloji and Makhubu, who appropriate colonial photography and bend its lightwave, so to speak, by mixing it with Black Anthropocenic understandings together with their own portraiture to come at the genre of photography anew or at least from Indigenous perspectives, Mujila employs Black Anthropocenic sensibilities to overhaul Western definitions of the novel. Ato Quayson ('Shakara Baby-chicks,' 2016: para. 4) writes that the musicality 'at the level of content' in *Tram 83* is:

> superbly augmented by rhythms at the level of narration and it is here that the experimental innovativeness of Mujila's narrative style must be recognized. The rhythmic character of the narration is systematically structured around a series of repeated sentences, phrases, and sequences in a sometimes harmonious but often dissonant distributional matrix. These provide the novel not only with the air of an improvisational jazz symphony, but, perhaps even more importantly, lend it a dramatic character, as if it was written to be visualized rather than just read, or for stage and screen, rather than just for the inert pages of the book we hold in our hands. There is something even danceable about it.

The book does this at the level of content, and it also places jazz in the narrative. In *Tram 83*, jazz is conceived of as a music of the bourgeoisie, a music that elevates all manner of persons. In the nightclub Tram 83, 'n'écoute pas le jazz pour l'odeur des cannes à sucre ou retrouver la conscience nègre' [you don't listen to jazz to get a whiff of sugar cane or reconnect with Negro consciousness] (2014: 18, 2015: 10). No, jazz is 'un signe de noblesse, c'est musique des riches et des nouveaux riches, de ceux qui construisent ce beau monde cassé' [a sign of nobility, it's the music of the rich and the newly rich, of those who build this beautiful broken world], and it is 'le seul levier dont se sert toute la racaille du Tram 83 pour changer de classe sociale comme on changerait de métro' [the only lever used by all the riffraff of Tram 83 to switch social class as one would subway cars]

(2014: 18–19, 2015: 11). Jazz functions as a railroad, a method of transportation elsewhere when stuck in an urban mining centre. It is the music that draws one into bars and speakeasies. Marabi music (an African keyboard style that borrows from jazz) functions similarly in Peter Abrahams's *Mine Boy*. Abrahams also relies on music and dance as a common denominator, as a reparative energy that pulses, enabling life to continue in Black urban settings despite the burdens of the Black Anthropocenic realities at work on Black bodies. The resistance to colonial structures, the attempt to make literature danceable and photography more Indigenous are self-determining, political endeavours. Material speaks. Genre matters.

Questions of immorality

Mujila has written a danceable immorality play in a globalised world within which the formation of western ethics (grounded in the encounter of the Other) is suspect at best, if not eco- and genocidal. Mujila, in *Tram 83*, exposes capitalism's underbelly, does not pretend morality, and slays heroes and anti-heroes alike. Those students who ride the trains through Ville-Pays may act entitled and better than the miner workers, but they are similarly reduced to passengers by Mujila when he writes:

> Les trains passaient sans s'arrêter. Quitte, pour les étudiants les plus rapides, à s'agripper à la ferraille, à la guerre comme à la guerre! Aux caprices de ces étudiants qui se croyaient tout permis s'opposait la bestialité des creuseurs qui partaient et revenaient par les mêmes engins. Les premiers reprochaient aux seconds de brader leur dignité aux exploitants et négociants miniers d'origines multiples. Les seconds s'en moquaient, démontrant avec leur poise et leur corps raidis à force de radioactivité qu'on n'a pas à passer sur les bancs de l'école pour baiser et trinquer, par la suite, avec une bière bien fraîche. D'ailleurs, certains étudiants butinaient dans les mines pour régler leurs dettes.
>
> [The trains passed without stopping, which meant the swiftest students had to grab ahold of the beat up railcars. All's fair in love and war! In stark contrast to these impulsive students with their sense of entitlement were the brutish diggers, who departed and returned on the same contraptions. The former reproached the latter for selling short their dignity to the mining operators and brokers of diverse origin. The latter couldn't care less, displaying, through their rotten luck and bodies stiffened with radioactivity, that you needn't spend time in the classroom to fuck and then clink glasses of ice cold beer after. Some students even scratched out a living in the mines to pay their debts.]
>
> (2014: 11, 2015: 3–4)

By book's end, no character is likeable; everyone is flawed. Quayson plumbs Lucien's idealism, which he maintains is retained 'only at the cost of not examining the compromised grounds upon which he is able to remain an intellectual in the first place' ('Shakara Baby-chicks,' 2016: para. 21). Lucien is a freeloader. He lives off Requiem, has run off with Requiem's wife, and plays with those other women who desire him, even as he appears to be the only male character able to keep his trousers on in the book. And it is this detail, or perhaps the silicone breasts of the novel's prostitutes, that seems to especially bother those, like Ikhide Ikheloa, who accuse Mujila of penning poverty porn (2017). As Mujila warns us in *Tram 83*, there are plenty of people out there who would like to tell Black African authors what African literature is supposed to do or be, but Black Anthropocenic life refuses to be hemmed in, just as Black African authors refuse to conform to certain politically inspired requests any longer. Mujila, like his fictional character Lucien, who is beaten repeatedly and jailed due to the political content of his writings, clamours after free speech.

This does not stop Ikheloa from accusing Mujila, in essence, of selling Africa's soul, panning the book on his blog: 'Africa's men doze, wake up, order dog meat and grilled rats and fuck more women, pretend humans with fake buttocks and "melon breasts" and return to sleep to continue with the misogyny and self-loathing. Africa has suffered' (2017). As if playing the role of those who tell Lucien in the novel how to write African literature, Ikheloa goes on to accuse Mujila of undoing the work that other Africans began when they 'themselves, in the middle of the twentieth century, took into their own hands the telling of their story' rather than accept the bad light cast by European literature on Africa (2017). He calls Mujila a self-exile, a diaspora writer 'lounging in alien cafes,' a person 'complicit in the mess' and potentially involved in restoking the West's claim that Africa is 'a space for caricatures, pretend-humans' (2017). The immorality tale that Mujila spins, however, is thick with irony, complete with sung praises to globalisation, no less:

> Alors pour sortir de la monotonie, fièvre, maladie du sommeil, séisme, cholera, éboulement, à l'exception de ceux qui traînent au Tram 83, tout le monde se met au porno américain ou au porno russe, vive la globalisation, vive le porno américain, vive le porno russe!
>
> [To escape the monotony, fever, sleeping sickness, earthquakes, cholera, and cave-ins, everyone, with the exception of those who hang out in Tram 83, gets into American porn or Russian porn. Long live globalization! Long live American porn! Long live Russian porn!]
>
> (2014: 43, 2015: 37)

And to the payouts of disaster: 'À dire vrai, toute la Ville-Pays se réjoissait lorsque la terre avalait . . . Les jours qui suivaient l'effondrement, les pierres se ramassaient partout, au Tram, à la gare' [Truth be told, the whole of the City-State rejoiced when the earth devoured . . . In the days following a cave-in, the stones could be gathered everywhere, from the Tram to the station'] (2014: 60, 2015: 57). Mujila's thick description even makes note of how 'des funérailles se transformaient en festivités après la découverte, au hasard, d'une pierre à teneur considérable' ['funerals would sometimes turn festive following the chance discovery of a high-grade stone'] (2014: 114, 2015: 116). Yet Mujila is not glorifying extraction; he is rather painting a realistic picture of it. This is also what Zukiswa Wanner, noted feminist and one of the African-based judges who awarded *Tram 83* the Etisalat Prize, argues when she writes about 'how accurate the novel is in capturing the [extractivist industry's imposed] misogynist reality of its setting' (Obi-Young, 2017a).

The extractivist industry's creation of prostitution, what with miners traditionally forced to labour away from their rural homes for months at a time if not longer, relies on disposability of women's bodies. Quayson ('Shakara Baby-chicks,' 2016: para. 20) may defend the idea that in Mujila's *Tram 83* sex workers control their own bodily economies ('in Mujila's novel sex . . . becomes first and last a transaction, in which the women have utter control on how the transaction is going to be conducted,' he writes), yet Mujila has no rosy idea of sex work. Baby-chicks end up dead in bathroom stalls in his novel. They, like so many others, are arguably just slaves who may think they are free, given come-on lines like: 'Je te donne mon corps, enchaîne-moi, fais de moi ton esclave, ta marchandise, ta chasse gardée' [I give you my body, chain me up, make me your slave, your property, your private hunting ground] (2014: 18, 2015: 10). Disposability in *Tram 83* knows no age limit. Child prostitutes are let into the club when it is realised that they can 'servir d'appât' [serve as bait] (2014: 79, 2015: 79) and 'engranger des recettes inespérées' [rack up unhoped-for revenue] (2014: 79, 2015: 79). As a result of living at a site where raw materials leave the earth and are rendered into value-added product elsewhere, the women in *Tram 83*, like the earth, are literally commodified objects, all but essentially non-human.

Are the men any better off? There are cave-ins, early deaths. The men in the novel, it is true, get to play bigger, more developed characters. *Tram 83*'s sex workers play bit parts if they even play parts at all. This said, we hear mention of Jacqueline, formerly Requiem's wife, now Lucien's, a woman held in high esteem by Lucien, if not also Requiem. Jacqueline may be physically absent (she remains throughout the novel in an unnamed location) and unvoiced, but she is nonetheless, one might argue, a central character, a Helen of Lubumbashi. Because of Jacqueline, it could be said that *all women* in *Tram 83* are not, as Petina Gappah would have us believe, 'disposable' (Obi-Young, 2017b). If Jacqueline were disposable, the chorus would not gibe: 'Lucien, égoïste, tu es là à t'occuper de ton ventre, tu ne cherches pas à savoir si Jacqueline parvient à vivre avec toutes ces crises boursières qui défraient la chronique' [Lucien, egoist, there you are focused on your belly, without even bothering to check if Jacqueline is able to survive, what with all the crashing stock markets that are hitting the headlines] (2014: 175, 2015: 183).

But whether *Tram 83* is misogynist porn or not is not particularly relevant here; what *is* relevant is the uncompromising nature of Mujila's swagger, his ability to blow his horn, to riff, to make the chorus the storyline. Mujila does not hold back. He will write his literature as he wishes, be it judged baseless or not. Here, for instance, is Ville-Pays hanging out all over itself:

> Des guimbardes en panne d'essence, des produits surgelés venus des îles Galápagos, des bibelots, des ventilateurs, des vidanges, des moutons, des sarcasmes, des corbillards sur le qui-vive, des œufs contaminés à la mélanine . . . des minarets à perte de vue . . . des moustiques résistants à tous les pesticides.
>
> [Jalopies out of gas, deep-frozen products from the Galapagos Islands, knickknacks, ceiling fans, oil changes, sheep, sarcastic remarks, hearses on alert, eggs contaminated with melamine . . . minarets as far as the eye can see . . . mosquitoes resistant to all pesticides.]
>
> (2014: 40, 2015: 34)

And just like the battered, undervalued Ville-Pays, Mujila's characters are, in effect, hustlers, bastards, and writers who understand their situation and make the best of it, all the while throwing wrenches into the workings of Ville-Pays, powered as it is by extractivism. And for this reason, Lucien and Requiem find themselves, by book's end, on the run.

Reclamation: a conclusion

Baloji's photographic series *The Beautiful Time* and Mujila's novel *Tram 83*, when read through a transcultural materialist lens, critique extraction and argue for a reclamation of what has been stolen. In Mujila's and Baloji's works, we sense the past's pulse. Colonial violence is a neck-chain, a slagheap, the 'construction métallique inachevée' [unfinished metal structure] (2014: 9, 2015: 1) of industry, recognisable in the present, too, as neocolonial violence. Here is the same slave (albeit believing himself emancipated and wanting to become like his master), the same slagheap, the same unfinished structure, the same capitalism creating *and* annulling time, or 'creating, accelerating, and exploding' it, as Mbembe argues (2016: 179–181). These works bother us with their refusal to look away from extractivist plunder. Black Anthropocenes are multitudinous and horrifying. As artisanal miners sicken with radiation poisoning, as humans, birds, fish, and multitudinous other poisoned species grow tumours and die, as the average human life expectancy plunges, as African nationals drown in the Mediterranean, as the global supply chain begets starvation, as climate chaos hits country after country with drought, fires, flooding, and death, *The Beautiful Time* and *Tram 83* bespeak amnesiac traces of Congolese cosmologies and Indigenous political arts. The works act upon the West by turning westerners (at least those who are, at bare minimum, consumers of minerals, literature, and art) to stone, and slowly chipping away at them, until they too recognise that their disposable lifestyles (increasingly wedded to the non-human machine – a tangle of artificial, organic, synthetic, and geological forces) connect them to Black and brown persons global capital marks as disposable.

As Mujoli and Baloji remind us, we all remain and must continue to remain in relationship with nature. And as Eze cautions: 'what affects nature or ecosystems also affects us' (2017: 629). We would do well to attend to an African cosmological view that ultimately translates into living as part of nature, our ethics not an ideology but 'a way of life' (Eze, 2017: 631). African cosmology allows that we live in relationships, that our relationships extend beyond the human realm, and that natural elements are sacred and should not be violated. Invoking Bantu cosmology and quoting the writings of Leopold Senghor, Eze posits that, according to African cosmology, every being and element contains a life force, a spirit (2017: 626). Illness is caused by disturbance to 'the cosmic world' and 'healing occurs at the metaphysical confluence between nature, intelligent beings, and the . . . ancestors' (2017: 626). The works by Baloji and Mujila would have us remember who we are *in relation* with each other, nature, and our ancestors, and urge us to choose to leave minerals in the ground, to acknowledge the agency of stone, and work to regain our self-determined sovereignty, all the while checking our idealism before even dreaming about moving on to 'forge memory, that is to say, build together spaces of collective decision-making' (Mbembe, 2016: 181) that might help establish equalities between all life forces.

Bibliography

Arndt, Lotte (2020), *Candice Lin: A Hard White Body* (Chicago: University of Chicago Press).

Baloji, Sammy (2015), 'Sammy Baloji Interview: The Past in Front of Us', *TheLouisianaChannel*, [Online], [Accessed 10 September 2022], available from: youtube.com/watch?v=jCS6RwnrEzE

Barad, Karen (2007), *Meeting the Universe Halfway: Quantum Physics and the Entanglement of Matter and Meaning* (Durham, NC: Duke University Press).

Bassey, Nimmo (2012), *To Cook a Continent: Destructive Extraction and Climate Crisis in Africa* (Nairobi: Pambazuka Press).
Campt, Tina (2017), *Listening to Images* (Durham, NC: Duke University Press).
Counihan, Clare (2016), 'Interview with Nomusa Makhubu', *Comparative Studies of South Asia, Africa and the Middle East*, 36:2, 307–319.
Eze, Michael (2017), 'Humanitatis-Eco (Eco-Humanism): An African Environmental Theory', in Adeshina Afolayan and Toyin Falola (eds.), *The Palgrave Handbook of African Philosophy* (Basingstoke: Palgrave Macmillan), 621–632.
Gioia, Ted (2021), *The History of Jazz* (New York: Oxford University Press).
Glasser, Roland (2016), 'Interview with Fiston Mwanza Mujila', *The White Review*, [Online], [Accessed 10 September 2022], available from: www.thewhitereview.org/feature/interview-with-fiston-mwanza-mujila/
Heise, Ursula K. (2016), 'The Environmental Humanities and the Futures of the Human', *New German Critique*, 128, 21–31.
Hoyos, Héctor (2019), *Things with a Past: Transcultural Materialism and the Literatures of Extraction in Contemporary Latin America* (New York: Columbia University Press).
Ikheloa, Ikhide (2017), 'Mujila Fiston Mwanza's *Tram 83*: Requiem for the African Writer, and Again, the Balance of Today's Stories', [Online], [Accessed 10 September 2022], available from: https://xokigbo.com/2017/04/16/mujila-fiston-mwanzas-tram-83-requiem-for-the-african-writer-and-again-the-balance-of-todays-stories/
Jewsiewicki, Bogumil (2010), *The Beautiful Time: Photography by Sammy Baloji* (Seattle, WA: University of Washington Press).
Lambert, Léopold and Sammy Baloji (2021), 'Colonial Extractivism and Epistemic Geologies in the Congo', *The Funambulist*, [Online], [Accessed 10 September 2022], available from: https://thefunambulist.net/magazine/decolonial-ecologies/colonial-extractivism-and-epistemic-geologies-in-the-congo
Madsen, Kristian Vistrup and Sammy Baloji (2019), 'Sammy Baloji – Interview: "I'm Not Interested in Colonialism as a Thing of the Past, But in the Continuation of That System"', *Studio International: Visual Arts, Design and Architecture*, [Online], [Accessed 10 September 2022], available from: www.studiointernational.com/index.php/sammy-baloji-im-not-interested-in-colonialism-as-a-thing-of-the-past-but-in-continuation-of-that-system
Makhubu, Nomusa (2014), 'Nomusa Makhubu', *Erdmann Contemporary*, [Online], [Accessed 10 September 2022], available from: https://issuu.com/erdmanncontemporary/docs/book_-_nomusa_makhubu_-_final
Mbembe, Achille (2016), *Necropolitics* (Durham, NC: Duke University Press).
Mbembe, Achille (2019), *Out of the Dark Night: Essays on Decolonization* (New York: Columbia University Press).
Mhute, Josiah Rungano (2001), 'Downcast: Mining, Men and the Camera in Colonial Zimbabwe, 1890–1930', *Kronos*, 27, 114–132.
Moten, Fred and Stefano Harney (2013), *The Undercommons: Fugitive Planning and Black Study* (London: Minor Compositions).
Mujila, Fiston Mwanza (2014), *Tram 83* (Paris: Éditions Métailié).
Mujila, Fiston Mwanza (2015), *Tram 83*, trans. Roland Glasser (Dallas, TX: Deep Velum Publishing).
Nakate, Vanessa (2021), *The Bigger Picture: My Fight to Bring a New African Voice to the Climate Crisis* (Boston, MA: Mariner Books).
'Nomusa Makhubu: Intertwined 2005–2015' (2016), *Contemporaryand*, [Online], [Accessed 10 September 2022], available from: https://contemporaryand.com/exhibition/nomusa-makhubu-intertwined-2005-2015/

Obi-Young, Otosirieze (2017a), 'Is Fiston Mujila's "Tram 83" Misogynist Poverty Porn? Zukiswa Wanner and Richard Oduku Lead Strong Reaction to Ikhide Ikheloa's Damning Criticism', *Brittle Paper*, [Online], [Accessed 10 September 2022], available from: https://brittlepaper.com/2017/04/fiston-mujilas-tram-83-misogynist-poverty-porn-zukiswa-wanner-richard-oduko-lead-strong-reaction-pa-ikhides-damning-criticism/

Obi-Young, Otosirieze (2017b), 'Is "Tram 83" Misogynist Poverty Porn? Petina Gappah, Tsitsi Dangarembga, Ainehi Edoro Deepen Conversation as Ikhide Ikheloa and Richard Oduku Publish New Essays', *Brittle Paper*, [Online], [Accessed 10 September 2022], available from: https://brittlepaper.com/2017/04/ikhide-tram-83/

Quayson, Ato (2016), 'Shakara Baby-chicks and Locomotive Tales: Fiston Mwanza Mujila's *Tram 83*', *Arcade*, [Online], [Accessed 10 September 2022], available from: https://arcade.stanford.edu/blogs/shakara-baby-chicks-and-locomotive-tales-fiston-mwanza-mujilas-tram-83

Samatar, Sofia (2015), 'Fiston Mwanza Mujila and Robert Glaser', *Bomb Magazine*, [Online], [Accessed 10 September 2022], available from: https://bombmagazine.org/articles/fiston-mwanza-mujila-roland-glasser/

Schuller, Gunther (1986), *Early Jazz: Its Roots and Musical Development* (New York: Oxford University Press).

Sontag, Susan (2004), 'Regarding the Torture of Others', *The New York Times Magazine*, [Online], [Accessed 10 September 2022], available from: www.nytimes.com/2004/05/23/magazine/regarding-the-torture-of-others.html

Yusoff, Kathryn (2018), *A Billion Black Anthropocenes or None* (Minneapolis: University of Minnesota Press).

Afterword

One must cultivate one's own garden

Helena Duffy

This volume has presented the partial findings of the research conducted by the study group 'Narrative and Violence.' Between 2020 and 2022, the group dedicated its energies to exploring and analysing the ways in which narrative symbolises violence with the aim of taming (or inflaming) it. 'Narrative and Violence' involved researchers at different stages of the academic career and adopted a multicultural and transgeneric approach to the analysis of narrative. Similarly, the present collection of essays has brought together the work of established, mid-career, and early-career scholars, as well as doctoral candidates. Their contributions have addressed texts produced in different cultural and linguistic contexts and representing a gamut of genres. However, in contrast to the research undertaken by 'Narrative and Violence,' which was concerned with topics as diverse as misogyny, racism, and slow violence, *Storying the Ecocatastrophe* has focused closely on cultural representations of the anthropogenic degradation of the natural environment. Its twelve core chapters have examined how this degradation has been articulated and interpreted by contemporary works of literary and cinematic fiction, both formally traditional and experimental, as well as by other genres such as poetry, visual arts, science books, and dance installations. The chapters have also interrogated the ethical engagement and political agendas of the texts they examine, and their potential mobilising impact on their audiences.

As we have seen, contemporary cultural productions hope to spur climate action with recourse to representational strategies that strikingly vary from those embraced by early climate fiction. More specifically, these productions have moved away from the apocalyptic scenarios in which humanity was obliterated (or nearly obliterated) by climate change, man-made pandemics, experimental accidents, or nuclear war. Though undoubtedly intended on spreading environmental awareness and encouraging readers and viewers to act more responsibly, these doomsday stories seem to have proven largely ineffective. This can be plainly seen from the rapidly

DOI: 10.4324/9781032726953-14

deteriorating environmental situation: air and water temperatures keep rising, oceans teem with plastic waste, between 0.01 per cent and 0.1 per cent of all species becomes extinct each year, and, as we have just learned by living through the subsequent waves of COVID-19, recurring pandemics are a very real threat. At the same time, western politicians, both on the right and on the left, continue to identify growth as the only antidote to the falling living standards and construe borders, barriers, and bans as the sole cure to a migrant crisis that has been caused by, among other factors, the disastrous consequences western economic growth has had on the less industrially developed countries. In any case, the coordinated efforts of politicians have been deemed inadequate by climate activists and scientists alike, and have now been additionally put in jeopardy by the food and energy crises exacerbated by Putin's attack on Ukraine. Indeed, for Russia, warfare has become the preferred solution to the prospected shortages of resources, which only gives credence to the prediction of some Holocaust analysts who see the impending scarceness of food and fuel as the origin of future genocidal conflicts.

In the face of the realisation that environmental damage cannot be reversed and that further degradation will not be avoided without a drastic – and therefore highly unlikely – policy change at a global level and an equally drastic shift in individual consciousness, writers and artists have co-opted new narrative strategies. To begin with, instead of showing eco-catastrophe as a future reality that can still, however narrowly, be avoided, the texts examined in this collection depict it as something that is already very much part of our everyday reality. Polluted air and water, coastal flooding, climate change-induced displacement, weather anomalies, depletion of natural resources, poverty, and loss of biodiversity are just some of the symptoms of the Anthropocene staged by the narratives analysed in this volume. These symptoms are not hard to imagine, as most of us have already witnessed or even experienced them first hand. Hence, instead of projecting future doom-and-gloom scenarios, contemporary narratives offer stories on how individual lives have been affected by the detrimental results of human activity and what can be done to mitigate these results.

Incredulous towards politicians, whatever their colour, as well as towards both science and organised climate action, these narratives would rather echo Candide's much commented upon solution to the myriad of ills described by Voltaire's satirical eponymous tale (1759). Under the current circumstances, to cultivate one's garden means to deal as best as one can with the ravages that bear witness to humanity's domination and destruction of the non-human world, which are the consequence of humanity's hubris and which have been sanctioned by dominant religious and philosophical systems. To tend responsibly a post-apocalyptic garden involves a (re)creation of a space where animals would be safe from slaughter, enslavement, and exploitation, and where plants could be restored to their subjectivity, agency, and cultural visibility. To build such a harmonious world, which, as the analysed narratives tell us, can only be achieved by embracing the ethics of difference and care, we require not only an interspecies solidarity, but also a collaboration across different ethnic, sexual, and social groups. As the twelve contributions have demonstrated, contemporary narratives indeed stress the connection between interhuman violence, as manifest in, for example, heteropatriarchy and (neo)colonialism, and our oppression of the animal, plant, and material world.

The twelve chapters in the present volume have also shown that representing the ecocatastrophe calls for a novel set of methods and tools, and that these will vary depending on the cultural context of the work. There is indeed an unease about traditional representational forms. One of these forms is the realist novel whose birth coincided with the onset of the Anthropocene, whose concerns are inevitably human-centred, and which, consequently, has been justifiably suspected of complicity in the very thinking that engendered the environmental collapse. Writers and artists have therefore reached for new idioms which are potentially capable of defamiliarising what is rapidly becoming all too familiar, and thus of resensitising readers and viewers to the urgency of the situation at hand. These new, future-oriented representational practices are designed to reinforce the message of hope, harmony, and solidarity that underpins the stories being told. They are meant to activate audiences, yet without being overtly didactic, making them anxious, or instilling in them a sense that all is lost and the world is beyond repair. On the contrary, as we have observed, many of the analysed texts strike a positive note and seek a silver lining to the environmental crisis. Namely, they hope for a change in our mindset regarding the environment, which would consist in an unsettling of the human/non-human divide and in seeing ourselves as an integral part of nature rather than its masters or stewards.

Index

Milton Keynes UK
Ingram Content Group UK Ltd.
UKHW052104130524
442568UK00005BA/66

9 781032 726946